8 Springer Series in Solid-State Sciences

Edited by Peter Fulde

Springer Series in Solid-State Sciences

Editors: M. Cardona P. Fulde H.-J. Queisser

Volume 1 **Principles of Magnetic Resonance** 2nd Edition
By C. P. Slichter

Volume 2 **Introduction to Solid-State Theory**
By O. Madelung

Volume 3 **Dynamical Scattering of X-Rays in Crystals**
By Z. G. Pinsker

Volume 4 **Inelastic Electron Tunneling Spectroscopy**
Editor: T. Wolfram

Volume 5 **Fundamentals of Crystal Growth I.** Macroscopic Equilibrium and Transport Concepts
By F. Rosenberger

Volume 6 **Magnetic Flux Structures in Superconductors**
By R. Hübener

Volume 7 **Green's Functions in Quantum Physics**
By E. N. Economou

Volume 8 **Solitons and Condensed Matter Physics**
Editors: A. R. Bishop and T. Schneider

Volume 9 **Photoferroelectrics**
By V. M. Fridkin

Volume 10 **Phonon Dispersion Curves in Insulators**
By H. Bilz and W. Kress

Solitons and Condensed Matter Physics

Proceedings of the Symposium on
Nonlinear (Soliton) Structure and Dynamics
in Condensed Matter
Oxford, England, June 27–29, 1978

Editors
A. R. Bishop T. Schneider

With 120 Figures

Springer-Verlag Berlin Heidelberg New York 1978

Alan R. Bishop, B.Sc. Ph.D.

Department of Physics, Queen Mary College, Mile End Road
London E1 4NS, England

Dr. Toni Schneider

IBM Zürich Research Laboratory, CH-8803 Rüschlikon, Switzerland

Series Editors:

Professor Dr. Manuel Cardona
Professor Dr. Peter Fulde
Professor Dr. Hans-Joachim Queisser

Max-Planck-Institut für Festkörperforschung
Büsnauer Strasse 171, D-7000 Stuttgart 80, Fed. Rep. of Germany

ISBN 3-540-09138-6 Springer-Verlag Berlin Heidelberg New York
ISBN 0-387-09138-6 Springer-Verlag New York Heidelberg Berlin

Offset printing: Zechnersche Buchdruckerei Speyer, Bookbinding: J. Schäffer oHG, Grünstadt.
2153/3130-543210

Preface

Nonlinear ideas of a "soliton" variety have been a unifying influence on the natural sciences for many decades. However, their universal appreciation in the physics community as a genuine paradigm is very much a current development. All of us who have been associated with this recent wave of enthusiasm were impressed with the variety of applications, their inevitability once the mental contraint of linear normal modes is removed, and above all by the *common* mathematical structures underpinning applications with quite different (and often novel) physical manifestations. This has certainly been the situation in condensed matter, and when, during the Paris Lattice Dynamics Conference (September 1977), one of us (T.S.) first suggested a condensed matter soliton Meeting, the idea was strongly encouraged. It would provide an opportunity to exhibit the common mathematical problems, illuminate the new contexts, and thereby focus the "subject" of nonlinear physics at this embryonic stage of its evolution.

The original conception was to achieve a balance of mathematicians and physicists such that each would benefit from the other's expertise and outlook. In contrast to many soliton Meetings, however, a deliberate attempt was made to emphasize physics contexts rather than mathematical details. The beautiful examples and techniques of rigorous soliton theory and algebraic topology developed so fully in the last few years have motivated the physicists' appreciation: we hope that this survey of condensed matter applications will be accepted as a tribute to the mathematics, and also serve to highlight some of the more pressing and universal mathematical problems now facing the nonlinear physicist.

These Proceedings should give a sufficiently coherent flavour of current activity to introduce nonlinear physics to the newcomer but they are certainly not exhaustive. Some omissions were deliberate (we considered nonlinear plasmas, optics, hydrodynamics, etc., as extremely important but sufficiently established in their own right); others were imposed by time limitations. We apologise to all those subjects and authors who were not represented, but feel sure that they too will benefit from this exposure of the greater nonlinear cause. Our main hope is that this contribution will mark a small step towards the ultimate percolation of the soliton paradigm throughout all levels of the physics community

The gratitude of the Organizing Committee:

S. Aubry, A.R. Bishop, R. Blinc, A. Bruce
R. Bullough, R.J. Elliott, J. Krumhansl, A. Luther
T. Schneider, R. Stinchcombe, H. Thomas and
S. Trullinger

goes to many individuals and organizations.

Without our Sponsors:

European Research Office (U.S. Army)
IBM Zurich Research Laboratory
National Science Foundation (U.S.A.)
RCA Laboratories (Zurich)

this venture would have been impossible. We are especially grateful for their encouragement and administrative cooperation which enabled us to meet optimistic deadlines. Queen Mary College (London University) (through A.R.B. and D.A.L. Jones) kindly agreed to administer the European grants and the University of Southern California (through S.E. Trullinger) acted as intermediary for the travel support from NSF.

We thank Springer-Verlag for undertaking such a rapid publication. This has necessarily placed uncomfortable burdens on authors; we apologize for any strong-arm tactics and trust they find the present outcome some compensation. Special thanks to the indefatigable E. Stoll and to the IBM Publications Department (Zurich): U. Bitterli, M. Grüneisen, P. Theus, K. Thoma, and R. Trachsler, for their unfailing patience and accommodations.

The Clarendon Laboratory and St. John's College, Oxford, provided splendid environments for a convivial and stimulating gathering. We are also deeply indebted to R.J. Elliott and his staff (especially Miss B. Green) for their hospitality and meticulous attention to local organization, both before and during the Symposium.

So onwards to the next Soliton Meeting. Meanwhile, it only remains for us to introduce the specific topics included in this one. Fortunately, we are relieved of this duty by the inventive mind of J.A. Krumhansl - we take great pleasure in reproducing "JAK's Odd Ode to Solitons"*, which accomplishes the task with eminent clarity and distinction.

Zürich, July 1978 A.R. Bishop T. Schneider

*Taken from Professor Krumhansl's presentation at the Symposium Dinner. The editors accept full responsibility for its punctuation and introduction!

Solitons in Oxford

First from Bullough we have heard,
of concepts wild and weird.
How to find waves solitary,
you must be bright and wary,
backing round an inverse path
à la Kruskal and his math.

Enter now the marching cordon,
flying banners of sine Gordon!

Then from Scott we hear the history,
how his namesake (hyphen Russell),
just to exercise his muscle,
rode his horse and set the course
we've followed hard these days.

Eilbeck, Gibbon, Kitchenside sing praise!

Now along comes Alan Newell,
and he's nobody's fool,
and tells us if we mess around with change
from systems thought conditioned well,
things will really go to hell.

In 1-d only can you freeze that pulse of K. DeVries!

Thence on we switch to physics statistical
and how those kinks somewhat mystical
jump through hoops in Bishop's rendition
of Gibbs' function of partition.

Through the saddle we must jump with the solitary hump!

Then shades of Alice's Cheshire cat,
as Wallace tells us where its at.
Through the complex plane, it's gone,
wondrous, wondrous instanton.

But now with dreaming call a truce,
listen to the words of Bruce.
Renormalize this group's odd missions
with rules well known for phase transitions.

Order, order, everywhere, but also lots of kinks!

And thence to Schneider's show with Stoll:
my how those IBM computers roll
with all the power they can muster
forming cluster after cluster.

1-d, 2-d, 3-d, 4, - now really is there need for more?

Thence, onward, onto wordly facts,
we hear from Messrs Bak and Axe
if there is sanity to save,
charge density's your wave.

And that is it so far this week,
but more is to come the morrow:
Solitons to be more truther
by the words of Alan Luther,
and perhaps sound neat and nice
in prose which rolls from Michael Rice.

More new words from Friend and Aubry
and finally, then, that liquid wacky
distorted by Kazumi Maki.
Thence to wind-up with Steve Trullinger -
(what rhymes with that?)

So farewell - and end!

These names were but a few of those
whose mystic, mathic, physic prose
has made these days true seem
Midsummer's solitonic dream.

Contents

I. Introduction

Solitons in Mathematics: Brief History

R.K. Bullough and R.K. Dodd

Department of Mathematics, U.M.I.S.T., P.O. Box 88
Manchester M60 1QD, Great Britain
and
School of Mathematics, Trinity College, Dublin 2, Ireland

1. Introduction

This article presents a brief history of the inverse scattering method for solving nonlinear evolution equations and the Hamiltonian structure associated with it. It is not a comprehensive survey of the different mathematics now concerned with soliton theory. To attempt the latter would be inappropriate for a meeting concerned with nonlinear dynamics and structure in condensed matter. In any case, soliton theory already ramifies into areas of mathematics, algebraic geometry, theory of Jacobian varieties, on the edge of the mathematical range of one of us (RKB).

The subject surely begins with the observation [1] of the single soliton solution of the Kortewes-de Vries (KdV) equation, in the month of August 1834, by JOHN SCOTT RUSSELL. Since then, and until quite recently, the physics and mathematics of the subject have evolved together, each influencing the other. At the present time there are signs of schism into distinct and disparate disciplines.

RUSSELL in effect established by experiment [1] that arbitrary initial data for the KdV equation

$$u_t + 6\,uu_x + u_{xxx} = 0 \tag{1}$$

broke up into solitons. The speed of the soliton depended on its amplitude. No serious mathematical developments followed this observation until the work of ZABUSKY and KRUSKAL [2,3]

In contrast, the sine-Gordon equation

$$u_{xx} - u_{tt} = \sin u \tag{2}$$

arose in a strictly mathematical context - in differential geometry in the theory of surfaces of constant curvature [4]. It is worth remarking that LIOUVILLE gave the general solution of the nonlinear Klein-Gordon equation

$$u_{xt} = \exp m\,u \tag{3}$$

(m = parameter) in 1853 [5]. Eq. (3) is a Klein-Gordon equation $u_{xt} = F(u)$ 'light cone' coordinates, but this is easily expressed in the Lorentz covariant form $u_{xx} - u_{tt} = F(u)$, the form of (2). LIOUVILLE's method was his own, but recently [6] we have used the general solution $u = f(x) + g(t)$ of $u_{xt} = 0$ and a Bäcklund transformation to rederive LIOUVILLE's solution in the form

$$\exp mu = \exp m(f-g)\{k\int^x \exp m f\, dx' + \frac{1}{2}k^{-1}m\int^t \exp(-mg)dt'\}^{-2} \qquad (4)$$

The Bäcklund transformation (BT) with free parameter k

$$\begin{aligned} u'_x &= u_x + 2k \sin\frac{1}{2}(u'+u) \\ u'_t &= -u_t + 2k^{-1}\sin\frac{1}{2}(u'-u) \end{aligned} \qquad (5)$$

for the sine-Gordon

$$u_{xt} = \sin u \qquad (6)$$

in light cone coordinates was known to Bäcklund before 1882. We now know that a BT is characteristic of the 'integrable systems'. This phrase is used loosely here to mean those systems of nonlinear evolution equations which have soliton solutions. We define 'integrability' more precisely later in terms of 'integrable Hamiltonian systems'. We shall show elsewhere [7] how these ideas relate to the classical integrability theorem of FROBENIUS.

Integrability is concerned ultimately with integrability conditions exemplified by (5). The BT (5) relates a solution u of the sine-Gordon (SG) (6) to another solution u' (it is strictly speaking an auto BT [6]). The integrability condition is equality of the second partial derivatives $u'_{xt} = u'_{tx}$. This means that u satisfies a partial differential equation which is (6). By adding u'_{xt} and u'_{tx} it also means that u' satisfies a partial differential equation which in this case is also (6). BT's are generalizations of contact transformations (canonical transformations) in which the derivatives of u' are expressed in terms of u, its derivatives, *and* u'. The structure has been extended by PIRANI [8]. The line of attack is to replace 'transformation' by 'map' in which x,t,u,u_x,u_t are independent variables. Introduce the idea of C^∞ (= infinitely differentiable) functions u equivalent to v if $u(x,t) = v(x,t)$ and all their derivatives at (x,t) are equal to order k. The k-jet of u at (x,t), $j^k_x u$, is the equivalence class thus defined. The k-jet bundle $J^k(x,t;u)$ is the set of all $j^k_x u$. This is a manifold coordinatized by, e.g., for $k = 1$, just (x,t,u,u_x,u_t). In this language, the Bäcklund map is a map $\psi : J^1(x,t,u,u_x,u_t) \times N(u') \to J^1(x,t,u',u'_x,u'_t)$ in which $N(u')$ is a C^∞ manifold coordinatized by u'. Thid definition can be generalized [8]. In essence J^1 is the natural manifold on which to calculate BT's like (5): we have used this covertly in finding BT's for the Klein-Gordon equations [6].

Not all equations with BT's are 'integrable' in the sense of having soliton solutions: in particular LIOUVILLE's equation (3) does not because u does not vanish as $|x| \to \infty$ (it is plain that $u = 0$ is not a solution: the soliton solutions decay exponentially to zero at $|x| \to \infty$ so there can be none of these).

LUND [9] recently extended the early work on the SG in differential geometry. We use his work here to exemplify the mathematics. Consider the problem of *embedding* the n-dimensional surface V_n in the n+1 dimensional Euclidean space E

$$E : x^i \quad (i = 1,2,\dots, n+1)$$

$$V_n : y^\mu \quad (\mu = 1,2,\dots, n), \text{with metric}$$

$$ds^2 = g_{\mu\nu}\, dy^\mu\, dy^\nu .$$

The embedding is isometric (preserves scalar products) if V_n can be defined through

$$x^i = x^i(y^1,\dots, y^n)$$

such that $g_{\mu\nu} = \partial x^i/\partial y^\mu,\ \partial x^i/\partial y^\nu$. Vectors X_μ in the tangent space of V_n, and the surface normal X_{n+1} define a basis in E, $X_\mu \equiv \partial\vec{X}/\partial y^\mu$, X_{n+1}. The expression of the vectors $\partial X_\mu/\partial y^\nu$, $\partial X_{n+1}/\partial y^\nu$ in terms of this basis is the (linear) system of Gauss-Weingarten equations. The integrability conditions $\partial^2 X_\mu/\partial y^\nu \partial y^0 = \partial^2 X_\mu/\partial y^0 \partial y^\nu$ (for all pairs y^0, y^ν) lead to a nonlinear system, the Gauss-Codazzi equations.

For a two-dimensional surface in three-dimensional Euclidean space, LUND [9] finds Gauss-Codazzi integrability conditions which ultimately take the form

$$\begin{aligned} &\frac{\partial}{\partial\xi}(\cot^2\theta\, \lambda_\eta) + \frac{\partial}{\partial\eta}(\cot^2\theta\, \lambda_\xi) = 0 \\ &\theta_{\xi\eta} - \frac{1}{2}\sin 2\theta + \frac{\cos\theta}{\sin^3\theta}\lambda_\xi \lambda_\xi = 0 \end{aligned} \tag{7}$$

for the Gauss-Weingarten system

$$\begin{aligned} \begin{pmatrix} v_{1,\xi} \\ v_{2,\xi} \end{pmatrix} &= \begin{pmatrix} -i\zeta + ip & q \\ -q^* & i\zeta - ip \end{pmatrix} \begin{pmatrix} v_1 \\ v_2 \end{pmatrix} \\ \begin{pmatrix} v_{1,\eta} \\ v_{2,\eta} \end{pmatrix} &= i \begin{pmatrix} r & s \\ s & -r \end{pmatrix} \begin{pmatrix} v_1 \\ v_2 \end{pmatrix} \end{aligned} \tag{8}$$

in which $p = (\cos 2\theta/2\sin^2\theta)\lambda_\xi$, $q = \theta_\xi + i\cot\theta\lambda_\xi$, $r = (1/4\zeta)\cos 2\theta - (2\sin^2\theta)^{-1}\lambda_\zeta$, $s = (1/4\zeta)\sin 2\theta$. An eigenvalue has been introduced by appealing to Lie invariance (essentially invariance under interchange of ξ,η) of the pair of coupled evolution equations (7) in the two fields θ,λ. The conditions (7) are sufficient and *necessary* for the integration of (8). The system (8) is however a generalized ZAKHAROV-SHABAT [10] scattering solution problem as we shall see shortly. Integration of this linear system is equivalent to solving the nonlinear system (7). We thus have a natural geometrical basis to an inverse scattering problem.

However, the BT (5) already contains a similar scattering problem for the SG. Set $\Gamma = \tan[(u+u')/4]$. Then

$$\Gamma_{1x} = \frac{1}{2} u_x (1 + \Gamma^2) + k \Gamma \tag{9}$$

$$\Gamma_{1t} = k^{-1} \Gamma \cos u - (2k)^{-1} (1 - \Gamma^2) \sin u.$$

The Ricatti transformation $\Gamma = v_2/v_1$ now yields

$$\begin{pmatrix} v_{1,x} \\ v_{2,x} \end{pmatrix} = \begin{pmatrix} -\frac{1}{2} k & \frac{1}{2} u_x \\ -\frac{1}{2} u_x & +\frac{1}{2} k \end{pmatrix} \begin{pmatrix} v_1 \\ v_2 \end{pmatrix}$$

$$\begin{pmatrix} v_{1,t} \\ v_{2,t} \end{pmatrix} = -\frac{1}{2k} \begin{pmatrix} +\cos u & \sin u \\ \sin u & -\cos u \end{pmatrix} \begin{pmatrix} v_1 \\ v_2 \end{pmatrix} . \tag{10}$$

This is the same as (8) for $\lambda = 0$ and $u = 2\theta$ (corresponding to a surface of constant intrinsic curvature) if $i\zeta \equiv k/2$. POHLMEYER [11] has obtained a BT for the nonlinear sigma models in n fields $u_1,\ldots,u_n$ coupled only by the constraint

$$\sum_{i=1}^{n} u_i^2 = 1.$$

For $n = 4$, the 0_4 invariant model, the system reduced to LUND's. For $n = 3$, the 0_3 invariant case, the system is the SG. POHLMEYER [11] has shown that the results (10) can be extended to (8) by using his BT. The BT may be much more fundamental than the idea that inverse scattering transforms like (8) or (10) are geometrical in origin. However, BT's can be derived for example, by the method of prolongation structures due to WAHLQUIST and ESTABROOK [12]. It is plain that geometrical ideas underlie this theory although the precise nature of these is not yet fully understood.

The BT can generate infinite sets of conserved quantities. We have demonstrated this elsewhere [6,13] and remark only that by using the BT between the SG and

$$u'_{xt} = \{1 - k^2 u_x^{\,2}\}^{1/2} \sin u' \tag{11}$$

which is

$$u'_x = k^{-1} \sin(u' - u)$$
$$u'_t = u_t + k \sin u' \tag{12}$$

one obtains the conserved densities of the SG

$$T^2 = u_x^{\,2},\ T^4 = u_{xx}^{\,2} - \frac{1}{4} u_x^{\,2},\ T^6 = u_{xxx}^{\,2} - \frac{5}{2} u_{xx}^{\,2} u_x^{\,2} + \frac{1}{8} u_x^{\,6}$$

$$T^8 = u_{xxxx}^{\,2} - \frac{7}{2} u_{xxx}^{\,2} u_x^{\,2} + \frac{7}{4} u_{xx}^{\,4} + \frac{35}{8} u_{xx}^{\,2} u_x^{\,4} + \frac{5}{64} u_x^{\,8},$$

$$T^{10} = \text{etc.} \tag{13}$$

Note that T^r has the rank r in that any term in $T^r, \Pi u_{nx}^{a_n}$, has rank $\sum n a_n = r$. The odd rank densities have been removed [13] because they are trivial. Note that, because of the assumed boundary conditions $u, u_x, u_{xx} \to 0$, etc., $d/dt \int_{-\infty}^{\infty} T\,dx = -\int_{-\infty}^{\infty} X_x\,dx = X(+\infty) - X(-\infty) = 0$ since a density T satisfies a conservation law $T_t + X_x = 0$. Thus $\int_{-\infty}^{\infty} T^n dx$ is a constant of the motion I_n (say). The existence of an infinite set of constants of the motion I_n is necesary but not sufficient for the SG to be a completely integrable (inifinite dimensional) Hamiltonian system. We return to this later.

A third equation, the nonlinear Schrödinger equation (NLS) [10]

$$i\,u_t + u_{xx} + 2u\,|u|^2 = 0 \tag{14}$$

has played an important role in the history of the subject. The equation of a 'simple wave' $u_t + uu_x = 0$ describes a nonlinear system without dispersion; the KdV describes a weakly dispersive, weakly nonlinear system; the NLS describes a strongly dispersive, weakly nonlinear system. The equation is of relatively recent origin [14]. Its role [10] in the development of the inverse scattering method is described below.

2. Discovery of the Inverse Spectral Transform

In §1 we mentioned three equations, the KdV, the SG and the NLS, important to the development of the inverse scattering method. The critical step was taken by KRUSKAL and co-workers [2,3] for the KdV. It seems worthwhile sketching the argumentation adopted by these authors as we have heard it [15] since it throws up a number of points. The argument is physically based and the mathematical content has emerged later following the work of LAX [16]. We indicated the nature of this in §3 following where we also connect the theory with the inverse scattering methods associated with Eqs. (10) and (8) introduced without further comment in §1.

KRUSKAL in the period 1955-65 was apparently concerned with the periodicity exhibited by one-dimensional lattices - the FERMI-PASTA-ULAM problem [17]. First results obtained with ZABUSKY appeared in 1965 [2]. The two authors were concerned with a continuum limit of a 1D-lattice which leads to

$$y_{tt} = y_{xx}\,(1 + \epsilon\, y_x). \tag{15}$$

The Riemannian invariants of the linear problem $u = 1/2(y_x + y_t) = 0(1)$, $v = 1/2(y_x - y_t) = 0(\epsilon)$ lead, for $\epsilon \ll 1$, to the one-way going wave $[\tau \equiv 4\epsilon(t-x)]$

$$u_\tau + u\,u_x = 0, \tag{16}$$

the simple wave. Characteristics of this equation are [18] $dx/d\tau = u =$ constant. These lines intersect and the system shocks (the equation can be integrated by the hodograph transformation [18]). It is interesting that (16) has an infinite set of polynominal conserved densities $T^n = u^n/n$: it has some of the structure of the integrable systems like the KdV equation (the $I_n \equiv \int T^n dx$ are the involution - see below). The addition of the dispersive term $[(\Delta x)^2/12]y_{xxxx}$, a further term in the Taylor expansion used to convert the discrete lattice model to a continuum, balances dispersion against nonlinearity and leads to $u + uu_x + (\delta^2/2\epsilon)u_{xxx} = 0$, the KdV. [For the record $\Delta x \sim 1/64$, $\epsilon \sim 1/20$ and $\delta^2 \equiv \Delta x^2/12\epsilon \sim (0.022)^2$.] Numerical solutions under periodic boundary conditions with a harmonic initial condition lead to periodic behavior in time. It is now known that more generally, under these conditions, the KdV is periodic in space and 'almost periodic' in time.

The simple wave shocks and with the small dispersive term added, the resultant KdV equation also develops a (smooth) jump. KRUSKAL and co-workers looked for conservation laws across the jump and found the polynominal conserved densities u [from the equation in the form $u_\tau + (1/2\ u^2 + \delta^2 u_{xx})_x = 0$] u^2 and $2u^3 - u_x^2$. They subsequently found ten polynominal conserved densities T^r: the rank of $\Pi u_{nx}^{a_n}$ in this case is $r = \sum(1 + n/2)a_n$. T^{10} has 32 terms $(1/10)u^{10} - 36u^7u_{1x}^2 - 630\ u^4u_{1x}^4 \ldots (419904/12155)\ u_{8x}^2$ [19].

FPU also looked at a lattice with quartic rather than cubic anharmonicity. This leads to $y_{tt} = y_{xx}(1 + \epsilon y_x^2)$ and thence to the modified KdV

$$v_t + v^2 v_x + v_{xxx} = 0. \tag{17}$$

This also has conservation laws. MIURA found the transformation $u = v^2 + \sqrt{-6}\ v_x$ connecting these. The remarkable $\sqrt{-6}$ is actually innocuous: the MIURA transformation [20] $u = v^2 + v_x$ is a BT connecting

$$\begin{aligned} u_t - 6\ uu_x + u_{xxx} &= 0 \\ v_t - 6\ v^2 v_x + v_{xxx} &= 0. \end{aligned} \tag{18}$$

The Ricatti transformation $v = \partial(\ell n\ \psi)/\partial\psi = \psi_x/\psi$ linearizes $v^2 + v_x$ and $u = \psi_{xx}/\psi$. The Galilean transformation

$$x' = x - Vt, \quad t' \to t, \quad u \to u + V \tag{19}$$

leaves $u_t - uu_x + u_{xxx} = 0$ invariant. Hence, $u \to u + \lambda = \psi_{xx}/\psi$ and the Schrödinger eigenvalue problem has emerged! Furthermore, $u = \psi_{xx}/\psi - \lambda$ into the KdV yields

$$\begin{aligned} &\text{(a)} \qquad \lambda_t = 0 \\ &\text{(b) a functional } B[\psi,u] \text{ such that } B[\psi,u] = \psi_t. \end{aligned} \tag{20}$$

GARDNER *et al.* [3] used these results and the property $u, u_x, u_{xx} \to 0$ as $|x| \to \infty$ to solve the KdV by the route $u(x,0)$ (initial data) $\to$ scattering data at $t = 0$ (by $B[\psi,u] = \psi_t$) scattering data at time $t \to$ [via the Gel'fand-Levitan-Marchenko linear integral equation (see, e.g., [13])] $u(x,t)$, the solution of the KdV. The method is now called the 'inverse spectral transform' (IST) because (a) $\lambda_t = 0$, i.e., the spectrum of the Schrödinger operator is invariant under the KdV flow, and (b) the method generalizes the Fourier transform for linear systems [21]. The remarkable discovery by KRUSKAL and co-workers has been the source of all the developments since, with the single exception of the direct methods developed by

HIROTA [22] and CAUDREY [22]. The precise connection of these with the IST is still to be established.

3. Operator Pair Formulation of Nonlinear Evolution Equations

LAX [16] stimulated an important development and added mathematical understanding to the IST. Given the relatively prime differential operators $u \to \hat{L}_u \equiv (-\partial^2/\partial x^2 + u)$, $\hat{B}_u \equiv \partial^{2n+1}/\partial x^{2n+1}$ + lower degree ($\hat{B}$ is skew symmetric, $\hat{B}^* = -\hat{B}$) such that there is a one-parameter family of unitary operators $\hat{U}$ satisfying $\hat{U}_t = \hat{B}\,\hat{U}$; and $\hat{L}_u$ is unitary equivalent under $\hat{U}$, i.e., $\hat{U}^{-1}(t)\,\hat{B}(t)\,\hat{U}(t)$ is independent of t:.

(i) The eigenvalues λ_u of $\hat{L}_u$ are integrals of the motion, i.e., $\lambda_{u,t} = 0$. [The proof is trivial: $\hat{L}_u(0) = \hat{U}^{-1}(t)\,\hat{L}_u(t)\hat{U}(t)$, $\hat{L}_u(0)\psi(0) = \lambda_u\psi(0)$. Then $\hat{L}_u(t)\psi(t) = \lambda_u\psi(t)$ with $\psi(t) = \hat{U}(t)\psi(0)$, i.e., $\hat{U}(t)$ is the evolution operator.]

(ii) $\frac{\partial}{\partial t}\,\hat{U}^{-1}(t)\,\hat{L}_u(t)\,\hat{U}(t) = 0 \Rightarrow$

$$-\hat{U}^{-1}\,\hat{U}_t\,\hat{U}^{-1}\,\hat{L}_u\,\hat{U} + \hat{U}^{-1}\,\hat{L}_{u,t}\,\hat{U} + \hat{U}^{-1}\,\hat{L}_u\,\hat{B}\,\hat{U} = 0.$$

Then $\hat{U}_t = \hat{B}\,\hat{U} \Rightarrow -\hat{U}^{-1}\,\hat{B}\,\hat{L}_u\,\hat{U} + \hat{U}^{-1}\,\hat{L}_{u,t}\,\hat{U} + \hat{U}^{-1}\,\hat{L}\,\hat{B}\,\hat{U} = 0$ and

$$\hat{L}_t = [\hat{B}, \hat{L}] \,. \tag{21}$$

(ii) Since $\hat{L}_t = u_t$, the operator equation (21) is an evolution equation $u_t = K[u]$ in which $K[u]$ is a functional of u. By trying for the skew symmetric operator $\hat{B} = \partial^3 + b\partial + \partial b$ ($\partial \equiv \partial/\partial x$), LAX found that with $b = -3u/4$ (our scaling is different from that used by LAX [16]) he regained from (21) the KdV equation. Further, the method generalizes to yield an infinite hierarchy of KdV equations of degrees 3 (the KdV), 5, 7, etc.

It is remarkable that LAX's operators can be found by defining the square root $\hat{R}$ of the operator $-\hat{L}$ by $\hat{R}^2 = -\hat{L}$ and $R = \partial + c_0 + c_1\partial^{-1} + c_2\partial^{-2} + \ldots$ with $\partial \equiv \partial/\partial x$). Then $-\hat{L}_u = \partial^2 - u$ and $R^2 = -L_u$ is $\partial^2 + 2c_0\partial + (c_{0,x} + c_0^2 + 2c_1) + (c_{1,x} + 2c_2 + 2c_0c_1)\,\partial^{-1} + \ldots$, so that $c_0 = 0$, $c_1 = -1/2\,u$, $c_2 = 1/4\,u_x$. Evidently, the principal part of $\hat{R}$ (that excluding inverse powers of ∂) is ∂ and $[\hat{R},\hat{L}] = L_t$ is

$$u_x = u_t. \tag{22}$$

However, the principal part of $\hat{R}^{3/2} = (-\hat{L}_u)\,\hat{R}$ is

$$\partial^3 - \partial u - \frac{1}{2}u\,\partial + \frac{1}{4}\,u_x = \partial^3 - (3/4)\,\partial u - (3/4)\,u\partial \tag{23}$$

and this is LAX's operator for the KdV. Other fractional powers $\hat{R}^{5/2}$, etc., generate LAX's hierarchy.

GELFAND and DIKII [23], and other workers (NOVIKOV, DRINFELD) referenced by MANIN [24], in particular show how to solve the KdV for its multi-soliton solutions on the real line $-\infty < x < \infty$ by reducing the problem to a system of algebraic equations. Analogous but more difficult methods are used by NOVIKOV [24,25] to solve the KdV under periodic boundary conditions. MANIN [24] notices interesting new properties in the multi-soliton solutions on the real line when the operators L, B are not relatively prime, i.e., are of degrees d (the order of the highest derivative) which are not relatively prime, 2 and 4 for example. He obtains remarkable 'geyseron' solutions in this case [24]!

4. Symplectic Structure

GARDNER [26] introduced the idea of a bracket (a Poisson bracket) but worked under periodic boundary conditions. For present purposes, define the functional or Frechet derivative $\delta F/\delta u$ of the functional $F[u] = \int_{\infty}^{\infty}\mathcal{F}[u]\,dx$ by

$$\frac{dF}{dt} = \int_{-\infty}^{\infty} \frac{\delta F}{\delta u}\frac{\partial u}{\partial t}\,dx \tag{24}$$

where t, usually the time, parametrizes u. Alternatively,

$$\lim_{\epsilon\to 0}\frac{\partial}{\partial\epsilon}\int_{-\infty}^{\infty}\mathcal{F}[u+\epsilon\delta u]\,dx = \int_{-\infty}^{\infty}\frac{\delta F}{\delta u}\,\delta u\,dx. \tag{25}$$

The bracket corresponding to that found by GARDNER is the unusual one

$$\{F, G\} = \int_{-\infty}^{\infty}\frac{\delta F}{\delta u}\frac{\partial}{\partial x}\frac{\delta G}{\delta u}\,dx. \tag{26}$$

GARDNER in effect proves

$$0 = \frac{dI_n}{dt} = \int_{-\infty}^{\infty}\frac{\delta I_n}{\delta u}\frac{\partial u}{\partial t}\,dx = \{I_n, I_m\} \tag{27}$$

if $u_t = \partial/\partial x\,(\delta I_m/\delta u)$. This evolution equation is a Hamiltonian flow with Hamiltonian the mth constant of the motion I_m of the KdV equation. Since $I_m = \int_{-\infty}^{\infty}T^m dx$ and $T^3 = 1/2(2u^3 - u_x^2)$, one finds that the KdV is a Hamiltonian flow. The constants of the motion I_m are in involution, there are precisely the right number of these, and the system constitutes a completely integrable infinite dimensional Hamiltonian system - the first example of this class.

To expand this a little more: with $\mathcal{H}_3 = 1/2(2u^3 - u_x^2) \equiv T^3$ and $p_x = u$

$$\begin{aligned}\frac{\partial u}{\partial t} &= \frac{\delta H_3}{\delta p} = 6uu_x - u_{xxx}\\ \frac{-\partial u}{\partial t} &= \frac{\partial}{\partial x}\left(\frac{\delta H_3}{\delta u}\right) = -6uu_x + u_{xxx}\end{aligned} \tag{28}$$

can be compared with the classical prescription for a Hamiltonian flow

$$\begin{pmatrix}\dot q\\ \dot p\end{pmatrix} = \begin{pmatrix}0 & 1\\ -1 & 0\end{pmatrix}\begin{pmatrix}\partial H/\partial q\\ \partial H/\partial p\end{pmatrix} = J\ \mathrm{grad}\ H. \tag{29}$$

In this, J is a linear map taking phase space to phase space. It is isometric (i.e., preserves inner products) and, and this is the key point, is *skew*. It connects the contravariant vector field $(\dot{q},\dot{p})$ with a covariant one (grad H). Phase space is symplectic: it carries a closed skew symmetric differential two-form ω [27]. For (29)

$$\begin{aligned}\omega &= \omega_{ij}\ \delta p^i \wedge \delta q^j \qquad (i,j = 1,2)\\ &= \delta p^1\ \delta q^2 - \delta q^1\ \delta p^2 \equiv \delta p \wedge \delta q.\end{aligned} \tag{30}$$

δp^1 and δp^2 (δq^1 and δq^2) are independent variations in p (q). The matrix J is ω_{ij} which is skew. The wedge product ($\wedge$) is (by definition [27]) a skew symmetric tensor $\delta p \wedge \delta q = - \delta q \wedge \delta p$.

The symplectic form ω is invariant (coordinate free) and closed, $d\omega = 0$. From Hamilton's principle if

$$\delta \int (p\ dq - H\ dt) = 0, \qquad \delta \int (\bar{p}\ d\bar{q} - \bar{H}\ dt) = 0 \tag{31}$$

describe the same trajectories in phase space,

$$\delta \int (p\ dq - Hdt - \bar{p}\ d\bar{q} + \bar{H}\ dt + dF) = 0$$

and

$$\delta(p\ dq) - \delta(\bar{p}\ d\bar{q}) + \delta\ dF = 0$$

so that

$$d(p\ \delta q) - d(\bar{p}\ \delta\bar{q}) + d\ \delta F = 0$$

and

$$dp\ \delta\ q - d\bar{p}\ \delta\bar{q} = \delta p\ dq - \delta\bar{p}\ d\bar{q}$$

or

$$dp\ \delta q - \delta p\ dq = d\bar{p}\ \delta\bar{q} - \delta\bar{p}\ d\bar{q} \tag{32}$$

i.e., $\omega = \bar{\omega}$ under the canonical transformation $p,q \to \bar{p},\bar{q}$, $H \to \bar{H}$. In this somewhat heuristic demonstration, we have used that $H(p,q) = \bar{H}(\bar{p},\bar{q})$, that is take the same values although their functional forms are different: hence $H\ dt = \bar{H}\ dt$ above.

In (28) the discrete variables q,p in (29) are replaced by the running set $u, p = \int_{-\infty}^{x} u(x')dx'$, for each x, and the symplectic form matrix ω_{ij} is replaced by the skew symmetric operator $\partial/\partial x$. This suggests the symplectic form

$$\omega = \int_{-\infty}^{\infty} dx \int_{-\infty}^{x} dy\ [\delta_1 u(x)\ \delta_2 u(y) - \delta_1 u(y)\ \delta_2 u(x)] \tag{33}$$

and the Poisson bracket (26). This bracket satisfies Jacobi's identity

$$\{\{A, B\}, C\} + \{\{C, A\}, B\} + \{\{B, C\}, A\} = 0, \tag{34}$$

a result which is conveniently proved via the identity

$$\{\{A, B\}, C\} = \int \left[\frac{\delta^2 A}{\delta u^2} \frac{\partial}{\partial x} \frac{\delta B}{\delta u} - \frac{\delta^2 B}{\delta u^2} \frac{\partial}{\partial x} \left(\frac{\delta A}{\delta u}\right)\right] \frac{\partial}{\partial x} \left(\frac{\delta C}{\delta u}\right) dx\ . \tag{35}$$

To prove involution of the constants of the KdV motion I_n it is sufficient to know [28] that there is a skew symmetric operator $K = - \partial^3 + 2u\partial + 2\partial u$ such that the vector field $u_t \equiv \partial_n u$ for the nth KdV flow (namely $u_t =$

$K_n[u] = [\hat{L}, \hat{B}_n]$ with $\hat{B}_n$ LAX's nth *skew* symmetric operator $\hat{B}$) satisfies $\partial^n_{n+1} u = K\, \delta H_n/\delta u$. For then

$$\{I_n, I_m\} = \int \frac{\delta I_n}{\delta u} \frac{\partial}{\partial x} \frac{\delta I_m}{\delta u}\, dx = \int \frac{\delta I_n}{\delta u} \partial_m u \; dx$$

$$= \int \frac{\delta I_n}{\delta u} K \frac{\delta I_{m-1}}{\delta u}\, dx = - \int K \frac{\delta I_n}{\delta u} \frac{\delta I_{m-1}}{\delta u}\, dx = - \int \partial_{n+1} u \frac{\partial I_{m-1}}{\partial u}\, dx$$

$$= - \int \frac{\partial}{\partial x} \frac{\delta I_{n+1}}{\delta u} \frac{\delta I_{m-1}}{\delta u}\, dx = \int \frac{\delta I_{n+1}}{\delta u} \frac{\partial}{\partial x} \frac{\delta I_{m-1}}{\delta u}\, dx = \{I_{n+1}, I_{m-1}\}, \quad (36)$$

and continuation down to $I_1 = u$ yields zero [28].

We shall take as our definition of a completely integrable (or integrable) system with $2n$ degrees of freedom such that (a) the system has n constants of the motion I_n, and (b) these are in involution, i.e., $\{I_n, I_m\} = 0$. Integrable systems can be given a more fundamental description and we provide this elsewhere [7]. For the KdV the *infinite* set of constants I_n in involution is not enough for complete integrability since *precisely* the right number of constants $n = \infty$ is required. However, the question of integrability was settled by ZAKHAROV and FADEEV [29] who explicitly and independently enunciated the fact that the KdV was an infinitely dimensional, completely integrable Hamiltonian system proved the symplectic form (35) is closed under the transformation which maps $u(x,t)$ via the Schrödinger eigenvalue problem, to the scattering data $w(t)$, ζ_ℓ and c_ℓ, and integrated the equations. The scattering data $S = \{w(t), \zeta_\ell, c_\ell\}$ are constituted as follows: $w(t)$ is the reflection coefficient for Jost functions [13] $\psi \sim e^{+ikx}$, $x \to -\infty$; ζ_ℓ is the ℓth bound state eigenvalue (which proves to lie on the positive imaginary axis in the ζ-plane) and c_ℓ is the normalization of the eigenfunction. This scattering data is sufficient to reproduce the potential $u(x,t)$ via the Gel'fand-Levitan-Marchenko equation. The Hamiltonian in terms of the scattering data proves to be of action angle type:

$$H = - \frac{8}{\pi} \int_{-\infty}^{\infty} k \log (1 - |w|^2)\, dk - \frac{32}{5} \sum_{\ell=1}^{m} \zeta_\ell^5$$

$$= 8 \int_{-\infty}^{\infty} k^3 P(k)\, dk - \frac{32}{5} \sum_{\ell=1}^{m} p_\ell^{5/2}. \quad (37)$$

This Hamiltonian reproduces the time evolution of the scattering data:

$$P(k)_{1t} = 0, \quad p_{\ell,t} = 0; \quad Q(k)_{1t} = 8 k^3, \quad q_{\ell,t} = -8 \zeta_\ell^3. \quad (38)$$

These results are not peculiar to the KdV (and the TODA lattice [30]). ZAKHAROV and SHABAT almost immediately [10] integrated the NLS equation $iu_t + u_{xx} + \chi u|u|^2 = 0$ in the LAX form $\hat{L}_t = i [\hat{L}, \hat{A}]$ with

$$\hat{L} = i \begin{bmatrix} 1+p & 0 \\ 0 & 1-p \end{bmatrix} \frac{\partial}{\partial x} + \begin{bmatrix} 0 & u^* \\ u & 0 \end{bmatrix}, \quad \chi = \frac{2}{1-p^2}$$

$$\hat{A} = -p\begin{bmatrix}1 & 0\\ 0 & 1\end{bmatrix}\frac{\partial^2}{\partial x^2} + \begin{bmatrix}|u|/(1+p) & iu_x^*\\ -iu_x & -|u|^2/(1-p)\end{bmatrix}. \tag{39}$$

The scattering problem $\hat{L}v = \zeta v$ with $v = [v_1, v_2]$ yields an infinite set of conserved densities. The NLS is also a Hamiltonian flow and proves to be completely integrable.

ABLOWITZ, KAUP, NEWELL and SEGUR [31] then provided a natural generalization: namely find $\hat{L}$, $\hat{A}$ such that

$$\hat{L}\,v = \zeta v, \qquad \hat{L} = \begin{bmatrix} i\partial/\partial x & -iq\\ ir & -i\partial/\partial x\end{bmatrix}$$

$$\hat{A}\,v = v_t, \qquad \hat{A} = \begin{bmatrix} A & B\\ C & -A\end{bmatrix}. \tag{40}$$

Choose A,B,C,q,r such that, under $\zeta_t = 0$, $\hat{L}_t = [\hat{A},L]$ is the required *pair* of evolution equations $q_t = K_1\,[q,r]$, $r_t = K_2\,[q,r]$ (here, $K_i\,[q,r]$ means functional of q and r). The potentials q,r form a natural canonical pair and the appropriate bracket is

$$\{F, G\} = \int_{-\infty}^{\infty}\left(\frac{\delta F}{\delta q}\frac{\delta G}{\delta r} - \frac{\delta F}{\delta p}\frac{\delta G}{\delta q}\right) dx\,. \tag{41}$$

An extension essentially due to CALOGERO [32] yields [33]

$$\frac{\partial}{\partial t}\,\hat{\sigma}\,s(x,\underset{\sim}{y},t) + h(\hat{L}^{\dagger},\underset{\sim}{y},t)\cdot\frac{\partial}{\partial \underset{\sim}{y}}\,\hat{\sigma}\,s(x,\underset{\sim}{y},t)$$
$$+ 2\,\Omega\,(\hat{L}^{\dagger},\underset{\sim}{y},t)\;s(x,\underset{\sim}{y},t) = 0 \tag{42a}$$

where

$$s = \begin{pmatrix} r\\ q\end{pmatrix}, \qquad \hat{\sigma} = \begin{bmatrix}1 & 0\\ 0 & -1\end{bmatrix}$$

$$\hat{L}^{\dagger} = \frac{1}{2i}\begin{bmatrix}\partial/\partial x - 2r\int_{-\infty}^{x} dx'q & 2r\int_{-\infty}^{x} dx'r\\ -2q\int_{-\infty}^{x} dx'q & -\frac{\partial}{\partial x} + 2q\int_{-\infty}^{x} dx'r\end{bmatrix}, \tag{42b}$$

as an integrable pair of evolution equations in x,t and any number of 'time-like' variables $\underset{\sim}{y}$. The extended LAX-AKNS operator pair formulation is

$$\Delta v\,(x,\underset{\sim}{y},t,\zeta) = \hat{A}\,v(x,\underset{\sim}{y},t,\zeta)$$
$$\Delta = \frac{\partial}{\partial t} + \underset{\sim}{h}\,(\zeta,\underset{\sim}{y},t)\cdot\frac{\partial}{\partial \underset{\sim}{y}}$$
$$\hat{L}\,v \equiv \begin{bmatrix} i\partial/\partial x & -iq\\ ir & -i\partial/\partial x\end{bmatrix} v = \zeta\,v \tag{43}$$

and $\Delta\hat{L} = [\hat{L},\hat{A}]$ with $\Delta\zeta = 0$. We have reported a BT for this system [34] and its canonical structure is established [33]. No physical application of this extended formalism is known to us. An extension to bigger $N \times N$

matrices has brought some mathematical solutions - the boomerons - with remarkable properties. This ingenious extension is due to CALOGERO and DEGASPERIS [32] who exploit Wronskian relations for example an $N \times N$ Schrödinger problem (in which u becomes an $N \times N$ matrix) to obtain linear equations of motion for the matrix reflection coefficients and to which there corresponds a nonlinear evolution equation for u. NEWELL [35] has recently extended the AKNS [31] scheme to $N \times N$ matrices and generalized the canonical structure to this case.

As an early example of the use of the AKNS 2×2 inverse scattering scheme consider the 'reduced MAXWELL-BLOCH' system of equations [36]

$$E_t = -s, \qquad r_x = -\mu s$$
$$s_x = Eu + \mu r, \qquad u_x = -Es. \tag{44}$$

It is easy to check that

$$A = \frac{-i\zeta}{4\zeta^2-\mu^2}\,u, \qquad B = \frac{-i\zeta}{4\zeta^2-\mu^2}\left[s + i\,\frac{\mu r}{2\zeta}\right]$$

$$C = \frac{-i\zeta}{4\zeta^2-\mu^2}\left[s - i\,\frac{\mu r}{2\zeta}\right], \qquad r = \frac{1}{2}E = -q = -q^* \tag{45}$$

with $r_x = -\mu s$ [do not confuse the q,r potentials with the three dependent variables r,s,u in (44)], reproduces (44) in the form $\hat{L}_t = [\hat{A},\hat{L}]$ under the condition $\zeta_t = 0$. The N-soliton solution is [36]

$$E^2 = 4\,\frac{\partial^2}{\partial x^2}\,\ln\det|M|$$

$$M_{nm} = \frac{2}{E_m+E_n}\cosh\left[\frac{1}{2}\,(\theta_n+\theta_m)\right]$$

$$\theta_n = \frac{1}{2}\,E_n\,(x - 4\,[E_n^2 + 4\mu^2]^{-1}\,t + \delta_n)\,. \tag{46}$$

Note that when $\mu = 0$ the system is the SG: equations (44) are $E_t = -s$, $s_x = Eu$, $u_x = -Es$. Put $s = -\sin\sigma$, $u = -\cos\sigma$, $E = \sigma_x$; the $\sigma_{xt} = \sin\sigma$. In this case

$$\hat{A} = \frac{i}{4\zeta}\begin{bmatrix}\cos\sigma & \sin\sigma\\ \sin\sigma & -\cos\sigma\end{bmatrix} \tag{47}$$

with q and r in the scattering problem (40) given by $q = -\sigma_x/2 = -r$, This is the form obtained from LUND's result (8) when $\lambda \equiv 0$. The $\hat{L},\hat{A}$ pair agree with the results obtained from the BT (5) quoted in (10). This pair was given first by ABLOWITZ, KAUP, NEWELL and SEGUR [37].

Whilst the BT thus shows that the $\hat{L},\hat{A}$ pair for the SG is a natural pair it is important to notice that this extension of the LAX pair formation by AKNS is critically different from LAX's original formulation and from its extensions by ZAKHAROV and SHABAT [10] and those reported by MANIN [24] .

For the sine-Gordon equation, a LAX pair of differential operators is a pair of 4×4 first-order (= first degree) differential operators [38]. In the 2×2 formulation in the AKNS scheme $\hat{A}$ is not a differential operator [see (47)] and it involves the eigenvalues of the scattering problem $\hat{L}\,v = \zeta\,v$. Thus, if (i) $[\hat{A},\hat{L}] = \hat{L}_t$, (ii) $\hat{L}v = \zeta v$, (iii) $\zeta_t = 0$, then $\hat{L}(Av - v_t) = \zeta\,(\hat{A}v - v_t)$ and $(\hat{A}v - v_t)$ is an eigenvalue belonging to ζ ,i.e, $\hat{A}v - v_t = \lambda(\zeta)v$ or $\lambda(\zeta) = 0$. It is easy to see that ζ^{-1} in $\hat{A}$ acts as an integral operator by calculating $[\hat{L},\hat{A}]$ from (47) explicitly. One finds

$$[\hat{L},\hat{A}] = \frac{-\sin\sigma}{2\zeta}\begin{bmatrix}0 & 1\\ 1 & 0\end{bmatrix}\begin{bmatrix}\partial/\partial x & \frac{1}{2}\sigma_x\\ \frac{1}{2}\sigma_x & -\partial/\partial x\end{bmatrix}$$

$$= i\,\frac{\sin\sigma}{2}\begin{bmatrix}0 & 1\\ 1 & 0\end{bmatrix} = i\begin{bmatrix}0 & \frac{1}{2}\sigma_{xt}\\ \frac{1}{2}\sigma_{xt} & 0\end{bmatrix} = \hat{L}_{1t} \tag{48}$$

The first step uses $Lv = \zeta v$ to cancel the ζ^{-1}.

5. Theory of the SG

The rest of this article is concerned with developing the theory of the SG in detail since the result is relevant, in particular,to the work of BISHOP [39] and TRULLINGER [40] and the work of this meeting.

The SG is a Hamiltonian flow: the two potentials q,r in the AKNS generalization (40) are $q = -\sigma_x/2$ $r = \sigma_x/2$ and $q = -r = -r^*$. The pair cannot constitute an independent pair of canonical variables. However, in this case, one finds quite generally [33] that

$$H(r,q) = H(q) - \frac{1}{2}\int \frac{\delta H}{\delta q}(q - r_{1x})\,dx.$$

Then

$$q_t = \frac{\delta H}{\delta r} = -\frac{1}{2}\frac{\partial}{\partial x}\frac{\delta H}{\delta q}$$

$$r_t = -\frac{1}{2}\frac{\delta H}{\delta q} + \frac{1}{2}\frac{\delta^2 H}{\delta q^2}(q - r_{1x}). \tag{49}$$

Thus, if $q = r_{1x}$, i.e., $r = \int_{-\infty}^{x} q\,dx'$ initially, then it is true for all time. This again indicates in a natural way that the correct choice for the Poisson bracket is GARDNER's choice (26) and that for the symplectic form is (33).

With this in mind, a symmetrized Hamiltonian for the SG is [13,33]

$$H = \frac{1}{4}\gamma_0^{-1}\int_{-\infty}^{\infty} dx\,[\cos 2\gamma_0\,p + \cos\{-2\int_{-\infty}^{x} q(x',t)dx'\} - 2]$$

with $p = 1/2\,\gamma_0\sigma$, $q = -1/2\,\sigma_x$. From this

$$-\frac{1}{2}\sigma_{xt} = q_t = \frac{\delta H}{\delta p} = -\frac{1}{2}\sin 2\gamma_0\,p = -\frac{1}{2}\sin\sigma$$

$$\frac{1}{2}\gamma_0^{-1}\sigma_t = p_t = -\frac{\delta H}{\delta q} = -\frac{1}{2}\gamma_0^{-1}\int_x^{\infty}\sin'\{-2\int_{-\infty}^{x} q(x'')dx''\}dx''$$

or
$$\frac{1}{2}\gamma_0^{-1}\sigma_{tx} = \frac{1}{2}\gamma_0^{-1}\sin\sigma \ . \tag{50}$$

The dimensionless number γ_0 is a coupling constant: note that it does not play a role in the equations of motion. It does play a role in the definition of the momentum conjugate to q. Thus, it plays a role in the Poisson brackets and is fundamental to the canonical quantization of the SG equation.

The scattering problem (40) for the SG is studied in terms of Jost functions. Consider the Jost function solutions for v, φ and $\bar{\varphi}$, such that

$$\varphi \sim \begin{bmatrix}1\\0\end{bmatrix} e^{-i\xi x} \ , \qquad \bar{\varphi} \sim \begin{bmatrix}0\\1\end{bmatrix} e^{i\xi x} \ , \quad x \to -\infty \ ;$$

$$\psi \sim \begin{bmatrix}0\\1\end{bmatrix} e^{i\xi x} \ , \qquad \bar{\psi} \sim \begin{bmatrix}1\\0\end{bmatrix} e^{-i\xi x} \ , \quad x \to +\infty \ . \tag{51}$$

The eigenvalue ξ is real. Since there can be only two linearly independent functions belonging to ξ

$$\varphi = a(\xi)\,\bar{\psi}\,(x,\xi) + b(\xi)\,\psi(x,\xi)$$
$$\bar{\varphi} = \bar{b}(\xi)\,\bar{\psi}\,(x,\xi) + \bar{a}(\xi)\,\psi(x,\xi) . \tag{52}$$

The quantities $w \equiv b(\xi)/a(\xi)$ and $1/a(\xi)$ are the reflexion coefficient and transmission coefficient for functions

$$\bar{\psi} \sim \begin{bmatrix}1\\0\end{bmatrix} e^{-i\xi x}, \text{ as } x \to \infty \ .$$

Both w and a^{-1} can be continued into the upper half ζ-plane: the zeros of $a(\zeta)$ there define the bound states of the scattering problem (40). The key result is that $\hat{A}[(1/a)\varphi] = \partial/\partial t\ [(1/a)\varphi] \Rightarrow a$ is a constant of the motion. We demonstrate this explicitly to show what is going on: we have for $x \to +\infty$ that

$$\frac{\partial}{\partial t}\frac{1}{a}\begin{bmatrix}1\\0\end{bmatrix} = (A_0 + f)\,\frac{1}{a}\begin{bmatrix}1\\0\end{bmatrix}$$

$$\frac{\partial}{\partial t}\frac{b}{a}\begin{bmatrix}0\\1\end{bmatrix} e^{i\xi x} = \begin{bmatrix}A_0 + f & 0\\ 0 & -A_0 + f\end{bmatrix}\left\{\begin{bmatrix}0\\1\end{bmatrix} e^{-i\xi x} + \frac{b}{a}\begin{bmatrix}1\\0\end{bmatrix} e^{+i\xi x}\right\} \tag{53}$$

from which

$$A_0 = -f, \quad a(\xi,t) = a(\xi,0)$$
$$b(\xi,t) = b(\xi,0)\, e^{-2A_0 t} \tag{54}$$

and these results can be continued for $\text{Im}\,\zeta > 0$. Notice that $\hat{A}$ is replaced by $\hat{A} + f(\zeta,t)\hat{I}$, but $[f\ \hat{I},\hat{L}] = 0$ in agreement with the analysis above (48). Further, only $\lim x \to +\infty$ of $\hat{A}$ is needed. Since the potential $\sigma_x/2$ vanishes for the SG there, $\lim_{x\to\infty}\hat{A} = \hat{A}_0$ is essentially the linearized dispersion relation. Indeed from (47), since $\sigma \to 0(\text{mod}2\pi), |x|\to\infty$,

$A_0 = (i/4\zeta)$ diag [1, -1], whilst the linearized dispersion relation of the SG (6) is $\omega = k^{-1} \to A_0 = i/4\zeta$ [see (10)]. The situation is quite general: for example, for the RMB equations (44), $B,C \to 0$ as $x \to \infty$ since $r,s \to 0$ as $x \to \infty$. Then $u \to -1$ and $A_0 = i\zeta\,[4\zeta^2 - \mu^2]^{-1} \equiv \Omega(\zeta)$. In general $\Omega(\zeta) = -1/2\, i\omega_q(-2\zeta) = 1/2\, i\omega_r(2\zeta)$ in terms of the linear dispersion relations for q and r of (40) [33]. It is easily checked that $\omega_q(k) = \omega_r(k) = k[k^2-\mu^2]^{-1}$ for the RMB equations. It is clear that $\Omega(L^{\dagger},\underset{\sim}{\chi},t)$ is the natural extension of $\Omega(L^{\dagger})$ and hence of the linearized dispersion relation in (42).

With $a(\zeta)$ a constant of the motion established by (54), it is expressed as the Wronskian $a(\zeta) = W(\ ,\psi) \equiv \varphi_1\psi_2 - \varphi_2\psi_1 \sim \varphi_1 e^{i\zeta x}_{\sim}$ for $x \to \infty$. In $W(\varphi,\psi)$, φ_1 and φ_2 are the *components* of φ, i.e., $\underset{\sim}{\varphi} = [\varphi_1,\varphi_2]$ and similarly for ψ. From $Lv = \zeta v$, the AKNS scattering problem (40) in terms of the potentials q,r, one finds that $\Phi \equiv \ln(\varphi_1 \exp i\zeta x)$ satisfies

$$2\,i\zeta\,\Phi_{1x} = \Phi_{1x}^{\;2} - q\,r + q\,\frac{\partial}{\partial x}\,(q^{-1}\,\Phi_{1x}). \tag{55}$$

As $|\zeta| \to \infty$, $\Phi \to 1$ so

$$\Phi_{1x} \sim \sum_{n=1}^{\infty} \mathscr{H}_n\,(2i\zeta)^{-n} \tag{56}$$

(say). Then from (55)

$$\mathscr{H}_{n+1} = q\,\frac{\partial}{\partial x}\,(q^{-1}\,\mathscr{H}_n) + \sum_{j+k=n} \mathscr{H}_j\,\mathscr{H}_k, \quad n = 1,2,\ldots.$$

$$\mathscr{H}_0 = -\,q\,r\,. \tag{57}$$

As $x \to \infty$, $\ln a(\zeta) \sim \Phi$ and since $a(\zeta) \sim 1$ as $|\zeta| \to \infty$

$$\ln a(\zeta) \sim \sum_{n=1}^{\infty} \zeta^{-n}\,(2i)^{-n} \int_{-\infty}^{\infty} \mathscr{H}_n\,dx = \sum_{n=1}^{\infty} \zeta^{-n}\,H_n. \tag{58}$$

Then $\ln a(\zeta)$ a constant of the motion yields $H_n = (2i)^{-n}\int_{-\infty}^{\infty}\mathscr{H}_n dx$ are constants of the motion and the $\mathscr{H}_n$ are conserved densities. For the SG $q = -r = -\sigma_x/2$ and these densities are T^2, T^4, etc., in agreement with (13).

By expansion about the ordinary point $\zeta = 0$ one finds a second infinite set of conserved densities. The point $\zeta = 0$ is chosen since it is a pole of the linearized dispersion relation for the SG $\Omega(\zeta) = \zeta/4$. This choice makes $\Omega(\zeta)$ take its proper form in the evolution equation described by Hamilton's equations of motion. For the SG this is just the SG itself, Eq.(6), in light cone coordinates and evolution equation form. This analysis for the SG is given in Refs.[13] and [33]. However, there is a trivial route to the first conserved density. Observe that $T^2 \equiv \sigma_x^{\;2}/2$ satisfies

$$\frac{1}{4}\,\gamma_0^{-1}\,(\sigma_x^{\;2})_t + \frac{1}{2}\,\gamma_0^{-1}\,\{\cos\sigma - 1\}_x = 0 \tag{59}$$

because $\sigma_{xt} = \sin\sigma$. But this SG equation is invariant under interchange of x and t (Lie invariance [41]) and therefore

$$\frac{1}{2}\,\gamma_0^{-1}\,\{\cos\sigma - 1\}_t + \frac{1}{4}\,\gamma_0^{-1}(\sigma_t^{\;2})_x = 0. \tag{60}$$

The density $\frac{1}{2}\gamma_0^{-1}\{\cos\sigma - 1\}$ is the first of the sequence obtained by expansion of $\ell n\, a(\zeta)$ about $\zeta = 0$ [13,33]. We now have in $\frac{1}{4}\gamma_0^{-1}\sigma_x^2$ a momentum density and in $\frac{1}{2}\gamma_0^{-1}\{\cos\sigma - 1\}$ a Hamiltonian density. We wish to express these in terms of the scattering data for the following reason: the scattering problem $Lv = \zeta v$ in (40) is again a canonical transformation; and the Hamiltonian in terms of scattering data again takes the simpler form of action-angle type depending only on the momenta. This is exactly the situation found by ZAKHAROV and FADEEV for the KdV [29] [compare (37)]. The expression of the Hamiltonian H and the momentum P in terms of scattering data is achieved by expressing $\ell n\, a$ in terms of the scattering data. This is done through CAUCHY's theorem [33]. The result, which is (65) below, shows that the SG $\sigma_{xt} = \sin\sigma$ is a completely integrable infinite dimensional Hamiltonian system. Then by (66) the SG (2) is also.

The symplectic form (33) in the case $u = q$, $p = -\gamma_0 \int_{-\infty}^{x} q\, dx'$ transforms into [13,33]

$$\bar{\omega} = \sum dp_j \wedge dq_j \quad \int_{-\infty}^{\infty} d\xi [d\, P(\xi) \wedge dQ(\xi)]. \tag{61}$$

One set of complex canonical coordinates is defined in terms of the scattering data [13,33] by

$$\begin{aligned} p_j &= \gamma_0\, \ell n\, \zeta_j\,, \quad q_j = 2\, \ell n\, b_j\,, \quad j = 1,2,\ldots, M. \\ P(\xi) &= (2\pi\, \xi\gamma_0)^{-1}\, \ell n\, \{a(\xi)\bar{a}(\xi)\}, \quad Q(\xi) = \arg b(\xi). \end{aligned} \tag{62}$$

The symmetrized Hamiltonian and momentum derived from the densities $\frac{1}{2}\gamma_0^{-1}$ $(\cos\sigma - 1)$ and $\frac{1}{2}\gamma_0^{-1}\sigma_x^2$, respectively, can be put in the forms [compare (50) and note we have now introduced a mass m in H]

$$\begin{aligned} H(p,q) &= \frac{m^2}{4\gamma_0} \int_{-\infty}^{\infty} \left\{\cos(-2 \int_{-\infty}^{x} q\, dx') + \cos\, (2\gamma_0\, p) - 2\right\} dx \\ P(p,q) &= \frac{1}{2\gamma_0} \int_{-\infty}^{\infty} (q^2 + \gamma_0^2\, p_x^2)\, dx. \end{aligned} \tag{63}$$

By expressing these conserved quantities in terms of the scattering data [13,33] they can be expressed in the forms

$$\begin{aligned} H &= -\frac{m^2 i}{2\gamma_0} \left\{-2 \sum_{j=1}^{K} (\zeta_j^{-1} - \xi_j^{*-1}) + 2i \sum_{j=1}^{L} \eta_j^{-1} + i\gamma_0 \int_{-\infty}^{\infty} d\xi\, \xi^{-1} P(\xi)\right\} \\ \mathbf{P} &= + \frac{2i}{\gamma_0} \left\{-2 \sum_{j=1}^{K} (\zeta_j - \zeta_j^{*}) - 2i \sum_{j=1}^{L} \eta_j - i\, \gamma_0 \int_{-\infty}^{\infty} d\xi\, \xi\, P(\xi)\right\} \end{aligned} \tag{64}$$

in which $2K + L = M$ [compare (62)]. The result for P suggests new momenta different from those quoted in (62) for which

$$\begin{aligned} H &= \sum_{j=1}^{K} \hat{h}_j + \sum_{j=1}^{L} h_j + \int_{-\infty}^{\infty} d\xi\, h(\xi)\, P(\xi) \\ P &= \sum_{j=1}^{K} \hat{P}_j + \sum_{j=1}^{L} p_j + \int_{-\infty}^{\infty} d\xi\, p(\xi)\, P(\xi) \end{aligned} \tag{65}$$

where $h_j p_j = 4m^2\gamma_0^{-2}$, $\hat{h}_j\hat{p}_j = 64\, m^2\gamma_0^{-2}\sin^2\theta$, $h(\xi)p(\xi) = m^2$ and $\theta_j = \arg\zeta_j$. These momenta with corresponding coordinates $\hat{q}_j, q_j$, etc, also close (61). Define now still new energies and momenta by $h(\xi) = 1/2\{h'(\xi) + p'(\xi)\}$, $p(\xi) = 1/2\{h'(\xi) - p'(\xi)\}$ and define similar quantities for h_j, p_j and $\hat{h}_j, \hat{p}_j$. Define a new mass $m' = 2m$ and drop the primed notation. Then

$$H = \sum_{j=1}^{K} [256m^2\gamma_0^{-2}\sin^2\theta_j + \hat{p}_j^2]^{1/2} + \sum_{j=1}^{L} [64m^2\gamma_0^{-2} + p_j^2]^{1/2} + \int_{-\infty}^{\infty} [m^2 + p^2(\xi)]^{1/2}\, P(\xi)\, d\xi$$

$$P = \sum_{j=1}^{K} \hat{p}_j + \sum_{j=1}^{L} p_j + \int_{-\infty}^{\infty} d\xi\, p(\xi)\, P(\xi). \tag{66}$$

The expression for H in this form was first achieved by TAKTADJAN and FADEEV [38] who used the LAX pair of first-order 4×4 matrix operators mentioned earlier. The route used here is via the 2×2 ZS-AKNS scattering problem (40). We can follow it because in light cone coordinates the SG (2) is the evolution equation $u_t = \int_{-\infty}^{x} \sin u\, dx'$ in which the right side is a functional of u.

The expression (66) transforms the Hamiltonian of (63) to that for a collection of *free* relativistic particles of masses m, $8m\,\gamma_0^{-1}$ (the SG kinks) and $16m\,\gamma_0^{-1}\sin\theta$ (the SG breathers). The transformation is canonical and the result exact. It is an easy matter to canonically quantize it since the canonical coordinates q_j, $\hat{q}_j$, $q(\xi)$ can be found from the symplectic form. One finds the canonical pair $\gamma_0^{-1}\theta_j$, $4 \arg b_j$ for the internal co-ordinates of the breather. The phase space $0 \leq \theta_j \leq \pi/2$, $0 \leq 4 \arg b_j \leq 8\pi$ is compact with volume $4\pi^2\gamma_0^{-1}$. The corresponding breather spectrum in (65) is therefore discrete. One finds [13] a discrete set of breather masses for (66) given by

$$M_n = \frac{16m}{\gamma_0} \sin \frac{n\gamma_0}{16} \qquad n = 1, 2, \ldots, N$$

where N is the largest integer $\leq 8\pi\gamma_0^{-1}$. This beautiful result has been reached by a number of authors by very different routes [42].

For the purposes of this meeting on nonlinear structure and dynamics of condensed matter, the SG Hamiltonian in the form (66) provides an excellent model for the statistical mechanics of a soliton system. It is clear that the solitons and breathers are new elementary excitations; the continuum contribution is phonon-like. The complication in (66) is that although the Hamiltonian is separable, it depends on the initial data. In particular, the number of solitons L, of breathers K, and the occupation of phonon-like state $P(\xi)$ depends on the initial data. It is therefore necessary to trace over all sets of available initial data, namely over the set of C^∞ functions decaying exponentially at $x = \pm\infty$. BISHOP [39] takes the somewhat different view involving (x,t) space rather than momentum space where interactions between solitons, breathers and 'phonons' appear as phase shifts and emptied phonon states. It remains an interesting technical problem to ally the two different points of view.

Notice that, because the soliton and breather masses depend on γ_0^{-1}, perturbation theory in γ_0 must be singular perturbation theory above sine-Gordon solitons and breathers (see [43] and [44]).

Finally, we note that we cannot hope to indicate all the ways in which the inverse spectral transform method is now being extended. We simply refer the reader to the paper [35] by NEWELL, recent work on the prolongation structure method of WAHLQUIST and ESTABROOK by DODD [45] and others, the work quoted by MANIN [24] and two remarkable papers by ZAKHAROV [46 , 47], especially.

References

1. J.S. Russell: "Report on Waves", British Association Reports (1844)
2. N. Zabusky, M.D. Kruskal : Phys. Rev. Lett. 15, 240 (1965)
3. C.S. Gardner, J.M. Greene, M.D. Kruskal, R.M. Miura: Phys. Rev. Lett. 19, 1095 (1967)
4. G.L. Lamb: Rev. Mod. Phys. 43, 99 (1971), and references therein
5. J. Liouville: J. Mathematiques Pures et Appliquées (Paris) 18 (1) 71-72 (1853)
6. R.K. Dodd, R.K. Bullough: Proc. Roy. Soc. (London) A 352, 481 (1977)
7. R.K. Dodd, R.K. Bullough: "Integrability of nonlinear evolution equations: prolongations and solitons", to accompany Proc. of the Chalmers Symposium on *Solitons and their Applications in Science and Technology*, Göteborg June 1978 (to be published in Physica Scripta, 1979)
8. Details communicated by W. Shadwick, Warsaw Meeting, September (1977)
9. F. Lund: Phys. Rev. Lett. 38, 1175 (1977); Proc. of NATO Advanced Study Institute on *Nonlinear Problems in Physics and Mathematics* (Istanbul, August 1977), ed. by A.O. Barut (D. Reidel Publishing Co., Dordrecht, Holland 1978)
10. V.E. Zakharov, A.B. Shabat: Zh. Eksp. Teor. Fis. (Soviet) 61, 118 (1971); JETP (Soviet) 34, 62 (1972)
11. K. Pohlmeyer: Comm. Math. Phys. 46, 207 (1976); *New Developments in Quantum Field Theory and Statistical Mechs.*, ed. by M. Levy and P. Nitter (Plenum Press, New York 1977) p.339
12. H.D. Wahlquist, F.B. Estabrook: J. Math. Phys. 16, 1 (1975); 17, 1293 (1976)
13. R.K. Bullough, R.K. Dodd: In *Synergetics. A Workshop*, Proc. Int. Workshop on Synergetics at Schloss Elman, Bavaria, May 1977, ed. by H. Haken (Springer-Verlag, Berlin, Heidelberg, New York 1977) pp.92-119
14. V.L. Ginzburg, L.P. Pitaevskii: JETP 34, 1240 (1958); L.P. Pitaevskii: JETP 35, 408 (1968); R.Y. Chiao, E. Garmire, C.H. Townes: Phys. Rev. Lett. 13, 479 (1964); P.L. Kelley: Phys. Rev. Lett. 15, 1005 (1965); T.B. Benjamin, J.E. Feir: J. Fluid. Mech. 27, 417 (1966); D.J. Benney, A.C. Newell: J. Math. Phys. 46, 133 (1967); V.I. Bespalov, A.G. Litvak, V.I. Tulanov: *Nanka* , 2nd All-Union Symposium on Nonlinear Optics, Collection of Papers, (Russian) (Moscow, 1968)
15. M.D. Kruskal: In *Proc. of Symposium on Nonlinear Evolution Equations Solvable by the Inverse Spectral Transform*, Accademia dei Lincei, Rome June 1977, ed. by F. Calogero (Pittman, London 1978); Proc. of NATO Advanced Study Institute on *Nonlinear Problems in Physics and Mathematics* (Istanbul, August 1977), ed. by A.O. Barut (D. Reidel Publishing Co., Dordrecht, Holland 1978)

16. P.D. Lax, Comm. Pure Appl. Maths. 21, 467 (1968)
17. E. Fermi, J.R. Pasta, S.M. Ulam: *Studies of Nonlinear Problems*, Vol. 1, Los Alamos Rept. LA-1940, (May 1955); *Collected Works of E. Fermi*, Vol. 2, (Univ.of Chicago Press, 1965) pp.978-88
18. R.K. Bullough: In Proc. of NATO Advanced Study Institute on *Nonlinear Problems in Physics and Mathematics* (Istanbul, August 1977), ed. by A.O. Barut (D. Reidel Publishing Co., Dordrecht, Holland 1978)
19. R.M. Miura, C.S. Gardner, M.D. Kruskal: J. Math. Phys. 9, 1204 (1968)
20. R.M. Miura: J. Math. Phys. 9, 1202 (1968)
21. F. Calogero: "Nonlinear evolution equations solvable by the inverse spectral transform", invited lecture presented at Int. Conf. on Mathematical Problems in Theoretical Physics, Rome University (June 6-15, 1977); A.C. Newell:"The Inverse Scattering Transform" in *Solitons*, ed. by R.K. Bullough, P.J. Caudrey, Topics in Current Physics (Springer, Berlin, Heidelberg, New York 1979)
22. R. Hirota: In, for example,"Direct Methods in Soliton Theory" in *Solitons*, ed. by R.K. Bullough, P.J. Caudrey, Topics in Current Physics (Springer, Berlin, Heidelberg, New York 1979) ; P.J. Caudrey: Proc. of NATO Advanced Study Institute on *Nonlinear Problems in Physics and Mathematics* (Istanbul, August 1977), ed. by A.O. Barut (D. Reidel Publishing Co., Dordrecht, Holland 1978); R. Hirota: Phys. Rev. Lett, 27, 1192 (1971); P.J. Caudrey, J.C. Eilbeck, J.D. Gibbon: J. Inst. Maths. Applics. 14, 375 (1974)
23. I.M. Gel'fand, L. Dikii: Uspeki mat. nank. 30, 67 (1975); Russian Maths. Surveys 30, 77 (1975); Funkt, Anal. i Ego Prilog. 10, 18 (1976)
24. Yu, I. Manini: Itogi Nanki i Tekniki 11, 5 (1978)
25. S.P. Novikov, Funkt. Anal. i Ego Prilozh. 8:3, 54 (1974); B.A. Dubrovin, I.M. Kricheven, S.P. Novikov: Dokl. AN SSSR (1976)
26. C.S. Gardner: J. Math. Phys. 12, 1548 (1971)
27. H. Flanders: *Differential Forms with Applications to the Physical Sciences* (Academic Press, New York 1963)
28. This argument and some earlier remarks were stimulated by access to notes by Prof. David Simms on Lectures by H.P. McKean at Calgary, 1978. We have not been able to check original sources and rely on our recollections
29. V.E. Zakharov, L.D. Fadeev: Funkt. Anal. i Ego Prilozh. 5, 18 (1971)
30. M. Toda: In *Studies on a Nonlinear Lattice*, Ark. for Der Fysiske Sem., Trondheim 2 (1974); "On a Nonlinear Lattice - the Toda Lattice" in *Solitons* ed. by R.K. Bullough, P.J. Caudrey. Topics in Current Physics (Springer, Berlin, Heidelberg, New York 1979)
31. M.J. Ablowitz, D.J. Kaup, A.C. Newell, H. Segur: Phys. Rev. Lett. 31, 125 (1973); Studies in Appl. Maths. 53 No.4, 249 (1974)
32. F. Calogero, A. Degasperis: In *Solitons*, ed. by R.K. Bullough, P.J. Caudrey Topics in Current Physics (Springer, Berlin, Heidelberg, New York 1979) ; Proc. NATO Advanced Study Institute on *Nonlinear Problems in Physics and Mathematics* (Istanbul, August 1977), ed. by A.O. Barut (D. Reidel Publishing Co., Dordrecht, Holland 1978)
33. R.K. Dodd, R.K. Bullough: In *The Generalized Marchenko Equation and the Canonical Structure of the AKNS-ZS Inverse Method* (to be published 1978)
34. R.K. Dodd, R.K. Bullough: Phys. Lett. 62A, 70 (1977)
35. A.C. Newell: "The general structure of integrable evolution equations" (Preprint 1977)
36. J.D. Gibbon, P.J. Caudrey, R.K. Bullough, J.C. Eilbeck: Lett. al Nuovo Cimento 8, 775 (1973)

37. M.J. Ablowitz, D.J. Kaup, A.C. Newell, H. Segur: Phys. Rev. Lett. 30, 1262 (1973)
38. L.A. Taktadjan, L.D. Fadeev: Teor. Mat. Fis. 21, 160 (1974)
39. A.R. Bishop: these Proceedings
40. S. Trullinger: In discussion at this Symposium
41. R.K. Dodd, R.K. Bullough: Proc. Roy. Soc. (London) A 351, 499 (1976)
42. R.F. Dashen, B. Hasslacher, A. Neveu: Phys. Rev. D 11, 3424 (1975); V.E. Korepin, L.D. Fadeev: Teor. Mat. Fis. 25, 47 (1975); A. Luther: these Proceedings and published work
43. A.C. Newell: these Proceedings
44. P.W. Kitchenside, A.L. Mason, R.K. Bullough, P.J. Caudrey: these Proceedings
45. R.K. Dodd: Proc. NATO Advanced Study Institute on *Nonlinear Problems in Physics and Mathematics* (Istanbul, August 1977), ed. by A.O. Barut (D. Reidel Publishing Co., Dordrecht, Holland 1978)
46. V.E. Zakharov, A.B. Shabat: Funkt. Analiz. i Ego Prilozh. 8, 43 (1974)
47. V.E. Zakharov: In *Solitons*, ed. by R.K. Bullough, P.J. Caudrey, Topics in Current Physics (Springer, Berlin, Heidelberg, New York 1979)

Solitons in Physics

J.A. Krumhansl

Office of the Assistant Director for Mathematical and Physical Sciences, and Engineering, National Science Foundation Washington, DC 20550, USA

1. Introduction

Given Bullough's review of the mathematics of soliton solutions of nonlinear field equations, we see how distinctly different those are from our traditional linear wave experiences. Therein lies the most serious barrier which limits much of physics today in this regard: during this century, generations of physicists have trained in and practiced linear thought, normal mode decomposition or superposition, and nonlinearity introduced only by perturbative corrections was used to varying degrees to correct this limited base. But in no way does an extended wave turn into a soliton by perturbation theory. The other side of the coin, however, is that in strongly nonlinear problems, mathematicians had not generally provided us with a game plan for systematically approaching nonlinearity, in the sense that mode analysis was utilitarian. But that is just what is happening with regard to solitons. It is being discovered that there is a non-negligible class of nonlinear field equations which, if not bearing exact soliton solutions, clearly have both spatially extended and spatially compact solutions, that these solutions are functionally independent to a high degree (if not exactly so), and that they are stable with respect to small perturbations. Now the physicist can do something with these as an augmented set of ingredients in the analysis of nature - in statistical mechanics, quantum liquids, structural phase transitions, quantum field theory, epitaxy in surface physics, polymer science, etc., to name some major areas whose utilization is growing.

This conference is about that theme; the mathematical situation is far more advanced than its deployment in these areas, but more than that I think there enters another consideration as we address the marriage of the soliton lore to physics. Specifically, it isn't so much that physical situations will often conform exactly to the sine-Gordon, nonlinear Schrödinger equation, or the rest of the integrable arsenal now available, the important point is that the dichotomy of nonlinear behaviors to be realized leads us to look for new phenomena. In that sense, a definition:

A Soliton is a Paradigm.

This is meant exactly in the sense of T.S. Kuhn the philosopher of science, i.e., a fundamentally new pattern of thought for regarding both old and new problems. That, for the physicist and engineer, is what is responsible for the excitement in this subject, and brings us together to see to what extent the joining of this new thought pattern to experiment and theory of condensed matter can be carried out. In the rest of this paper I point to examples where progress is taking place. The literal term *soliton* is overused; the thought pattern is a different matter.

2. Excitation Modes in Classical Nonlinear Systems [1]

The soliton class of equations allows us to write a field solution as

$$\psi(x,t) = \psi_{SOL}(x,t) + \psi_{OSC.}(x,t) + \psi_{TRANS}.$$

This additive statement is utilitarian and totally surprising, indeed it looks like superposition all over again. Of course, it is not; in point of fact $\psi_{OSC.}$ depends implicitly on the form of ψ_{SOL}, whether that is a one-soliton, two-soliton etc. solution. But although not a mathematically linear superposition it is perfectly good for calling out two distinctly different behaviors. Details are well developed in several instances; for example, in sine-Gordon put in a single soliton, then:

$$\psi_{SOL} = \psi \tan^{-1} \exp \frac{(x - x_0)}{\xi_0\sqrt{1 - (v_s/c)^2}} \tag{1a}$$

$$\psi_{OSC} = (2\pi)^{-1/2}(i/\omega_k)\, e^{ik(x-x_0)} \quad k + i\,\frac{\omega_0}{c} \tanh \frac{\omega_0(x-x_0)}{c} \tag{1b}$$

$$\psi_{TRANS} = (d/dx)\,\psi_{SOL}. \tag{1c}$$

Indeed, in this case the separation is exact and a sum over $-\infty < k < \infty$ can be used to make up ψ_{OSC}. The sine-Gordon equation has many remarkable features, but among others note the perfect transmission (save for phase shift) of the plane waves through the soliton, a significant measure of independence which the physicist finds useful. Next, note that $\psi_{SOL} + \varepsilon\,\psi_{TRANS}$ boosts $\psi_{SOL}(x,t) \to \psi_{SOL}(x + \varepsilon,t)$, another useful operation physically (and to be expected from translational invariance). Finally, we note that the eigenfrequencies of the oscillatory modes are all real, $\omega_k^{\,2} = \omega_0^{\,2} + c^2k^2$; thus the soliton is stable against perturbations.

In summary: (a) Simultaneous presence of localized and extended fields, (b) recipe for translating the soliton, and (c) stability against fluctuations.

As it turns out this pattern of setting out the problem now has considerable practical utility [1] as a general recipe for perturbations about a fully nonlinear solution. That is:

(a) Write $\psi = \psi_{SOL} + \delta\psi$

(b) Substitute in the (fully nonlinear) equation of motion

(c) Collect terms linear in $\delta\psi$; the resulting equation is linear and self adjoint, for which a Green's function may be determined. The eigenfunctions and eigenvalues form a basis for stability analysis, and response to applied forces, etc.

Application of this method is straightforward; for example, FOGEL, TRULLINGER, BISHOP, and KRUMHANSL [2] have studied the motion of a soliton in the presence of impurities and applied fields.

3. An Extension to Quantum Mechanics

Luther will discuss fully quantum theoretical methods for an electron gas system and others equivalent by transformations to the *quantum* sine-Gordon system; these are exact models.

However, the method of the previous section can be extended to include quantum corrections to the classical field. Thus, JACKIW [3], and others have written $\hat{\Psi}(x,t) = \psi_{SOL}(\hat{x}_0,t) + \delta\hat{\Psi}$ where the tent indicates quantum variables; that is, we promote the coordinate x_0 of the soliton to a quantum variable $\hat{x}_0$, and eventually find a systematic expansion of the quantum Hamiltonian in $\hat{x}_0$, $\delta\hat{\Psi}$ and their canonical momenta. In lowest order, the soliton behaves as a massive free particle, and the $\delta\hat{\Psi}$ as a quantized oscillator field.

This forges one more link to workaday physics, i.e., quantum calculations. Significantly, the classical soliton solution is the basis for the method, and the perturbation about this develops into the quantum fluctuations from motion of the soliton as a classical particle.

4. Consequences in Statistical Mechanics

BISHOP will review this topic in some detail but to point the way from the discussion above, we again adopt the physical dichotomy of the two classes of excitation. The central quantity, in general, for statistical thermodynamics is the partition function $Z = \{\int\} e^{-\beta H}$ (or quantum mechanically an appropriate trace). It is phenomenologically tempting to hope that $Z = Z_{SOL} Z_{OSC}$, or that $F = -k_B T \ln Z = F_{SOL} + F_{OSC}$. In fact, KRUMHANSL and SCHRIEFFER [4] showed, and checked with an *exact* functional integral calculation in 1-d, that this separate indeed holds to significant accuracy, although much further work has been needed to understand the conditions of validity and determine higher-order corrections.

Thus we have proceeded one step further in the deployment of the solitary nonlinear excitations into physics: the counting of phase space in the Gibbs ensemble lends significant weight to solitonic excitations in many nonlinear systems.

A very important point to note here concerns the different requirements of quantum theory *vs.* statistical mechanics. To lead to a truly stationary quantum state the system must be integrable, whereas this is not necessary in order to appear in the partition function - where it is only necessary that extremal paths of the Hamiltonian phase space be dense for a solitary mode.

4. Topology, Dynamics, and Structural Phase Transitions

The spatial variation of the nonlinear fields frequently has a geometric character associated with it. For example, if the field variable in a sine-Gordon equation is the rotation of a spin in a plane perpendicular to a straight line of spins in an applied field, the single soliton is a 2π kink in the chain; if a uniaxial perpendicular anisotropy is introduced, π kinks can result.

The local potential $(-A\phi^2/2 + B\phi^4/4)$ is a widely used potential (called ϕ-*four*) to describe a field system having two possible minimum energy configurations locally - and if carried uniformly through the region, lead to long-range order. While ϕ-*four* is not truly integrable, it has kinks which are highly stable, and carry ϕ from one ordering orientation to the other. A consequence is the possibility that the field breaks up into locally-ordered microdomains, with nonlinear solitary wave configurations forming the domain walls. The ramifications of this insight are many - far too extensive to even begin to summarize here. SCHNEIDER and STOLL will report on several ongoing series of molecular dynamics studies which not only agree with known statistical mechanics predictions for phase transitions, but they also show graphically the presence of clusters; in addition BRUCE develops an analysis which goes far towards incorporating these phenomenological ideas into an analytical framework.

Moving on to ^{3}He we find an even richer topological ingredient, as will be evident in MAKI'S discussion. Again, solitons play an important role in breaking the symmetry of the ordered phase, as well as introducing dynamic effects on an entirely different time scale from the traditionally thought of spin-wave excitations.

Finally, I mention the fertile studies of charge-density wave, mass-density wave, and similar fields being studied in solids and reported here (RICE, BAK, FRIEND, AXE, HOROVITZ). The main physics is: (a) the materials are highly 1-d and 2-d in character, thus enhancing the stability of solitary structures, (b) structural phase changes can take place via propagation of nonlinear kinks, (c) frequently the displacing ions (electron gas) are loosely bound so that anharmonicity sets in easily.

Thus, in still further ways, we see how the localized and topological character of solitary nonlinear excitations can signficantly broaden the concepts we use in condensed matter.

5. Summary and Acknowledgements

By means of several general examples, I have tried to point out that the studies of solitons in mathematical physics present us with a paradigm for new perspectives on a wide class of phenomena in condensed matter. The ideas are exciting, and although physicists may not find many true solitons the hunt will reveal many common cousins.

My appreciation for discussions goes to many here, but my particular thanks go to my former colleagues: Alan Bishop, Steve Trullinger, and Baruch Horovitz.

References

1. A.C. Scott, F.Y.F. Chu, D.W. McLaughlin: Proc. IEEE 61, 1443 (1973)
2. M.B. Fogel, S.E. Trullinger, A.R. Bishop, J.A. Krumhansl: Phys. Rev. Lett. 36, 1411 (1976), 37, 314(E) (1976); Phys. Rev. B 15, 1578 (1977)
3. R. Jackiw: Rev. Mod. Phys. 49, 681 (1977)
4. J.A. Krumhansl, J.R. Schrieffer: Phys. Rev. B 11, 3535 (1975)

II. Mathematical Aspects

Numerical Studies of Solitons

J. Chris Eilbeck

Department of Mathematics, Heriot-Watt University
Edinburgh, EH14 4AS, Scotland

1. Introduction

It is now well-known that the first quantitative study of soliton-like behaviour was made by SCOTT-RUSSELL [1,2] in 1834. Using what we would call today an analogue computer, but refered to in Victorian parlance as a wave tank, he observed the remarkable stability of the solitary wave in shallow water and the breakup of an initial pulse into two solitons. The delightful description of his first encounter with the 'Wave of Translation' has been reprinted in the survey by SCOTT et al [3] : it seems likely that the canal involved was the Grand Union Canal which by happy coincidence passes close to the Riccarton Campus of Heriot-Watt University.

The work of KORTEWEG and DE VRIES [4] lead to an analytic treatment of the solitary wave in 1895, but the full richness of the nonlinear phenomena involved was ignored until interest was revived by the numerical experiments of FERMI PASTA and ULAM [5] in an apparently unrelated field. In 1965 ZABUSKY and KRUSKAL [6,7] derived the KdV equation as an anharmonic crystal model and showed by a series of classical numerical experiments that solitary wave solutions were extremely stable on collision. Independently, and three years earlier, PERRING and SKRYME [8] had investigated a nonlinear Klein-Gordon equation later to be christened the Sine-Gordon (SG) equation, and had shown, first numerically and then analytically, that a two soliton (kink) solution of this equation existed. However the world of high energy physics was at that time unready for nonlinear field theories, and this work received little attention. In contrast the work of Zabusky and Kruskal fell on fertile ground and quickly let to the explosion of theoretical work which is still growing today.

Once the full richness of the analytic results became available, the numerical work took second place for some years. Indeed it seemed at one time that almost all equations with solitary wave solutions would be shown to exhibit exact multisoliton behaviour, if only the right scattering problem could be found. It is now becoming more and more apparent that there are many examples of inexact, quasi-soliton behaviour. In addition 'exact' soliton equations extended to two or more space dimensions have localized solutions which exhibit facinating features but which are not solitons in the classical sense [9]. For these problems little or no analytic results are known and numerical studies are essential for a quantitative and qualitative understanding of the phenomena.

In a paper on numerical studies of solitons it is necessary to say a little about numerical methods for nonlinear dispersive partial differential equations especially since several recent papers are guilty of exhibiting numerical results with no mention of the associated numerical analysis. Section 2 is devoted to a lightning survey of some basic methods commonly used in the literature. In Section 3 studies of nonlinear Klein-Gordon type equations are reported, together with a discussion of the appropriate numerical techniques. The rather different problems associated with 'long wave' equations such as the KdV and the Regularized Long Wave equation are discussed in Section 4, followed by a brief and rather random survey of other single and coupled equations in one space dimension in Section 5. Finally in Section 6 some recent work on equations in two and three space dimensions is reviewed.

2. Basic Numerical Methods for Nonlinear Dispersive Wave Equations

In the field of partial differential equations, numerical analysis is still almost as much an art as a science. For any specific equation, the question of the 'best' numerical method is an extremely complicated one. Even a partial answer to this question depends on many factors: for example the final accuracy required for a specific range of the independent variables, the limitations of time and storage space, and machine word length. Equally important is the amount of time and effort available for the development of the appropriate software. Most computational physicists or applied mathematicians are content to derive one reasonably efficient method and few detailed comparisons are made with other methods. For this reason no outstanding claims are made for the methods described below except in the cases where meaningful tests have been carried out. Much further work is required, not just in theoretical analysis of the accuracy and efficiency of different methods, but in practical field tests involving near-optimum codes.

In this brief section I can do no more than outline a very basic set of numerical tools: a full description will be found in the books by RICHTMYER and MORTON [10], MITCHELL [11], MITCHELL and WAIT [12] and AMES [13], and in the review talks in the Conference on the State of the Art in Numerical Analysis held in York in 1976 [14]. For definiteness I shall discuss methods as applied to the general evolution equation

$$u_t = L(u) \tag{2.1}$$

Where $L(u)$ is some general nonlinear differential operator in the space variable x, and finite boundaries are prescribed with appropriate boundary conditions.

Two basic approaches are used to reduce the problem to one involving only a finite number of parameters: the function approximation approach and the finite difference approach.

The function approximation method, as its name implies, approximates the exact solution $u(x,t)$ by an approximate solution defined on a finite dimension subspace (usually in the x variable only):

$$u(x,t) \approx \tilde{u}(x,t) = \sum_{i=1}^{N} c_i(t)\,\phi_i(x) \tag{2.2}$$

The $\phi_i(x)$ are appropriately chosen basis functions. Common choices for these are the trigometric functions, leading to a finite Fourier Transform or pseudo spectral method and piecewise polynominal functions with a local basis, giving the Finite Element Method. Assuming the ϕ_i have been chosen to satisfy the boundary conditions, (2.1) becomes

$$r(x,t) = \tilde{u}_t - L(\tilde{u}) = \sum_1^N \dot{c}_i(t)\phi_i(x) - L(\tilde{u}) \tag{2.3}$$

where the dots represent differentiation with respect to t.
The residual r is to be made small in some sense: the most common criteria is the Galerkin approach

$$\int_0^1 r(x,t)\phi_j(x)dx = 0 \qquad j = 1,\ldots,N \tag{2.4}$$

leading to a system of ordinary differential equations.

An alternative approach is to set the residual to be zero at a given set of points $x_1,\ldots,x_N$

$$r(x_j,t) = 0 \qquad j = 1,\ldots,N \tag{2.5}$$

This is the collocation method: although little used at the present in partial differential equations, it is gaining popularity in the field of ordinary differential equations [15].

Function approximation methods are relatively new, and for many applications the more familiar finite difference methods are still the most popular. For this approach we seek approximations u_m^n to the original function u(x,t) at a set of points x_m, t_n on a rectangular grid in the x,t plane, where $x_m = hm$ and $t_n = kn$. By expanding function values at grid points in a Taylor series, approximations to the differential equation involving algebraic relations between grid point values can be obtained. Assuming some suitable approximation $L_h(u_m^n)$ for L(u) has been derived, various types of scheme can be constructed according to the type of approximation for the time derivative. Some common choices, together with the appropriate computational molecule [13] are shown below.

$$u_m^{n+1} - u_m^n = k\,L_h\,(u_m^n) \tag{2.6}$$

$$u_m^n - u_m^{n-1} = k\,L_h\,(u_m^n) \tag{2.7}$$

$$u_m^{n+1} - u_m^{n-1} = 2k\,L_h(u_m^n) \tag{2.8}$$

Eqns (2.6-8) are known as the simple explicit, simple implicit, and leapfrog schemes respectively. Two further permutations are important, the Crank-Nicholson scheme obtained by adding (2.6) and (2.7), and the Hopscotch scheme [16] which applies (2.6) at odd values of (n+m) and (2.7) at even values: this cunning combination can result in (2.7) becoming explicit, as the following diagram shows.

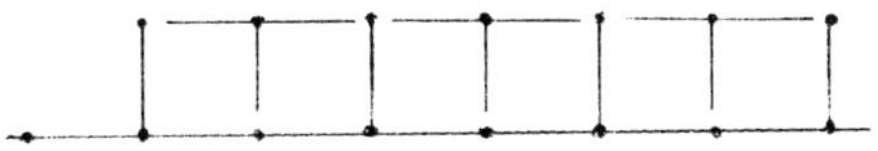

For (2.7) to become explicit, the nonlinear part of $L_h(u_m^n)$ must be replaced by a space average $L_h(\frac{1}{2}(u_{m+1}^n + u_{m-1}^n))$. The computational molecules shown above are for the case where L is a second order operator and the simplest difference schemes are used: for more complicated cases (e.g. the KdV equation) some modification will be necessary.

All methods must satisfy the basic stability requirement, that is $|u(x_m,t_n) - u_m^n|$ must remain bounded as $n \to \infty$. In addition we would like this error term to remain as small as possible. Stability requirements often lead to a limit on the time step k as a function of h, and accuracy requirements require both these to be small. The local truncation error is defined to be (p.d.e - f.d. approx) and the method is said to be of order l in time and j in space if it is $O(k^l + h^j)$.

Comparing these finite difference methods, for linear equations the Crank-Nicholson scheme is probably the most efficient of those discussed here, but in nonlinear cases it results in a set of nonlinear simultaneous equations at each time step, requiring much more work. The leapfrog scheme (2.8) is unstable for all choices of the timestep k for the linear heat equation, but a similar version for hyperbolic second order equations is an extremely practical method, as we shall see in Section 3 below. Finally the Hopscotch scheme [16] is extremely simple, fast, and stable but the step length k in the parabolic case is constrained to $k \approx h^2$ to preserve reasonable accuracy. Little work has been done on comparisons of finite difference and function approximation methods for nonlinear problems.

With this simple survey of numerical methods we can now discuss some numerical studies of solitons, starting with a review of work on nonlinear Klein-Gordon equations.

3. Nonlinear Klein-Gordon Equations

The equations

$$u_{xx} - u_{tt} = F(u) \tag{3.1}$$

for various choices of F, have played an important part in the study of solitons. The most common are

$$F(u) = \sin u \qquad \text{(Sine-Gordon)} \tag{3.2}$$

$$F(u) = -u + u^3 \qquad (\phi^4_-) \tag{3.3}$$

$$F(u) = u - u^3 \qquad (\phi^4_+) \tag{3.4}$$

$$F(u) = \sin u + \lambda \sin 2u \qquad \text{(Double Sine-Gordon)} \tag{3.5}$$

The ϕ-four equations (3.3) and (3.4), so named from the Lagrangian from which they are derived have kink and soliton solutions respectively, whereas the Sine-Gordon equations have kink solutions (different asymtotic values as $x \to \pm\infty$).

As mentioned in the introduction, PERRING and SKYRME [8] were the first to investigate the SG equation numerically. They considered two simple numerical methods. First was a simple leapfrog scheme, equivalent to

$$u_m^{n+1} = -u_m^{n-1} + \frac{k^2}{h^2}\left[u_{m+1}^n + u_{m-1}^n\right] + 2\left[1 - \frac{k^2}{h^2}\right]u_m^n - k^2 \sin u_m^n \tag{3.6}$$

Numerical tests, and a linear stability analysis, showed that this scheme was unstable for $k = h$, but it was found that reducing k to $0.95\,h$ was sufficient to remove this instability. The second scheme involved re-writing the equation as a pair of first order equations, probably in the form

$$u_x + u_t = v \tag{3.7}$$
$$v_x - v_t = \sin u$$

and introducing new variables $\eta,\xi = t \pm x$ so that (3.7) become

$$u_\eta = \tfrac{1}{2} v \;,\; v_\xi = -\tfrac{1}{2}\sin u \tag{3.8}$$

the socalled characteristic form. Since the characteristics in this case are straight lines, (3.8) can be solved by standard predictor-corrector techniques for ordinary differential equations [13]. This method in [8] was probably based on the familiar trapezium rule approximation which gives a second order accurate method. Although slightly more accurate than (3.6) it involves more work since an iteration of the corrector step is required at each time step.

Using both these schemes the kink-kink interaction was investigated and an analytic formula was derived following the numerical results. The bound state kink-antikink pair (meson/bion/breather) was also investigated and a numerical study showed that a kink-bion interaction was also stable.

Following the well-known analytic discoveries concerning the solutions of the SG equation, little progress was made on numerical methods until ABLOWITZ, KRUSKAL and LADIK [17] showed that the leapfrog scheme (3.6) can be stabilized for $k = h$ by using a space average $\frac{1}{2}(u_{m+1}^n + u_{m-1}^n)$ instead uf u_m^n in the final term of (3.6). This gives, for $k = h$

$$u_m^{n+1} = -u_m^{n-1} + u_{m+1}^n + u_{m-1}^n - k^2 F(\tfrac{1}{2}(u_{m+1}^n + u_{m-1}^n)) \tag{3.9}$$

This modification reduces the number of multiplications by one, but more importantly the term in u_m^n is now absent, and calculations can be performed on a diagonal grid involving only even values of $(n+m)$. This reduces the running time by a factor of 2. In practice only two time levels need to be stored and the final code is extremely simple and efficient. Numerical tests carried out by the author have confirmed that at least for medium size step lengths ($h = 0.1$) the scheme (3.9) is more efficient than both second and fourth order predictor-corrector methods based on characteristics. The numerical tests have also shown no sign of the nonlinear instability of the kind exhibited by NEWELL [18] in schemes for nonlinear diffusion equations and the KdV equation, though this possibility is still under investigation.

Using scheme (3.9), ABLOWITZ et al [17] investigated the collisions of two solitary wave solutions of both the ϕ^4_- equation (3.3) and the Double Sine-Gordon equation (3.5) with $0 \leqslant \lambda \leqslant 10$. For the DSG equation they showed that when the two pulses were moving very slowly with respect to each other there was a significant amount of radiation during the collision process, but at higher relative velocities very little, if any, radiation was present. Independent calculations by DUCKWORTH et al [19,20] confirm that at moderate relative velocities the DSG kinks show near-exact soliton behaviour. It seems likely that the inelasticity on collision is present at all relative velocities, but is so small at high velocities that more accurate calculations would be required to observe it. However the possibility of some sort of inverted threshold effect cannot be ruled out. The work by DUCKWORTH et al [19,20] also reveals a large amount of interesting phenomena in solutions of the DSG equation which are discussed in detail in the references.

For the ϕ^4_- equation ABLOWITZ et al [17] found evidence of inelastic behaviour at all relative velocities, a result obtained independently by KUDRYAVTSEV [21]. Similar results were found for a square wave potential version of F(u). If the relative velocities of the kinks is small enough, their identity is lost on collision, and a relatively long lived oscillating state is produced which behaves much like the Sine-Gordon bion.

It is interesting to note that analytic multisoliton-like solutions in two space dimensions have been found for the DSG and the ϕ^4 equations [22]. These solutions collapse into a single kink in one space dimension, so they are of little help in understanding the results discussed above.

Figure 1 shows some frames from a computer film in preparation showing solutions of the Sine-Gordon and ϕ^4_- equations. Although the sequence depicted, a two bion collision, was obtained from the exact analytic solution of the SG equation, it would probably have been more efficient to generate it numerically. The film also shows the collision of two kinks in both the SG and ϕ^4_- models.

The bion solution of the SG equation is an exact solution formed by taking a two-kink solution with complex conjugate energies for each kink, in other words a bound state of two kinks. Since kink collisions in the SG equation are elastic, there is no way that the bion SG solution can be formed from two free kinks. However in the ϕ^4_- model, two kinks travelling at small relative velocities can lose enough energy during collision to fall into the bound-state potential well. These ϕ^4_- bions are a single oscillating hump [9] and it is possible that solutions with more structure as in the SG bions shown in Fig.1 remain to be discovered.

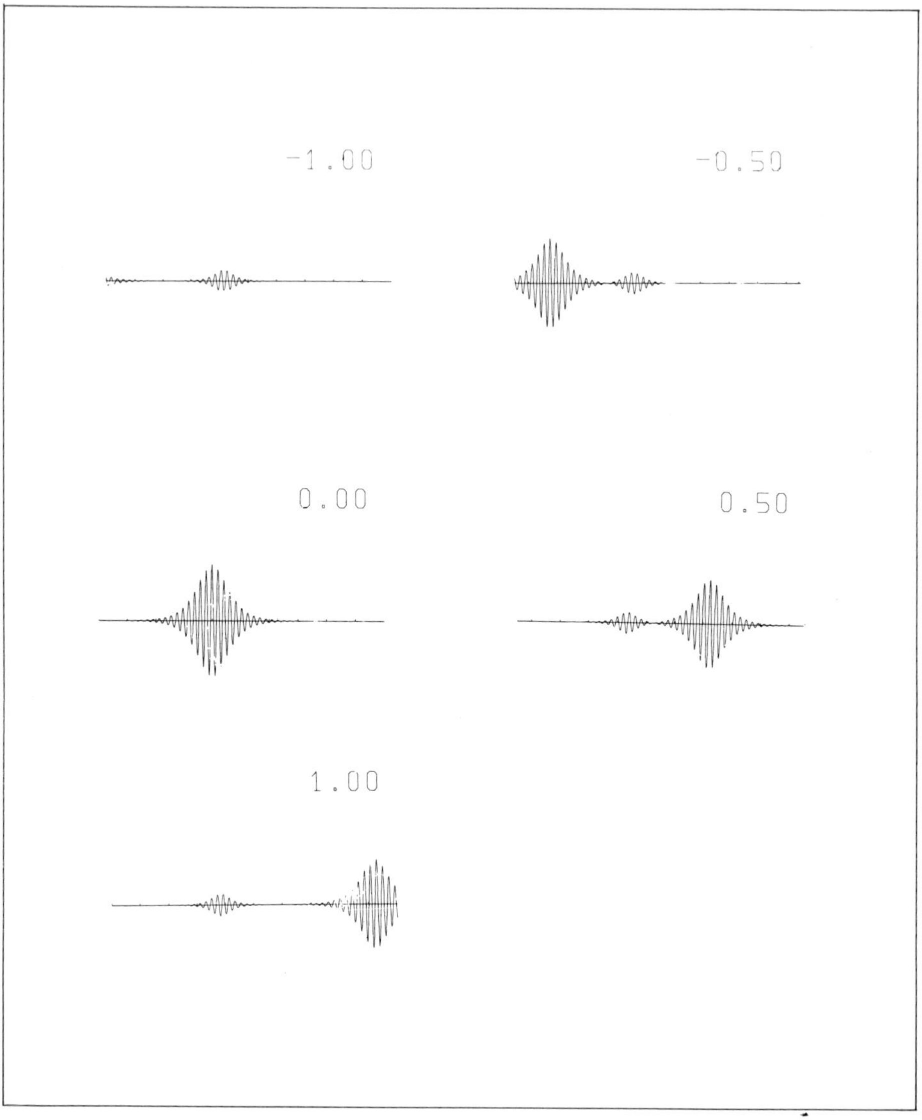

Fig.1 Frames from a computer produced 16 mm film showing analytic and numerical solutions of the Sine-Gordon and other nonlinear Klein-Gordon equations. This sequence shows the analytic two-bion solution of the SG equation.

4. 'Long Wave' Equations

The most familiar equation for long water waves is that due to KORTEWEG and DE VRIES [4] in 1895.

$$u_t + u_x + uu_x + u_{xxx} = 0 \tag{4.1}$$

The linear term in u_x can be removed by a transformation to a frame moving with unit velocity to give the more familiar form

$$u_t + uu_x + u_{xxx} = 0 \tag{4.2}$$

This equation played a fundamental role in the understanding of soliton phenomena: it was first studied numerically by Zabusky and Kruskal [6,7]. They used a leapfrog finite difference scheme.

$$u_m^{n+1} = u_m^{n-1} - \tfrac{1}{3}(k/h)(u_{m+1}^n + u_m^n + u_{m-1}^n)(u_{m+1}^n - u_{m-1}^n) \tag{4.3}$$

$$-(k/h^3)(u_{m+2}^n - 2u_{m+1}^n + 2u_{m-1}^n - u_{m-2}^n)$$

The space average term for u in the finite difference approximation for uu_x is a device to conserve the energy $\frac{1}{2}\Sigma(u_m^n)^2$ to within terms of order $O(k^2)$. The linear stability requirement for this scheme is that [23]

$$k/h^3 \leqslant \frac{1}{4+h^2|u_0|} \tag{4.4}$$

which means that a very small time step must be used to preserve stability (here u_0 is the maximum value of u in the range of interest). The nonlinear stability properties of this scheme have been extensively studied by NEWELL [18]. A survey of some finite difference schemes for the KdV equation appeared in the paper by VLIEGENT HART [24] in 1971 and in 1976 GREIG and MORRIS [23] proposed a Hopscotch scheme for this equation. This scheme has a considerably less stringent stability requirement.

$$k/h^3 \leqslant \frac{1}{|2-h^2u_0|} \tag{4.5}$$

The Hopscotch scheme requires less storage than the scheme (4.3) and can be shown to possess only a small phase error in its Fourier modes. However the small time step required even with (4.5) is apparent: a two soliton collision described in [23] required 4800 time steps to achieve an accuracy of 1% in the final amplitudes.

Very recently, FORNBERG and WHITHAM [25] have published details of numerical studies of the KdV equation and related equations, especially of the type

$$u_t + uu_x + \int_{-\infty}^{\infty} K(x-\xi)u_\xi(\xi,t)d\xi = 0 \tag{4.6}$$

where $K(x)$ can be chosen to give various dispersive effects. These authors used a pseudo spectral method for the x variable together with a leapfrog method in t. This approach is ideally suited to equations with linear integrals terms such as (4.6). The scheme used in [25] has the stability requirement

$$k/h^3 \leqslant \frac{3}{2\pi^2} \approx 0.1520 \tag{4.7}$$

The accuracy of this method is claimed to be high, but unfortunately there is no direct comparison with the Hopscotch scheme in [23]. Ref.[25] also considers the higher order KdV equations

$$u_t + u^p u_x + u_{xxx} = 0 \tag{4.8}$$

and numerical results show that with $p \geqslant 3$ the soliton collision is inelastic. However, results for (4.6) with $K(x) = \frac{\pi}{4}\exp(-|x|\pi/2)$ show some evidence for an exact two-soliton solution.

As an alternative model to the KdV equation, PEREGRINE [26], and BENJAMIN et al [27] have proposed the socalled regularized long-wave equation

$$u_t + u_x + uu_x - u_{xxt} = 0 \tag{4.9}$$

Numerical schemes for this equation were studied by EILBECK and MCGUIRE [28, 29]. They derived a three level finite difference scheme which was stable for all practical values of k, and in fact works best if $k = h$. This means that a considerably greater time step can be taken in the numerical study of (4.9) as compared to the KdV equation (4.1). The scheme in [29] can be made conservative by a space-averaged nonlinear term as in the scheme (4.3).

The numerical study in [29] showed that after a collision of two solitary waves the pulses re-appeared with amplitudes within 0.3% of their original amplitude. On the basis of this evidence, and on the evidence of a computer produced film showing the two and three solitary wave collision, it was concluded that the RLW equation (4.9) had exact multisoliton solutions. A sequence of frames showing the three 'soliton' solution is shown in Fig 2.

In addition to the two and three-solitary wave collision the film shows the breakup of an arbitrary initial pulse, in this case chosen to be a square wave of height 0.9 and length ≈ 12.17. With these initial conditions in the KdV equation, theory tells us that two solitary waves plus a large oscillating tail (radiation) would be observed. The numerical results for the RLW equation show results in close agreement with the theoretical results for the KdV equation, even though the amplitudes involved are large enough for substantial differences between the two equations to manifest themselves. At present we have not even a qualitative theory for the evolution of one or more solitary waves from arbitrary initial data for the RLW equation, and this remains an outstanding problem for the analyst.

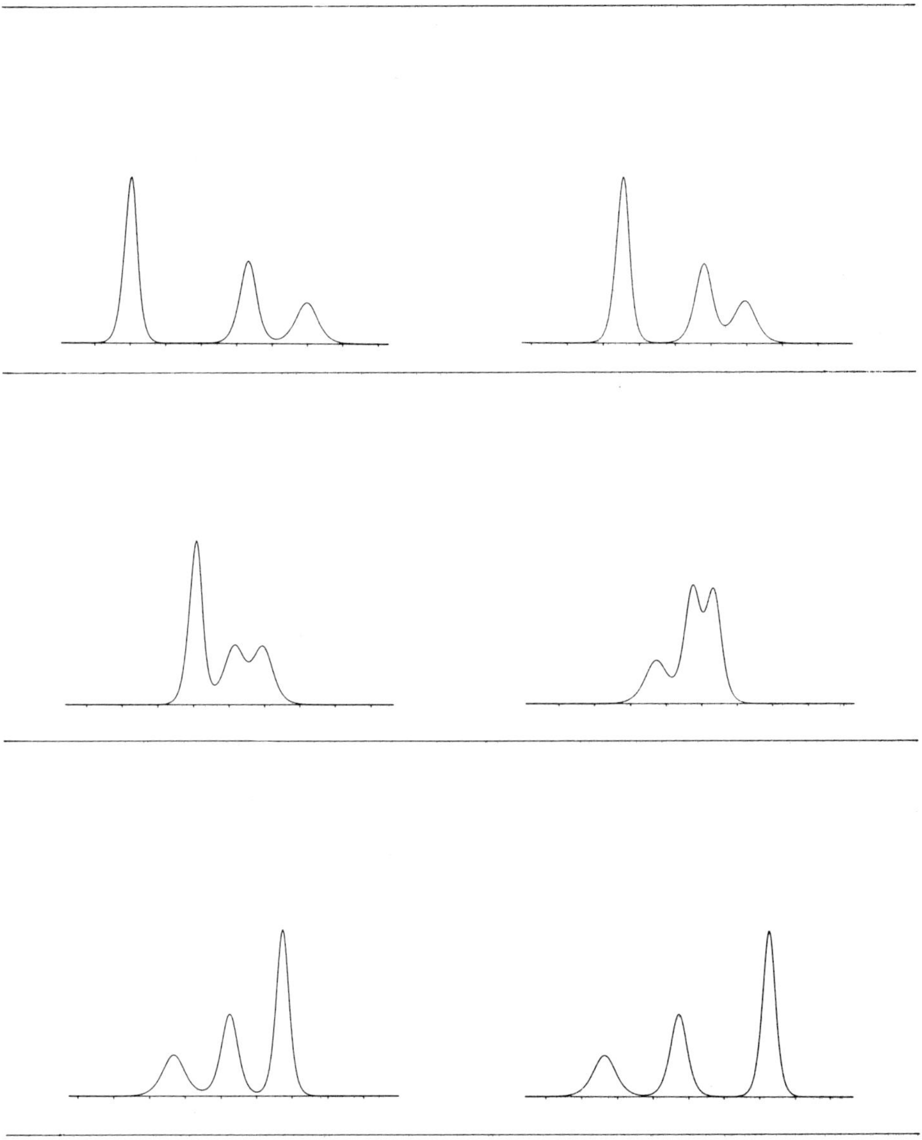

Fig.2 Frames from a computer produced film showing numerical solutions of the regularized Long-Wave equation (4.9). This sequence shows the collision and re-emergence of three 'solitons'.

Unfortunately the pictorial evidence was not sufficiently accurate to pick up the inelastic radiation occurring for the two solitary wave collision in this model. It was left to a Russian team, ABDULLOEV et al [30] to point out that when very large solitary waves ($\sim$10) collide in the RLW equation, a very small oscillating tail ($\sim 10^{-3}$) appeared, which could not be adequately explained as numerical error. These authors showed that for higher order RLW equations corresponding to (4.8) the inelasticity was even more pronounced. A more striking demonstration of this effect has been achieved by CALOGERO and SANTARELLI [31] who made use of the fact that (4.9) has solitary wave solutions travelling in both directions. When two travelling in opposite directions are allowed to collide the inelastic effect is much greater and extra solitary waves appear.

Finally, BONA, PRITCHARD and SCOTT [32] have recently derived a scheme which is fourth order in space and time for the RLW equation, and have demonstrated rigorously the existence of the small oscillating tail for a range of two solitary wave collisions.

Two important points arise from the study of the KdV and RLW equations. Firstly, if testing for exact soliton behaviour, a very careful analysis and an accurate numerical scheme are required. Secondly, in deriving an efficient numerical scheme to model a physical situation, it may be necessary to return to the original mathematical model and modify it in a way which will lead to well-behaved numerical schemes. It was precisely for this reason that PEREGRINE [26] first proposed the RLW equation.

Exactly why numerical schemes for the KdV equation should require such a small time step is unclear, but one possible explanation is that for large wave numbers the linear Fourier modes of the KdV equation have a large negative velocity, whereas the velocity of the corresponding modes of the RLW equation remain small and positive [27].

5. Other Equations in One Space Dimension

Nonlinear Optics has always provided a fertile area for interaction between experimental results, numerical studies and analytic work. The discovery of the phenomena of Self-Induced Transparency (SIT) by MCCALL and HAHN [33] followed from the numerical and analytic study of solitary wave solutions of the appropriate equations for the envelope of an intense coherent light pulse in a low density dielectric. Other numerical work in this field is reported in the excellent review by LAMB [34]. Although the envelope and approximate field equations have exact multisoliton solutions [35,36], numerical studies of the exact field equations [37,38] based on fourth-order predictor corrector methods have shown that exact soliton behaviour breaks down at high densities.

The field of plasma physics has also provided many nonlinear wave equations with soliton-like properties. Some important studies on KdV-type equations have been reported by TAPPERT [39,40]. A great deal of work on soliton phenomena in plasma physics has been performed in the USSR, and the recent review by MAKHANKOV [9] is a valuable source of reference. In this work, numerical studies of various modifications of the Boussinesq equation and of the nonlinear Schrödinger equation with self consistent potential are made, with emphasis on soliton-like phenomena in plasma physics. In all cases the modifications to the basic equations introduces some degree of inelasticity into the soliton interaction. Details of the numerical scheme used in the case of the modified Boussinesq equation are given in the paper by BOGOLUBSKY [41].

Some interesting work on nonlinear partial difference equations approximating the nonlinear Schrödinger equation has been described by ABLOWITZ and LADIK [42]. By construction these difference schemes can be solved by the inverse scattering method and have multi-soliton solutions. The resulting scheme is rather inefficient and slow, but studies of this kind may be useful in "shedding light on the difficult questions of fully nonlinear analysis".

The wide-ranging review by MAKHANKOV [9] also deals with realistic field theory models. In one space dimension, in addition to the SG and ϕ^4 equations described in section 3, he also reports work on various coupled relativistic field equations. However the most interesting results in this review occur in three-space dimentional calculations, which will be discussed in the following section.

6. Numerical Studies of Solitons in Higher Space Dimensions

Studies of nonlinear wave motion in more than one space dimension are more restricted than the more familiar results in one dimension. In general, most papers are devoted solely to a study of a single 'solitary wave' solution, and the goal is to find such solutions which remain localized for all time. Such solutions are commonly called 'solitons', especially in the particle physics literature, although no collisional stability of the sort exhibited by one dimensional 'classical' solitons are demonstrated. Some multisoliton solutions of equations in several space dimensions have been derived, but these are almost always plane wave solitons: one rather complicated exception with localized multisoliton solutions is given by NEWTON [43].

MAXON and VIECELLI [44,45] have derived spherically and cylindrically symmetric versions of the KdV equation for small amplitude acoustic waves in a collisionless plasma of warm electrons and cold ions. The equations are

$$u_\eta + (\nu\eta)^{-1}u + uu_\xi + \tfrac{1}{2}u_{\xi\xi\xi} = 0 \qquad (6.1)$$

where η is the time coordinate, ξ the radial coordinate, and $\nu = 1,2$ for speerical and cylindrical symmetry respectively. The authors examined the evolution of a single solitary wave using a leapfrog finite difference scheme based on that used by ZABUSKY and KRUSKAL [7]. It was found that the pulse steepened and narrowed as time increased, a result confirmed recently by the analytic work of CUMBERBATCH [46].

Similar work on the collapse of plasmons in the dynamics of Langmuir turbulence in plasmas is described in the review by MAKHANKOV [9]. The problem is quite analogous to the self-focusing effect of a laser pulse in matter and plasmas, reviewed in the paper by AKHMANOV et al [47] and recently discussed, with appropriate numerical schemes, by MATTAR [48], and MATTAR and NEWSTEIN [49].

Perhaps the most interesting results on soliton-like solutions in more than one dimension are the discovery of long-lived spherically symmetric pulsating solutions of the ϕ^4 and SG equations by BOGOLUBSKY and MAKHANKOV [7,50-54]. These solutions, called pulsons, are seen when the radially symmetric equations are numerically integrated with initial conditions corresponding to a spherical kink. For example for the ϕ^4_- equation initial values of u(r,t) were taken to be

$$u(r,t) = \tanh\left[(r-R_0)/\sqrt{2}\right],\ R_0 \gg 1 \tag{6.2}$$

together with unspecified boundary values for u_t (presumably $u_t = 0$, although the numerical results were not sensitive to small changes in the initial conditions). Initially this spherical kink collapsed towards the origin, and close to the origin the solution oscillates wildly and is reflected, in much the same way as a wave reaching the unsupported end of an elastic string or bar. After reflection the wave returns to a state close to its initial one, losing kinetic energy in the process, then the whole cycle is repeated. During each cycle energy is gradually leaked away to infinity, but, depending on the initial value of R_0, the pulson can maintain its identity for several cycles until a final outburst of energy leads to its collapse.

Figure 3 shows the results of a similar calculation made by the present author during preliminary work on a computer film on pulson solutions of non-linear Klein-Gordon (NLKG) equations. Unfortunately no details of the numerical methods used by BOGOLUBSKY and MAKHANKOV have been made available. For the calculations presented here and in the accompanying cine film extracts I used a simple leapfrog scheme analogous to that used in one space dimension for the NLKG equations (3.9), applied to the spherically symmetric NLKG equation

$$u_{rr} + 2r^{-1}u_r - u_{tt} = F(u) \tag{6.3}$$

At $r = 0$ this becomes, assuming $u_r(0) = 0$ [55]

$$3u_{rr} - u_{tt} = F(u) \tag{6.4}$$

A leapfrog scheme for (6.4) at the $r = 0$ gridpoint would give rise to large oscillations in the case $\Delta r = \Delta t$, so at this point an implicit scheme based on a time averaged approximation to the space derivative was used. A fixed grid spacing of $\Delta r = 0.025$ was taken, but in view of the large gradients during the reflection process it would be more satisfactory to introduce a finer grid near the origin or to stretch the variables in this region to give higher accuracy [56].

BOGOLUBSKY and MAKHANKOV studied pulson solutions to the SG equation and the ϕ^4_+ equations (3.2,3.4) in addition to the ϕ^4_- equation, and found similar results. For these three equations, near-exact small amplitude pulson solutions were found analytically. In addition BOGOLUBSKY [53] has studied 'heavy' pulson solutions of the SG equation, in which the initial kink drops from 4π to 0. He showed that this pulson behaved like a '4π' pulse for some oscillations, then 'cascaded' into a '2π' pulson for further oscillations before its identity eventually disappeared. Finally in a recent letter BOGOLUBSKY et al [54] have studied the head-on collision of two small-amplitude pulsons in the ϕ^4_- case, and have shown that these pulsons regain some individual identity after the collision process, if relative velocities are large enough.

These initial results on multi-dimensional localized solutions are very interesting, and no doubt many more results will follow from this line of investigation. It will be interesting to see if any analogue of the 'resonance' effect observed by ABLOWITZ et al [17] in the low energy collision of the ϕ^4_- kinks in one space dimension is also seen in higher dimensions. Another area of interest is the effect of the departure from spherically

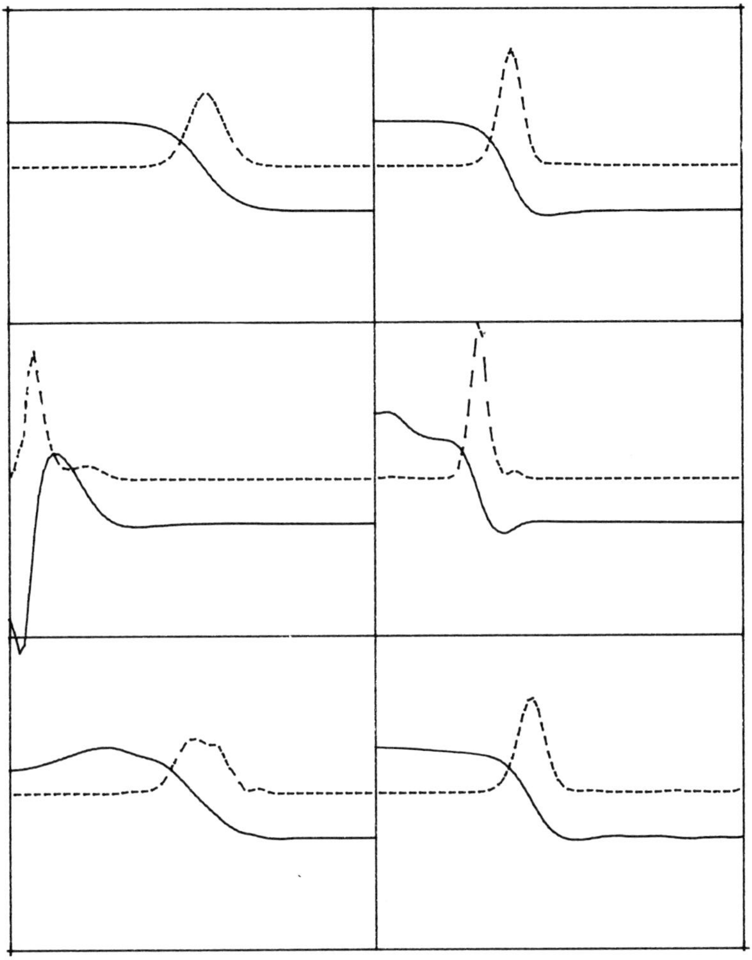

Fig.3 Frames from a computer cine film showing a pulson solution of the spherically symmetric ϕ^4_- equation. The solid curve is $u(r,t_i)$ and the dashed curve is the energy density $H(r,t_i)$. Each frame represents a change of 5 units in t, and the horizontal axis range is $0 \leqslant r \leqslant 15$, the vertical axis $-3.5 \leqslant u \leqslant 3.5$ or $-875 \leqslant H \leqslant 875$. The pulson is initially centred at $r = 8.0$.

symmetry on the evolution of pulsons. SCOTT [57] has suggested that some amount of angular momentum in nonspherical initial conditions may offset the pulson collapse to some extent and prolong the pulson lifetime. The present author is currently investigating the behaviour of nonspherically symmetric pulsons in the NLKG models, using a two-space dimensional version of the scheme (3.9) with $k = h/\sqrt{2}$ for stability.

Although these pulson solutions are of interest in particle physics [9], it would be nice to know if these results have any applications in other areas where the SG and ϕ^4_- equations are physical models. It is to be hoped that one success of this present symposium will be to spotlight areas in solid-state physics where these computational methods and computational studies can be usefully applied.

Acknowledgements

I apologise to all the people whose work I have failed to recognise in these pages. In particular I have said nothing about soliton-like behaviour in discrete systems, a topic which will be discussed in other papers at this symposium. Any other ommissions arise from ignorance rather than deliberate intent. If all those working in this area send me preprints I will have no excuse in future!

The work described here on the RLW and Maxwell-Bloch equations was joint work with Drs G. R. McGuire and P. J. Caudrey respectively, and I am most grateful for many helpful conversations. In addition it is a pleasure to acknowledge help and encouragement from many of my colleagues referred to in this paper. Finally, I would like to acknowledge travel grants from the Royal Society of London, the National Science Foundation and the Carnegie Trust for research visits to Italy and the U.S.A. during the last two years, and to thank the Science Research Council for financial and computational support.

References

1. J. Scott Russell, Report of the 14th Meeting of the British Association, 311 (1845)
2. G. S. Emmerson, in 'Scott Russell', London: Murray (1977)
3. A. C. Scott, F. Y. F. Chu, and D. W. McLaughlin, Proce. IEEE 61 1443(1973)
4. D. J. Korteweg and G. DeVries, Philos. Mag. 39 422 (1895)
5. E. Fermi, J. Pasta and S. Ulam, Los Alamos Report LA 1940, 143 (1955)
6. N. J. Zabusky, in Proc. Symp. on Nonlinear Partial Differential; W.F. Ames, ed., Academic Press, New York, 223 (1967)
7. N. J. Zabusky and M. D. Kruskal, Phys. Rev. Lett. 15 240 (1965)
8. J. K. Perring and T. H. R. Skyrme, Nucl. Phys. 31 550 (1962)
9. V. G. Makhankov, Phys. Lett. C35 1 (1978)
10. R. D. Richtmyer and K. W. Morton, Difference Methods for Initial Value Problems, 2nd Ed. Interscience, London (1967)
11. A. R. Mitchell, Computational Methods in Partial Differential Equations, J. Wiley, London (1969)
12. A. R. Mitchell and R. Wait, The Finite Element Method in Partial Differential Equations, J. Wiley, London (1977)
13. W. F. Ames, Numerical Methods for Partial Differential Equations 2nd Ed. Nelson, London (1977)
14. D. Jacobs, ed., The State of the Art in Numerical Analysis, Academic Press, London (1977)
15. J. H. Ahlberg and T. Ito, Math. Comp 29 761 (1975)

16. A. R. Gourlay, Proc. Roy. Soc. Lond. A 323 219 (1971)
17. M. J. Ablowitz, M. D. Kruskal and J. F. Ladik, 'Solitary Wave Collisions', preprint, to be published in SIAM J. Appl. Math.
18. A. C. Newell, SIAM J. Appl. Math. 33 133 (1977)
19. S. Duckworth, R. K. Bullough, P. J. Caudrey and J. D. Gibbon, Phys. Lett. 52A 19 (1976)
20. R. K. Bullough, in "Nonlinear Evolution Equations Solvable by the Spectral Transform", ed. F. Calogero, Pitman, London (1978)
21. A. Kudryavtsev, JETP Lett. 22 82 (1975)
22. J. D. Gibbon, N. C. Freeman and R. S. Johnson, Phys. Lett. 65A 380 (1978)
23. I. S. Greig and J. Ll. Morris, J. Comp. Phys., 20 64 (1976)
24. A. C. Viliegenthart, J. Engrg. Math., 51 137 (1971)
25. B. Fornberg and G. B. Whitham, Phil. Trans. Roy. Soc. 289 373 (1978)
26. D. H. Peregrine, J. Fluid Mech. 25 321 (1966)
27. T. B. Benjamin, J. L. Bona, and J. J. Mahony, Phil. Trans. Roy. Soc. 272 47 (1972)
28. J. C. Eilbeck and G. R. McGuire, J. Comp. Phys. 19 43 (1975)
29. J. C. Eilbeck and G. R. McGuire, J. Comp. Phys. 23 63 (1977)
30. Kh. O. Abdulloev, I. L. Bogolubsky and V. G. Makhankov, Phys. Lett. 56A 427 (1976)
31. F. Calagero and A. R. Santarelli, private communication, (1978)
32. J. L. Bona, W. G. Pritchard and L R Scott, University of Essex Fluid Mechanics Research Institute Report (1978)
33. S. L. McCall and E. L. Hahn, Phys. Rev. Lett. 18 908 (1967)
34. G. L. Lamb Jr., Rev. Mod. Phys. 43 99 (1971)
35. G. L. Lamb Jr., Physica 66 298 (1973)
36. J. C. Eilbeck, J. D. Gibbon, P. J. Caudrey and R. K. Bullough, J. Phys. A. 6 1337 (1973)
37. J. C. Eilbeck and R. K. Bullough, J. Phys. A. 5 820 (1972)
38. P. J. Caudrey and J. C. Eilbeck, Phys. Lett. 62A 65 (1977)
39. F. D. Tappert and C. N. Judice, Phys. Rev. Lett 29 1308 (1972)
40. F. D. Tappert, Lect. appl. math. Am. Math. Soc. 15 215 (1974)
41. I. L. Bogolubsky, Comp. Phys. Comm. 13 149 (1977)
42. M. J. Ablowitz and J. F. Ladik, Studies in Appl. Math. 55 213 (1976)
43. R. G. Newton, J. Math. Phys. 19 1068 (1978)
44. S. Maxon and J. Viecelli, Phys. Rev. Lett. 32 4 (1974)
45. S. Maxon and J. Viecelli, Phys. Fluids. 17 1614 (1974)
46. E. Cumberbatch, Phys. Fluids 21 375 (1978)
47. S. A. Akhmanor, A. P. Sukhorukov and R. V. Khokhlov, Soviet Phys. Uspekhi 10 609 (1968)
48. F. P. Mattar, 'Transient Propagation of Optical Beams in Active Media', preprint, Department of Physics, University of Rochester (1978)
49. F. P. Mattar and M. C. Newstein, in "Cooperative Effects in Matter and Radiation", eds. C. M. Bowden and D. W. Howgate, Plenum Press, N.Y. (1977)
50. I. L. Bogolubsky and V. G. Makhankov, JEPT Lett. 24 12 (1976)
51. I. L. Bogulubsky, JETP Lett. 24 535 (1976)
52. I. L. Bogulubsky and V. G. Makhankov, JETP Lett. 25 107 (1977)
53. I. L. Bogolubsky, Phys. Lett. 61A 205 (1977)
54. I. L. Bogolubsky, V. G. Makhankov, and A. B. Shvachka, Phys. Lett. 63A 225 (1977)
55. G. D. Smith, "Numerical Solution of Partial Differential Equations", Oxford U.P. (1969)
56. P. J. Roache, "Computational Fluid Dynamics", Hermosa Publishers, Albuquerque, New Mexico (1972)
57. A. C. Scott, private communication.

Poles of the Toda Lattice

John Gibbon*

Department of Mathematics, U.M.I.S.T., P.O. Box 88
Manchester M60 1QD, Great Britain

Introduction

The Toda lattice is a simple nonlinear differential-difference equation which was derived by TODA [1] as a model describing the propagation of waves along a chain of particles of equal mass with an exponential restoring force between them. If Q_n is the displacement from equilibrium of the nth particle then the equation of motion is

$$\ddot{Q}_n = \exp(Q_{n-1} - Q_n) - \exp(Q_n - Q_{n+1}) \tag{1}$$

Eq. (1) is also a model for a nonlinear electric filter [2]. The great interest in (1) lies in the fact that it is a completely integrable Hamiltonian system [3,4]. Here it is our intention to firstly show that there is a connection between (1) and other interesting Hamiltonian N-body systems and secondly to investigate the equivalent p.d.e.'s obtained in the continuum limit. In both cases there is a simple connection with the KdV/Boussinesq equations.

N-Body Systems

In 1971 CALOGERO [5] showed that the quantum mechanical problem of a set of N particles on a line interacting via inverse square potentials is a solvable one. MOSER then showed [6] that the equivalent classical problem is also completely integrable. This system can be written in the Hamiltonian formulation

$$H = \tfrac{1}{2}\sum_j \dot{x}_j^2 + \sum_{k\neq j} V(x_j - x_k) \tag{2}$$

where $V(x) = x^{-2}$ for the CALOGERO-MOSER problem. The equivalent equation of motion is

$$\ddot{x}_j = 2\sum_{k\neq j} (x_j - x_k)^{-3} \tag{3}$$

It has been shown subsequently by various workers [7,8,9] that several other N-body problems of type (2) are of great physical interest some of which are integrable and some not: (i) Sutherland's problem: $V(x) = 1/\sin^2 x$ [6] (ii) $V(x) = x^{-2} + ax^2$ [7] (iii) $V(x) = \wp(x)$ [8] where $\wp$ is the doubly periodic Weierstrassian elliptic function (iv) $V(x) = -\log|\sin x|$ [9]. The first three are integrable by an extension of the method used for (3). Case (iv) has been considered by CALOGERO and PERELOMOV [9] although a method for its complete integrability is not yet known.

Connection with the Toda Lattice and Evolution Equations

The connection between N-body systems such as (3) and nonlinear evolution

---

* Address after 1st Oct. 1978: Dept. of Mathematical Physics, University College, Belfield, Dublin 4, Eire.

equations was pointed out by AIRHAULT, MCKEAN and MOSER (AMM) [10]. Consider the Boussinesq equation

$$(u_{xx} + 6u^2 + u)_{xx} = u_{tt} = 0 \tag{4}$$

AMM chose u to be in the form of an expansion about poles $x_j(t)$ such that

$$u = \sum_j (x - x_j)^{-2} \tag{5}$$

The choice of (5) seems a little odd but if we take $x_j = (2j + 1)i\pi + ct$ then (5) is just the Laurent expansion of a squared hyperbolic secant about its poles and hence (5) in this particular case represents an expansion about the poles of a soliton. Using (5) in (4) yields two equations of motion

$$\ddot{x}_j = 2\sum_{k\neq j}(x_j - x_k)^{-3} \tag{6}$$

$$\dot{x}_j^{\,2} = 1 - \sum_{k\neq j}(x_j - x_k)^{-2} \tag{7}$$

Taken together (6) and (7) show that the velocities and positions of the poles in the complex plane must satisfy

$$\sum_{k\neq j}(\dot{x}_k + \dot{x}_j)(x_k - x_j)^{-3} = 0 \tag{8}$$

Eq. (6) is exactly the inverse square potential problem except that this problem is in the complex plane whilst the N-body problems previously mentioned are on the real axis.

Where does the Toda Lattice fit into this set up? It is different to the Hamiltonian system (2) in the sense that the equation of motion only contains nearest neighbour interactions while (2) contains all possible pair interactions. It is possible to show that, similar to the Boussinesq equation the Toda lattice can be transformed into a complex pair-potential Hamiltonian system. A more convenient form of (1) can be obtained by taking a new variable $Q_n = S_{n-1} - S_n$ yielding an equation of motion

$$1 + \ddot{S}_n = \exp(S_{n+1} + S_{n-1} - 2S_n) \tag{9}$$

A travelling wave solution of (9) is (equivalent to the one soliton solution)

$$S_n = \log\cosh\tfrac{1}{2}(an - wt) \qquad w = \pm 2\sinh(\tfrac{1}{2}a) \tag{10}$$

If we appeal to the usual product expansion for $\cosh\frac{1}{2}z$ then this indicates that the equivalent expansion for S_n in analogy to (5) is

$$S_n = \sum_j \log(n - x_j) \tag{11}$$

Using (11) in (9) yields (for an infinite lattice only) the equations of motion

$$\ddot{x}_j = 2\sum_\ell (x_j - x_\ell)^{-3} \prod_{k\neq j\neq \ell}[1 - (x_j - x_k)^{-2}] \tag{12}$$

$$\dot{x}_j^{\,2} = \prod_{k\neq j}[1 - (x_j - x_k)^{-2}] \tag{13}$$

The combination of the two shows that in analogy to (8)

$$\sum_{k\neq j}(\dot{x}_j + \dot{x}_k)(x_j - x_k)^{-3}[1 - (x_j - x_k)^{-2}] = 0 \tag{14}$$

The system can be more tidily expressed by the Hamiltonian

$$H = \tfrac{1}{2}\sum_j \dot{x}_j^{\,2} - \prod_{k\neq j}[1 - (x_j - x_k)^{-2}] \tag{15}$$

with canonical variables x_j and $\dot{x}_j$. The Hamiltonian (15) and (14) are obviously generalisations of the results that AMM obtained for the Boussinesq equation since the first term in the product expansion in (15) yields the inverse square potential.

We have yet to demonstrate complete integrability for this system on the real axis as MOSER has done for the inverse square potential. However since it is a generalisation of the inverse square case it may prove in the future to be tractable.

Lack of space precludes us from giving a fuller analysis of the motion but some information about the motion can be obtained very quickly. Briefly if we set $q = n^{-1}$ and recursively differentiate w.r.t. q and set $q = 0$ at each step, we obtain a set of relations the first of which is $\sum_j \ddot{x}_j = 0$. A combination of the rest gives the result

$$\sum_j x_j^{\,m} = \sum_{p=0}^{m} a_{mp} t^p \qquad (m = 1,2,\ldots\ldots..) \tag{16}$$

where the a's are constants. The first result simply means that we have a closed system whose centre of mass is stationary and is confirmed by summing(12). Equation (16) tells us that asymptotically the x_j approach a limit which is no faster than $x_j \sim c_j t$. Two examples of the motion can be considered. Firstly, if we choose the x_j to be the poles of a soliton: $x_j = ct \pm (2j + 1)i\pi$, then the equation of motion and (14) are satisfied as we would expect. Secondly, to show that not all motions are linear in time, a two particle solution is

$$x_1 = \alpha + [(t - \beta)^2 + \tfrac{1}{4}]^{\frac{1}{2}} \; ; \; x_2 = \alpha - [(t - \beta)^2 + \tfrac{1}{4}]^{\frac{1}{2}} \tag{17}$$

This means that $\dot{x}_1 + \dot{x}_2 = 0$ and so (14) is satisfied.

The Continuous System

In order to go into the continuum we shall use the shift operator [1] $\exp(\pm\partial/\partial n)f(n) = f(n \pm 1)$ and consider S_n to be a function of a continuous variable. Further, it is convenient to introduce a small scaling parameter ε and a new variable x such that $x = \varepsilon^{\frac{1}{2}}n$. Equation (9) now turns out to be

$$1 + \ddot{S} = \exp\left\{[2\cosh(\varepsilon^{\frac{1}{2}}\frac{\partial}{\partial x}) - 1]S\right\} \tag{18}$$

Expanding (18) in powers of ε and introducing a new variable $u = S_{xx}$, we find that

$$u_{tt} = 2\sum_{j=0}^{\infty} \frac{\varepsilon^{j+1}}{(2j + 2)!} A_{xx}^{(j)} \tag{19}$$

$$A^{(0)} = u \qquad\qquad A^{(1)} = u_{2x} + 6u^2$$

$$A^{(2)} = u_{4x} + 30uu_{2x} + 60u^3 \qquad\qquad u_{nx} = \partial_x^n u$$

To order ε^2 therefore, the Toda lattice is equivalent to the Boussinesq equation with contributions from $A^{(0)}$ and $A^{(1)}$. The contributions from the other $A^{(j)}$ $(j>2)$ are higher nonlinear dispersive corrections to the Boussinesq equation. It is these higher terms which contribute to the higher power terms in the product expansion in the potential in the Hamiltonian (15). A set of higher KdV equations can be obtained if we ignore leftward travelling waves:

$$u_t + 2\sum_{j=0} \frac{\varepsilon^{j+1}}{(2j + 2)!} A_x^{(j)} = 0 \tag{20}$$

After the KdV equation itself $(A^{(1)})$ the next operator $A^{(2)}$ is an evolution equation on its own (inconvenient numbers have been scaled out)

$$u_t + A_x^{(2)} = 0 \tag{21}$$

has been shown to be solvable by an inverse scattering operator and Bäcklund transformations and N-soliton solutions exist [11,12]. The set of $A^{(j)}$ form a ladder of KdV equations which are different to those found by LAX. In [12] it was shown that the scattering operator is of the third order:

$$L = \partial^3 + 6u\partial \qquad B = 9\partial^5 + 90u\partial^3 + 90u_x\partial^2 + 60(u_{2x} + 3u^2)\partial \tag{22}$$

such that $\dot{L} = [B,L]$ yields the evolution equations (21). Simple recursion relations also exist between the $A^{(j)}$:

$$\frac{\partial A^{(j)}}{\partial u_{2nx}} = \frac{(2j+2)!}{(2j-2n)!} \cdot \frac{2}{(2n+2)!} A^{(j-n-1)} \tag{23}$$

which enables the complete ladder to be constructed [11]. Quite a lot of other information can be found which we do not have room for here. However, in conclusion, it is interesting that the Toda lattice generates at higher orders of perturbation, integrable nonlinear evolution equations such as (21). It seems to be these which account for the simplicity of its pole structure. This is certainly not always the case for discrete integrable equations. For instance, KAC and VAN MOERBEKE's equation [6]

$$\dot{u}_n = \tfrac{1}{2}(\exp u_{n+1} - \exp u_{n-1}) \tag{24}$$

is integrable in the same way as the Toda lattice and to first order in the continuum yields the KdV equation. However no "sensible" or solvable equations are generated at higher orders of perturbation nor does it have a simple pole structure.

References

1. M. Toda: Prog. Theoret. Phys. Suppl. 45, 174, 1970
2. R. Hirota and K. Suzuki: J. Phys. Soc. Japan 28, 1366, 1970
3. M. Henon: Phys. Rev. B: 9, 1921, 1974
4. H. Flaschka: Phys. Rev. B: 9, 1924, 1974
5. F. Calogero: J. Math. Phys. 12, 419, 1971
6. J. Moser: Advances in Mathematics 16, 197, 1975
7. M. Adler: Comm. Math. Physics 55, 195, 1977
8. F. Calogero: Lett. Nuovo Cimento 13, 411, 1975
9. F. Calogero and A. Perelomov: Comm. Math. Phys. 59, 109, 1978
10. H. Airhault, H. McKean and J. Moser: Comm. Pure Appl. Math. 30, 95, 1977
11. P. Caudrey, R. Dodd and J. Gibbon: Proc. Royal Soc. Lond. A 351, 407, 1976
12. R.K. Dodd and J.D. Gibbon: Proc. Royal Soc. Lond. A 358, 287, 1977

Perturbation Theory of the Double Sine-Gordon Equation

P.W. Kitchenside, A.L. Mason, R.K. Bullough, and P.J. Caudrey

Department of Mathematics, U.M.I.S.T., P.O. Box 88
Manchester M60 1QD, Great Britain

A general perturbation theory has recently been developed by KAUP and NEWELL [1] for dealing with almost completely integrable nonlinear partial differential equations, for which conventional inverse scattering transform (IST) techniques are inapplicable. One such equation is the double sine-Gordon equation (DSGE).

$$u_{xx} - u_{tt} = \sin u + \frac{\lambda}{2} \sin 1/2 u. \qquad (1)$$

Eq. 1 arises in nonlinear optics in the theory of resonant pulse propagation in degenerate media (DUCKWORTH [2]). A solution of particular interest here, with boundary conditions $u \to 2\pi$ when $|x| \to \infty$, is given by

$$u = 4\tan^{-1}\left(\sqrt{\frac{\lambda}{4-\lambda}}\ \cosh\frac{1}{2}\sqrt{4-\eta}\ x\right) \qquad (2)$$

in its rest frame. This represents an unstable 2π-kink-antikink bound pair. In the optical context such a solution is known as a '0π' pulse.

For initial data corresponding to a compression of the spacing of the kink and antikink, it oscillates (Fig. 1) much as the breather solution of the sine-Gordon equation ($\lambda = 0$): for data corresponding to an extension it unbinds as shown in Fig. 2.

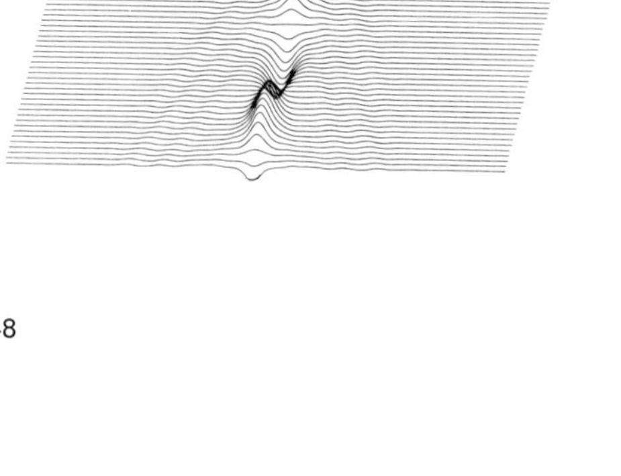

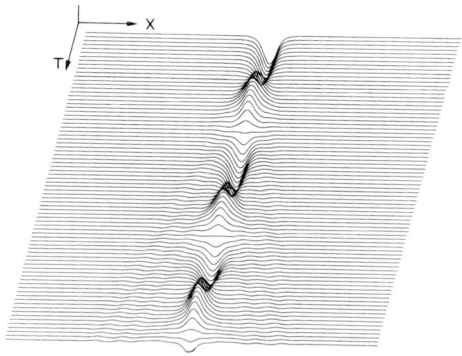

Fig.1 A numerical integration of (1) for the oscillatory regime

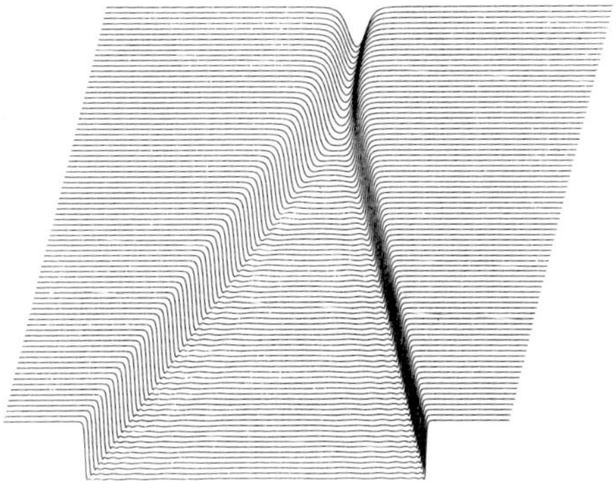

Fig.2 The break-up regime

The SGE is completely solvable by the IST and it is natural to regard the DSGE as a perturbation, for small λ. A convenient starting point is a two-kink solution of the SGE, which may be written using standard notation, as

$$u = 4\tan^{-1}\left|\frac{1 - \left(\frac{\zeta_1-\zeta_2}{\zeta_1+\zeta_2}\right)^2 \frac{\gamma_1}{2\zeta_1}\frac{\gamma_2}{2\zeta_2}\, e^{i2(k_1+k_2)X}}{\frac{\gamma_1}{2\zeta_1}\, e^{i2k_1K} + \frac{\gamma_2}{2\zeta_2}\, e^{i2k_2X}}\right|, \quad k_i = \frac{1}{2}\left(\zeta_i - \frac{1}{4\zeta_i}\right) \tag{3}$$

in terms of the scattering data. Here γ_1 and γ_2 are the residues of the reflection coefficient b/a, (when this quantity may be analytically continued into the upper half plane), evaluated at the complex eigenvalues ζ_1 and ζ_2 respectively. For the SGE, the time evolution of the scattering data is trivial. For (1), however, the appropriate equations are [1]

$$\zeta_{it} = -\frac{\lambda}{8\gamma_i \dot{a}_i^2}\int_{-\infty}^{\infty} \sin\frac{u}{2}\cdot[\phi_1^2 + \phi_2^2]_{\zeta_i}\, dX, \quad (i=1,2) \tag{4.1}$$

$$\gamma_{it} = i2\gamma_i\omega_i - \frac{\gamma_i \ddot{a}_i}{\dot{a}_i}\zeta_{it} - \frac{\lambda}{8\dot{a}_i^2}\int_{-\infty}^{\infty} \sin\frac{u}{2}\cdot\frac{\partial}{\partial\zeta}[\phi_1^2 + \phi_2^2]_{\zeta_i}\, dX \tag{4.2}$$

$$\left(\frac{b}{a}\right)_t = i2\omega\left(\frac{b}{a}\right) - \frac{\lambda}{8a^2}\int_{-\infty}^{\infty} \sin\frac{u}{2}\cdot[\phi_1^2 + \phi_2^2]_{\zeta}\, dX \tag{4.3}$$

$$\omega = \frac{1}{2}\left(\zeta + \frac{1}{4\zeta}\right), \quad \gamma_i = \frac{b_i}{\dot{a}_i}\ .$$

where $\dot{a}_i$ denotes differentiation of a with respect to ζ evaluated at $\zeta = \zeta_i$. The expression $(\phi_1^2 + \phi_2^2)_\zeta$ denotes the squared eigenfunction of the SGE IST evaluated at ζ . Eqs. (4) are simplified considerably if we discard the continuum components [and therefore (4.3)] on the assumption that they are negligible. This seems intuitively plausible if we start with the purely discrete (3) and λ is taken to be *sufficiently* small. Using the breather parametrization $\zeta_1 = -\zeta_2^*$ and $\gamma_1 = -\gamma_2^*$ the identifications

$$\zeta_1 = \frac{1}{2}e^{i\phi} , \quad \gamma_1 = \tan\phi e^{i\psi} \tag{5}$$

ensure that (3) is symmetrical about the origin in the rest frame. Putting $\phi = \cos^{-1}\sqrt{\lambda/4}$ and $\theta = \psi - \phi = 0$ in (3) give the required initial solution (2). The subsequent time evolution may not be determined from (4) neglecting the continuum. This has been done by MASON [3], and provided the breather solution remains valid the resulting equations are

$$C_t = \frac{\lambda}{2} \tan Q \cdot \frac{J(a)+a}{1+a} , \quad a = \cos^2\theta\tan^2\phi, \quad J = \frac{a}{\sqrt{1+a}}\tanh^{-1}\frac{a}{\sqrt{1+a}}$$

$$Q_t = \sin^2 Q - \left(\frac{\lambda}{4} - C\right) \cos^2 Q \tag{6}$$

$$Q = \tan^{-1}(\tan\theta\cos\phi), \quad C = \cos^2\phi .$$

Eqs. (6) show that for initial conditions $\theta = 0, 0 < \phi < \cos^{-1}\sqrt{\lambda/4}$ the resulting motion is oscillatory: for $\phi > \cos^{-1}\sqrt{\lambda/4}$, ϕ eventually reaches $\pi/2$ [3]. At this point, the solution becomes a rapidly separating kink and antikink. The breather form is now inappropriate and the analysis has to be repeated using kink parameters. A suitable parametrization is now

$$\zeta_1 = \frac{i}{2} e^{-q} , \quad \frac{\gamma_1}{2\zeta_1} = \coth q \, e^{p} ,$$

the equations of motion being

$$p_t = p^2 - C^2 - \frac{\lambda}{4} , \quad p = \coth p \sinh q, \quad C = -\sinh^2 q \tag{7}$$

$$C_t = \frac{\lambda}{4} P \frac{J(a)+a}{1+a} , \quad a = \coth^2 q \sinh^2 p.$$

We have numerically integrated (6) and (7) for the initial conditions

$$\theta = 0, \quad \cos^2\phi = \frac{1}{4} \lambda(1 - \varepsilon)^{-1}$$

with $\varepsilon = 0.1$ or -0.1 and $\lambda = 1.0$. When $\varepsilon = -0.1$ the motion is oscillatory as shown in Fig. 3. For $\varepsilon = 0.1$ break-up occurs (Fig. 4), requiring a change of parametrization when C becomes negative. A comparison of Fig. 4 with Fig. 2 shows good agreement for break-up in this *no-continuum* approximation. However, eventually truncation errors cause the parametrization to break down due to the asymptotic properties of the hyperbolic functions in (7). For the oscillatory regime, agreement is almost perfect until u first reaches its maximum amplitude. Subsequently, a slight periodic emission of ripples may be discerned representing energy being pumped into the continuum. The net result is to shorten the period of the breather-like oscillation.

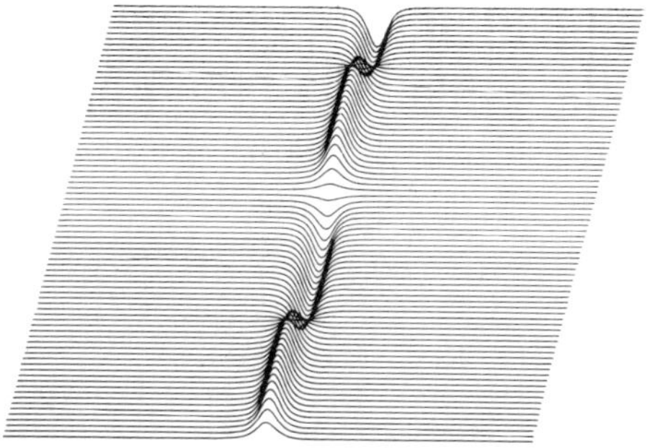

Fig.3 The oscillatory regime described by (6)

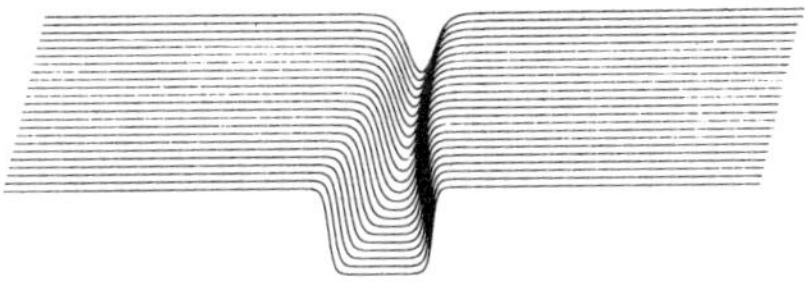

Fig.4 Break-up described by (6) and (7)

We are currently investigating this discrepancy by trying to account for the growth of the continuum. The procedure is however extremely complicated and seems unlikely to yield a set of equations with the relative simplicity of (6) or (7). Despite this anomaly there seems to be good agreement even for $\lambda = 1.0$.

References

1. D.J. Kaup, A.C. Newell: "Solitons as Particles, Oscillators and in Slowly-Varying Media: A Singular Perturbation Theory", Proc. Roy. Soc. (London, 1978) (to be published)
2. S. Duckworth: Ph.D. Thesis, University of Manchester (1976)
3. A.L. Mason: In *Proc. of the NATO Advanced Study Inst.*, "Nonlinear Equations in Physics and Mathematics", ed. by A.O. Barut (Istanbul, August 1977) (to be published)

Soliton Perturbations and Nonlinear Focussing

Alan C. Newell

Department of Mathematics, Clarkson College of Technology
Potsdam, NY 13676, USA

Abstract

A method for analyzing equations which are close to soliton equations is given and several physical examples are discussed in detail. The possible connection between certain focussing properties on nonlinear waves and turbulent bursts observed in many fluid dynamical situations is discussed.

0. Preamble

The first aim of these lectures is to survey the method by which one can analyze equations which are perturbations of soliton equations. The method is simply to apply the same transformation which carries the coordinates of the unperturbed equation to action-angle variables and then to compute the slow rate of change of the action variables and angular phases due to the cumulative effects of the perturbation. The approach we describe is completely equivalent to the apparently more straightforward but in fact more difficult method of perturbing the equation directly. The second aim is to use the method to examine the effects of applying typical types of perturbations to the canonical and ubiquitous Korteweg deVries (KdV), nonlinear Schrödinger (NLS) and sine-Gordon (SG) equations. In each case an attempt is made to show that these results can also be found by the judicious use of certain shortcuts. The particular problems we study are (a) the reversal in polarity due to depth changes of an internal solitary wave travelling on a thermocline, (b) the synchronous response of an envelope solitary wave to external forcing and the ramifications of the result to one-dimensional condensates, and (c) the tunnelling of envelope pulses through barriers created by regions of increased density. Along the way, several important points concerning the nature of NLS and SG solitons, which do not appear to be generally appreciated are brought out. Most of these results were obtained in collaboration with DAVID KAUP and are presented in more detail in references [1-4]. The final aim of this survey is to introduce some new ideas in connection with the behavior of waves which are created by the onset of an instability. The tendency of such a system to reach an ordered, cooperative state (the presence of a most unstable wave inhibits the growth of neighboring waves) on the one hand is counterbalanced on the other by the tendency of the modulational instability to spread the energy in wavenumber space and therefore cause focussing to occur in real space. A criterion which expresses this balance is given. It is suggested that when dispersive effects dominate and when the dimension of the system is more than one, the resulting focussing may be the mechanism for producing the intense irregular bursts often observed in these situations.

1. Introduction and General Discussion

The discovery of the soliton [5,6,7] has had a profound impact in the mathematical sciences. Quite simply it is a new type of normal mode of a special class of nonlinear, infinite dimensional, mechanical systems, systems often described by partial differential equations of evolution. The special property of these equations is that each is integrable; namely, one can find a canonical transformation, the inverse scattering transform or IST which carries the coordinates in which the equation is originally written to new coordinates which are action-angle variables. The IST is the nonlinear analogue of the Fourier transform. Consider the KdV equation,

$$q_t+6qq_x+q_{xxx} = 0, \qquad -\infty<x<\infty \tag{1.1}$$

and associate with a real $q(x,t)$ the eigenvalue problem

$$v_{xx} + (\zeta^2+q(x,t))v=0, \qquad -\infty<x<\infty. \tag{1.2}$$

Define for real ζ, the solutions $\phi(x,t;\zeta)$ and $\psi(x,t;\zeta)$ with the following asymptotic properties:

	ϕ	ψ
$x = -\infty$	$e^{-i\zeta x}$	$a(\zeta,t)e^{i\zeta x}-b^*(\zeta,t)e^{-i\zeta x}$
$x = +\infty$	$a(\zeta,t)e^{-i\zeta x}+b(\zeta,t)e^{i\zeta x}$	$e^{i\zeta x}$

(1.3)

We call $a(\zeta,t)$ $(a(-\zeta,t)=a^*(\zeta,t))$ and $b(\zeta,t)$ $(b(-\zeta,t)=b^*(\zeta,t))$ the scattering functions. In the linear limit a is 1 and

$\zeta b(\zeta,t) = \frac{i}{2}\int_{-\infty}^{\infty} q(y)e^{-2i\zeta y}d\zeta$ is simply the ordinary Fourier transform of $q(x,t)$. In the physicist's terminology, a^{-1} is the transmission coefficient and ba^{-1} the reflection coefficient. If $q(x,t)$ decays sufficiently rapidly as $x\to\pm\infty$, $a(\zeta,t)$ may be extended to $\mathrm{Im}\zeta>0$ and its zeros there, $\zeta_k = i\eta_k$, $k=1,\ldots,N$, are the discrete eigenvalues of (1.2) and at $\zeta=\zeta_k$, $\phi_k\to b_k e^{-\eta_k x}$, $\psi_k\to -b_k^* e^{\eta_k x}$ as $x\pm\infty$ respectively. The IST is a mapping from $q(x,t)$ to either of the two sets.

$$S_+\{\frac{b}{a},\ \zeta \text{ real};\ \{\zeta_k = i\eta_k,\ \gamma_k = \frac{b_k}{a'_k} = 2i\eta_k e^{2\eta_k \bar{x}_k}\}_1^N\} \text{ or}$$

$$S_-\{\frac{b^*}{a},\ \zeta \text{ real};\{\zeta_k = i\eta_k,\ \beta_k = -\frac{1}{b_k a'_k} = -2i\eta_k e^{-2\eta_k \bar{x}_k}\}_1^N\}, \tag{1.4}$$

called the scattering data, and from a knowledge of S one can reconstruct $q(x,t)$. In particular if $\frac{b^*}{a} \equiv 0$, $N=1$, $\zeta=i\eta_1$, $\beta_1=-2i\eta_1 e^{-2\eta_1\bar{x}_1}$, then $q(x,t) = 2\eta_1^2 \operatorname{sech}^2\eta_1(x-\bar{x}_1)$ one of the reflectionless potentials of (1.2) and the soliton of (1.1).

As $q(x,t)$ evolves in time so does S, usually in a most complicated way, in which the time behavior of $\frac{b^*}{a}(\hat{\zeta},t)$ at one $\zeta=\hat{\zeta}$ depends on this quantity at all other values of ζ. The system remains strongly coupled. However for a very special, and fortunately for us, a very useful class of evolutions of $q(x,t)$, the time rate of change of S is simple and suitable functions of the quantities [8,9,10] $|\frac{b^*}{a}|$, η_k, $k=1,\ldots,N$, $\mathrm{Arg}(\frac{b^*}{a})$, $\bar{x}_k$ $k=1,\ldots,N$ are the <u>action</u> and <u>angle</u> variables respectively. For (1.1),

$$|\frac{b^*}{a}|_t = 0, \quad (\mathrm{Arg}\,\frac{b^*}{a})_t = -8i\zeta^3, \tag{1.5a}$$

$$\eta_{kt} = 0, \quad \bar{x}_{kt} = 4\eta_k^2. \tag{1.5b}$$

It is very natural, then, to analyze the effects of perturbations to (1.1),

$$q_t+6qq_x+q_{xxx} = \varepsilon F, \tag{1.6}$$

by writing down expressions for the corresponding change in the action-angle variables. We find [1] that

$$\eta_{kt} = -\frac{i\varepsilon}{2\eta_k}\int_{-\infty}^{\infty} F\,\gamma_k\psi_k^2 dx \tag{1.7a}$$

$$\beta_{kt} + \left(\frac{a_k''}{a_k'} + \frac{1}{\zeta_k}\right)\beta_k\zeta_{kt} = -8i\zeta_k^3\beta_k + \frac{\varepsilon}{2i\zeta_k a_k'^2}\int_{-\infty}^{\infty} F\left(\frac{\partial\psi^2}{\partial\zeta}\right)_k dx \tag{1.7b}$$

$$\left(\frac{b^*}{a}\right)_t = -8i\zeta^3\,\frac{b^*}{a} + \frac{\varepsilon}{2i\zeta a^2}\int_{-\infty}^{\infty} F\,\psi^2 dx \tag{1.7c}$$

and

$$a_t = \frac{\varepsilon}{2i\zeta}\int_{-\infty}^{\infty} F\phi\psi dx. \tag{1.7d}$$

The program at this stage is to solve (1.7) iteratively and obtain asymptotic expansions in appropriate powers of ε for the quantities η_k, β_k, $\frac{b^*}{a}$. In general these expansions will be nonuniform in time but by choosing the zeroth order approximations to be slowly varying functions of time the nonuniformities can be removed. A long time $t\sim\varepsilon^{-\alpha}$, some $\alpha>0$, description of the system is thereby obtained.

Whereas in principle the method holds very generally, in practice it only works (at this time) when the zeroth order approximation is close to a multisoliton state. The reason for this is that although the squared eigenfunctions can be calculated in principle, their form does not facilitate the calculations in (1.7) unless the continuous spectrum is ignored to leading order. However the continuous spectrum still can be excited by the perturbation even though the initial state consists only of solitons. In certain of these cases, its presence can be handled and in fact we show that without it our understanding of the problem would be quite incomplete. The energy is transferred to the continuous spectrum by resonance. If the interaction term $\int F\psi^2 dx$ in (1.7c) contains a time dependence $e^{i\sigma(\xi)t}$, then it is possible that $\sigma(\xi)$ matches the natural frequency $-8i\xi^3$ of $\frac{b^*}{a}$ either (a) for no ξ, (b) for a particular $\xi=\hat{\xi}$ say or (c) for a continuous range (perhaps all) of ξ. If (a), then the continuous spectrum is not excited to any appreciable extent. If (b), then $\frac{b^*}{a}$ developes a Dirac delta function behavior about $\xi=\hat{\xi}$ in long time and results in an order ε nondecaying contribution from the continuous spectrum to $q(x,t)$. If (c), then the continuous spectrum will grow to order one on the $1/\varepsilon$ time scale and should therefore have been incorporated in the leading order approximation, a task, as we have mentioned, at present beyond the practical application of the theory.

It is natural to consider the alternative approach of perturbing (1.6) directly; if $q=q_0 + \varepsilon\delta q + \dots$,

$$q_{0t} + (3q_0^2+q_{0xx})_x = 0 \tag{1.8a}$$

$$(\delta q)_t + (6q_0\delta q + (\delta q)_{xx})_x = F(q_0). \tag{1.8b}$$

The disadvantage of this approach is that the problem of finding the appropriate basis in which to resolve δq and therefore separate the left hand side of (1.8b) is as difficult as (and in fact directly connected with) the problem of finding the appropriate eigenvalue problem to solve (1.8a) in the first place. From the scattering theory we know that [see ref. [10]] the appropriate basis for δq is

$E = \{\frac{\partial\phi^2}{\partial x}, \zeta \text{ real}; (\frac{\partial\phi^2}{\partial x}, \frac{\partial^2\phi^2}{\partial x\partial\zeta})_k\}$ which is adjoint to the set

$G = \{\psi^2, \zeta \text{ real}; (\psi^2, \frac{\partial\psi^2}{\partial\zeta})_k$. Now the squared eigenfunction function ψ^2 satisfies,

$$-\frac{1}{4}(\psi^2)_{xxx} - \frac{1}{2}q_x\psi^2 - q(\psi^2)_x = \zeta^2(\psi^2)_x \tag{1.9}$$

and if $q_0(x,t)$ satisfies (1.8a),

$$(\psi_0^2)_t = (2q_{0x}-8i\zeta^3)\psi_0^2 + (4\zeta^2-2q_0)(\psi_0^2)_x. \tag{1.10}$$

Multiply (1.8b) by ψ_0^2, use (1.9,10) and integrate to find,

$$\frac{\partial}{\partial t}\int_{-\infty}^{\infty}\delta q\,\psi_0^2 dx + 8i\zeta^3\int_{-\infty}^{\infty}\delta q\,\psi_0^2 dx = \int_{-\infty}^{\infty}F(q_0)\psi_0^2 dx. \tag{1.11}$$

But $\int_{-\infty}^{\infty}\delta q\,\psi_0^2 dx$ is simply the infinitesimal change $\delta(\frac{b^*}{a})$ in the reflection coefficient times $2i\zeta a^2$ (see (1.7c); there the infinitesimal change is $\frac{\partial}{\partial t}$) and thus (1.11) is precisely what one obtains at $O(\varepsilon)$ from (1.7c) by setting $\frac{b^*}{a} = (\frac{b^*}{a})_0 + \varepsilon\delta\ (\frac{b^*}{a})$. Therefore the two methods are precisely equivalent and in fact if one employs the second and direct method, one is led back to (1.7). A Green's function method suggested by KEENER AND MC LAUGHLIN [11] is very closely related to these ideas.

At this stage we also state two other useful facts. First even though the set G is not a basis (for L_2; $\frac{\partial\psi^2}{\partial\zeta}$ does not tend to zero as $x \to -\infty$), we can still expand the special function $q(x,t)$ as

$$q(x,t) = \frac{2}{i\pi}\int_{-\infty}^{\infty} \xi\, \frac{b}{a}\, \psi^2 d\xi - 4\sum_{k=1}^{N} \gamma_k \zeta_k \psi_k^2 \ . \tag{1.12}$$

Second we emphasize that the soliton shape assumes the form of the square of a hyperbolic secant i.e., $\psi_k = \frac{1}{2} e^{-\eta_k \bar{x}_k} \operatorname{sech}\eta_k(x-\bar{x}_k)$ only when it is separate from all the other normal modes. We will find that even the simplest of perturbations excites the $\xi=0$ wavenumber of the continuous spectrum and this contribution attaches itself to the soliton which as a result is distorted from the hyperbolic secant shape (see section 2).

Before we go on to discuss the results of perturbations we list and comment on some of the features of the soliton solutions for the other equations we shall study. The nonlinear Schrödinger (NLS) equation

$$q_t - iq_{xx} - 2iq^2q^* = 0 \tag{1.13a}$$

has the one soliton solution

$$q(x,t) = 2\eta \operatorname{sech} 2\eta(x-\bar{x}) \exp(-2i\xi x - 4i(\xi^2-\eta^2)t + i\phi),\ \bar{x}_t = -4\xi. \tag{1.13b}$$

The single kink (2π pulse) solution of the sine-Gordon equation

$$u_{TT} - c^2 u_{XX} = -u_{xt} = -n^2 \sin u,\ x = \frac{X+T}{2}\ ,\ t = \frac{X-T}{2} \tag{1.14a}$$

is

$$u(X,T) = 4\tan^{-1} \exp n\gamma(X-VT),\ \gamma = (1 - \frac{V^2}{c^2})^{-1/2}\ , \tag{1.14b}$$

and the breather (0π pulse) solution is

$$u(X,T) = 4\tan^{-1} \frac{2\eta c}{\Omega} \operatorname{sech} 2\eta\gamma(X-VT) \cos\Omega\gamma(T - \frac{VX}{c^2}) \tag{1.14c}$$

with frequency $(A = \frac{2\eta c}{\Omega})$

$$\Omega^2 = n^2 - 4\eta^2 c^2 \text{ or } \Omega^2 = \frac{n^2}{1+A^2}\ . \tag{1.14d}$$

The energies of (1.14b) and (1.14c) are Mc^2, $M=8\gamma$ and $2Mc^2(1 - \frac{\Omega^2}{n^2})^{1/2}$, respectively. We note that in the small amplitude limit, the solution to (1.14a) can be written as the product of a fast sinusoidal oscillation and an envelope function which satisfies the NLS equation (1.13a). There is a close connection between the solutions of (1.13a) and the small amplitude, almost monochromatic, solutions of equations like (1.14a) which in the weak nonlinear limit have a soft spring potential.

Let me make two very important points. Note first that the natural frequency of the breather is less than n (the "plasma" or "pinning" frequency). This distinguishes these normal modes from those associated with the continuous spectrum whose frequencies lie above n (cf. the linear limit $\omega^2 = n^2+c^2k^2$). Now there are travelling wave solutions of (1.14a) whose natural frequencies are reduced below n by the nonlinear modification. For example, in the small amplitude limit, $u=Ae^{i(kX-\omega T)}+(*)$ where $\omega^2=n^2(1-\frac{1}{2}AA^*)+c^2k^2$, is a solution and has a frequency, for sufficiently large A and small k, below n. The minus sign in front of AA^* comes from the fact that the force sinu derives from a soft spring potential. However this solution corresponds to the x independent solution of (1.13a) and is unstable in the BENJAMIN-FEIR [12] sense. Thus below the plasma frequency the only stable modes are the breathers, just as the only stable modes of positive frequency of (1.13a) are the solitons (1.13b).

The second point concerns the nature of the soliton (1.13b). Notice that whereas its frequency is modified by amplitude, the velocity of its envelope phase is not, a fact which does not faithfully reflect the nature of the equations such as (1.14a) from which it derives. In fact the soliton (1.14b) of the NLS equation moves with the linear group velocity $k-2\varepsilon\xi$ of the underlying carrier wave. This weakness can be corrected by including a term $i\varepsilon\delta q_{xt}$, usually neglected, on the R.H.S. of (1.13a) which leads to $\bar{x}_t = -4\xi+4\varepsilon\eta^2\delta + O(\varepsilon\xi^2)$. While this inclusion may not be important in many situations in which the NLS equation is used as a model, it is very important when we examine the tunnelling of soliton pulses.

2. A List of Results and Their Applications

1a. The effect of damping on the KdV equation. Consider

$$q_t+6qq_x+q_{xxx} = -\Gamma(t)q, \quad \Gamma(t)\ll 1, \tag{2.1}$$

where $\Gamma(t)\equiv 0$, $q(x,t) = 2\eta_0^2 \operatorname{sech}^2\eta_0(x-\bar{x})$, $\bar{x}_t= 4\eta^2$, $t<t_0$. Let q(x,t) be its subsequent evolution and let its scattering data be (1.4). Using (1.7) we find

$$\eta_t = -\frac{2}{3}\Gamma\eta, \quad \bar{x}_t = 4\eta^2 + \frac{\Gamma}{3\eta} \tag{2.2}$$

from which η and $\bar{x}$ can be found. Using (1.7c) we compute the excitation of the continuous spectrum and find that $\frac{b^*}{a}$ behaves like a Dirac delta function at $\xi=0$ and gives a nonvanishing $O(\Gamma)$ contribution $q_c(x,t)$ to q(x,t). Using formula (1.12),

$$q_c(x,t)= \frac{\Gamma}{6\pi\eta} \tanh^2\eta(x-\bar{x}) \left\{\int_{-\infty}^{\infty} \frac{\sin2\xi(x-\bar{x})-\sin2\xi(x-x_0+4\xi^2t)}{2\xi}\, d(2\xi)\right\}+O(\tfrac{1}{t}),$$

$$\simeq -\frac{\Gamma}{3\eta}, \quad x_0<x<\bar{x} \text{ and zero otherwise.} \tag{2.3}$$

Unable to balance both the total mass and energy flux relations $\frac{\partial}{\partial t}\int_{-\infty}^{\infty} q dx = -\Gamma\int_{-\infty}^{\infty} q dx$ and $\frac{\partial}{\partial t}\int_{-\infty}^{\infty} q^2 dx = -2\Gamma\int_{-\infty}^{\infty} q^2 dx$, the soliton changes η in such a way to satisfy the latter and develops a shelf in order to compensate for its inability to satisfy the former. Note that $\frac{\partial}{\partial t}\int_{-\infty}^{\infty} q dx = \frac{\partial}{\partial t}\int_{-\infty}^{\infty} q_s dx + \frac{\partial}{\partial t}\int_{-\infty}^{\infty} q_c dx = -\frac{2}{3}\Gamma(4\eta) - \frac{1}{3}\Gamma(4\eta) = -\Gamma(4\eta)$, the exact result. Thus the changing soliton accounts for $\frac{2}{3}$ of the rate of mass loss ($\Gamma>0$) while the shelf supplies the other $\frac{1}{3}$. When the soliton is at $\bar{x}=\bar{x}(\bar{t})$ at time $\bar{t}$, the shelf $q_c(x,\bar{t}) = -\frac{\Gamma}{3\eta}\exp(\int_{\bar{t}}^{t}\Gamma dt)$ which may be written as a function of x through $x=\bar{x}(t)$, $\bar{x}_t=4\eta^2$ the soliton position. The total area $A_c(t)$ in the shelf at time $\bar{t}$ is $4\eta_0\exp(-\int_{t_0}^{\bar{t}}\Gamma dt) - 4\eta_0\exp(-\frac{2}{3}\int_{t_0}^{\bar{t}}\Gamma dt)$ and in the soliton is $A_s(\bar{t}) = 4\eta_0\exp(-\frac{2}{3}\int_{t_0}^{\bar{t}}\Gamma dt)$, which agrees with the exact result $A(\bar{t}) = A(t_0)\exp(-\int_{t_0}^{\bar{t}}\Gamma dt)$. We can also compute the distortion of the soliton,

$$q_s(x,t) = -4\gamma_1\zeta_1\psi_1^2 = 2\eta^2\operatorname{sech}^2\eta(x-\bar{x})\left(1-\frac{\Gamma}{6\eta^2}(\bar{x}-x_0)\tanh\eta(x-\bar{x})\right), \qquad (2.4)$$

which means that the soliton is steeper in front if $\Gamma>0$. Hence it travels faster. We may also verify (2.2) by looking at the exact relation for the motion of the center of gravity $\frac{\partial}{\partial t}\int_{-\infty}^{\infty} xq dx = 3\int_{-\infty}^{\infty} q^2 dx - \Gamma\int_{-\infty}^{\infty} xq dx$. Using (2.4) and the expression q_c for the shelf height we find $\bar{x}_t = 4\eta^2 + \frac{\Gamma}{3\eta}$. Defining $\bar{y} = \int xq/\int q$ we find for small $t-t_0$, $\bar{y}\simeq\bar{x} - \Gamma\frac{\bar{x}}{12\eta^2} + O(\bar{x}^2)$ whence $\bar{y}_t \simeq 4\eta^2$ which agrees with the exact result $\frac{d\bar{y}}{dt} = \int q^2/\int q = \frac{3E}{M} = \frac{3E_0}{M_0}\exp(-\int_{t_0}^{t}\Gamma dt) \simeq 4\eta^2$ for $t\simeq t_0$. Then there is no sudden change in velocity of the total center of mass. In order to examine what happens on the time scales on which the perturbation theory breaks down $2\eta^2 = O(\frac{\Gamma}{\eta})$, we must take account of the change in the nature of $a(\zeta,t)$ which developes an irregular singular point at $\xi=0$, $a(\zeta,t) = \frac{\zeta-i\eta}{\zeta+i\eta}\exp\frac{2i}{3\zeta}\int\Gamma\eta dt$.

Note that by a judicious use of the exact "conservation" laws we would have been able to derive all the results (2.2) and (2.3). Could we have simply used these shortcuts initially? As it is with many expedient devices, the answer is yes (although we would have to calculate the soliton distortion by direct methods) with the proviso that one understands the nature of the problem. Certainly the range and role of the shelf was a major controversy until resolved in [1] by the "exact" perturbation theory. However once one understands the reason for the shelf we can use the idea directly. For example if we wish to perturb the soliton solution $q = 2\eta \text{sech} 2\eta(x-\bar{x})$, for the modified KdV equation $q_t + 6q^2 q_x + q_{xxx} = -\Gamma q$, then

$\frac{\partial}{\partial t} \int q^2 dx = -\Gamma \int q^2 dx$ implies $\eta_t = -2\Gamma\eta$, the shelf height when initially created is $-\frac{\pi\Gamma}{\bar{x}_t}$, $\bar{x}_t \simeq 4\eta^2$, (and then $\frac{\partial}{\partial t} \int_{-\infty}^{\infty} q dx \simeq -\Gamma\pi$) and at time $\bar{t}$ is $\frac{-\pi\Gamma}{4\eta^2} \exp \int_{\bar{t}}^{t} \Gamma dt$, $t=t(x)$. For the next correction to $\bar{x}_t$ (it turns out there is none [1]) we would need to calculate the soliton distortion.

1b. Application to Internal Waves. This work is being done in collaboration with COLLEN KNICKERBOCKER and more detailed results will be published shortly. We consider a lake model consisting of two layers of water with slightly different (approx. 6%) densities. Initially the bottom layer is fairly deep and thus long disturbances will travel along facing into the deeper water. However as the pulse enters the shallow portion of the lake, it reaches a point at which the deeper layer is the one on top. This situation is modelled by the equation $u_x + 6f(x)uu_t + u_{ttt} = 0$. The coefficient $f(x)$ is assumed to be a slowly varying function of position and changes sign at $x=x_1$ (e.g. $f(x) = \varepsilon x - 1$, $x_1 = \frac{1}{\varepsilon}$). If we set $u = \frac{+1}{f} q$, then we obtain (2.1) with x and t interchanged with $\Gamma(x) = \frac{-f_x}{f}$, and we may apply the previous results. If $f = \varepsilon x - 1$, then from (2.2) if $\eta_0 = 1$, $\eta = (-f)^{2/3}$, $q_{max} = 2(-f)^{4/3}$ and $u_{max} = -2(-f)^{1/3}$. The area $\int q_s dt$ under q_s is $4(-f)^{2/3}$ and under u_s, $\int u_s dt$, is $-4(-f)^{-1/3}$ and under the shelf $\int q_c dt$ is $4(-f) - 4(-f)^{2/3}$ or under u_c, $\int u_c dt$ is $-4+4(-f)^{-1/3}$. As a function of x, $u_c(x,\bar{t}) = \frac{\varepsilon}{3}(1 - \frac{7\varepsilon}{12}(x-x_0))^{8/7}$, $x_0 < x < \bar{x}$ a fact confirmed by numerical experiment. Furthermore, <u>the shelf faces into the top layer</u>. The theory breaks down when $2\eta^2 = O(\Gamma/\eta)$ or $(-f) = (\frac{\varepsilon}{6})^{1/3}$ at which point $u_s = O(\varepsilon^{1/9})$. However the shelf is small and slowly varying in x and thus propagates from $x_1(1-(\frac{\varepsilon}{6})^{1/3})$ to $x_1(1+(\frac{\varepsilon}{6})^{1/3})$ without significant distortion. At this point, it is in a region where it is a potential <u>well</u> and it has a positive area equal to that of the original soliton. Hence it creates solitons which face into the top and now deeper layer. The net result is that the original soliton appears to have changed its polarity at the turning point with (a) a change in phase, (b) slightly less amplitude and (c) trailing oscillations. Our numerical calculations agree closely with these predictions. We expect a similar qualitative behavior to obtain in many parallel situations.

2. The Synchronous Response of a NLS Soliton to the Combined Effects of Damping and Forcing.

Consider

$$q_t - iq_{xx} - 2iq^2q^* = -\Gamma q - Ee^{i\omega_0 t}, \tag{2.5}$$

for small Γ, E and suppose

$$q \simeq 2\eta \operatorname{sech} 2\eta(x-\bar{x})e^{-2i(\sigma+\pi/4)} \tag{2.6}$$

where $\sigma = \xi x + \bar{\sigma}$. The detailed rates of change of the action η, ξ and angle variables $\bar{x}$, $\frac{1}{\xi}\bar{\sigma}$ are written in [1]. Here we will assume the solution retains the form (2.6) and present a shortcut to the main result. From the exact law $\frac{\partial}{\partial t}\int_{-\infty}^{\infty}(q^*q_x - qq_x^*) = -2\Gamma\int_{-\infty}^{\infty}(q^*q_x - qq_x^*)dx$ we find straight away that $\frac{\partial}{\partial t}(\xi\eta) = -2\Gamma(\xi\eta)$ which implies that, depending on the initial ratio $\frac{\xi}{\eta}$, either ξ or η will tend to zero. If the former occurs, the forcing field loses the soliton. Let us assume then that $(\frac{\xi}{\eta})_0$ is such that $\xi \to 0$ and formally set $\xi=0$. Next use the exact relation $\frac{\partial}{\partial t}\int_{-\infty}^{\infty} qq^*dx = -2\Gamma\int_{-\infty}^{\infty} qq^*dx - [Ee^{i\omega_0 t}\int_{-\infty}^{\infty} q^*dx + (*)]$, from which we find $\eta_t = -2\Gamma\eta + \frac{\pi E}{2}\sin\chi$, $\chi = \omega_0 t + 2\bar{\sigma}$. Now to leading order $\bar{\sigma}_t = -2\eta^2$ and thus $\chi_t = \omega_0 - 4\eta^2$. The points $\eta = \frac{\sqrt{\omega_0}}{2}$, $\chi = 2n\pi + \chi_0$, $\sin\chi_0 = \frac{2\Gamma\sqrt{\omega_0}}{\pi E}$, are stable nodes or spirals. Thus a synchronized soliton $q = \sqrt{\omega_0}\operatorname{sech}\sqrt{\omega_0}(x-\bar{x})e^{i\omega_0 t - i\chi_0 - i\pi/2}$ is achieved provided the forcing frequency lies in the window $0 < \omega_0 < \frac{\pi^2E^2}{4\Gamma^2}$. Note that $\omega_0 > 0$. For $\omega_0 < 0$, only the continuous spectrum is excited. Conversely if $\omega_0 > 0$ is given, and the forcing amplitude slowly increased, the system does not respond until $E > \frac{2\Gamma}{\pi}\sqrt{\omega_0}$ and then once formed the soliton amplitude is independent of the forcing amplitude.

2b. Applications to one-dimensional condensates. In a recent letter [13], RICE, BISHOP, KRUMHANSEL AND TRULLINGER (RBKT) suggested that the solitons (1.14b) of (1.14a) may result in the condensate having a large d.c. conductivity at low temperatures. As a natural extension of these ideas, KAUP and NEWELL [14] have suggested that the breather solutions (1.14c) may contribute to the a.c. conductivity of the condensate. In the RBKT model, a breather would correspond to an oscillating electric dipole. It has the advantage over the kink that it needs no minimum activation energy. We expect these modes to be excited by an a.c. source with frequency below n, the pinning frequency of the condensate. Consider the RBKT model and include the effects of forcing and damping; here u(x,t) is the phase of the order parameter and satisfies

$$u_{tt} - c^2u_{xx} + n^2\frac{dV}{du} = -\Gamma u_t - 2n\varepsilon\cos\Omega t \tag{2.7}$$

where $\varepsilon = \frac{4e\bar{E}}{\pi Mc}$, e is electric charge, $M = 8n_s m^* n/cq_0^2$ and n_s, m^*, $q_0 = 2k_F$ are the density of condensed conduction electrons, the condensed electron effective mass and the fundamental periodicity of the undeformed condensate; $\Omega = n - \delta\omega$ and V is a soft spring periodic potential. Now look for solutions $u = \frac{1}{2}\psi e^{-int} + (*)$ and find (here we take $V = 1 - \cos u$), to a first approximation,

$$\psi_t - \frac{ic^2}{2n}\psi_{xx} - \frac{i}{16}n\psi^2\psi^* = -\frac{1}{2}\Gamma\psi - i\varepsilon e^{i\delta\omega t} \qquad (2.8)$$

to which we apply our earlier results. First we predict a significant absorption peak should occur for values of the external frequency just below the pinning frequency n of the condensate. Second we suggest that the width of the window $\Delta\omega = \frac{\pi^2}{32}\left(\frac{\varepsilon}{\Gamma}\right)^2 n$ may provide a means of measuring the phenomenological damping. Using representative values for TTF-TCNQ, We find $\frac{\Delta\omega}{n} \simeq \left(10^{-4}\frac{\bar{E}n}{\Gamma}\right)^2$, which for $\bar{E} = 10^3$ V/cm and $\Gamma \simeq 10^{-1}n$ would be order one. Third we find the average power absorbed from a single frequency $\langle p\rangle = -\langle\bar{E}\int_{-\infty}^{\infty} j(x,t)\cos\Omega t\,dx\rangle$, $j = n_s e q_0^{-1} u_t$, is $2\Gamma Mc^2\left(\frac{2\delta\omega}{n}\right)^{1/2}$, independent of the applied field. Finally, we calculate the power absorbed over a range of frequencies and using a breather distribution $n_b = \frac{n}{c}\left(\frac{kT}{Mc^2}\right)$ suggested by CURRIE [15] find the total power absorbed $\int n_b \langle p\rangle d(\delta\omega)$ is $n_s kT\left(\frac{n}{\Delta\omega_E}\right)\ (5.9\times10^{-5}\bar{E})^3$ where T is the temperature, k Boltzmann's constant and $\Delta\omega_E \gg \Delta\omega$ is the bandwidth of the applied field, a distinctly non-ohmic response.

3. NLS Solitons in Density Gradients.

3a. Consider

$$q_t - iq_{xx} - 2iq^2q^* = F, \quad F \ll 1. \qquad (2.9)$$

There are two cases of interest. First if $F = -\Gamma q$ we already have given the result. Second if $F = -2i\alpha xq - 2i\bar{\alpha}x^2 q$, write

$$X = x + a(t),\ T = t,\ q(x,t) = Q(X,T)e^{ixE + i(aE+F)}$$

where $a' + 2E = 0$, $(F+aE)' = -E^2 + 2\bar{\alpha}a^2$, $E' = -2\alpha + 4\bar{\alpha}a$ and find $Q_T - iQ_{XX} - 2iQ^2Q^* = -2i\bar{\alpha}X^2Q$. Then if $Q(X,T) \simeq 2\hat{\eta}\,\mathrm{sech}\,2\hat{\eta}(X-\hat{X})\ e^{-2i\hat{\xi}X - 4i\int\hat{\xi}^2 dT + 4i\hat{\eta}^2 T + i\hat{\phi}}$ and $\bar{\alpha}$ is small, the application of perturbation theory shows that $\hat{\eta}_T = 0$, $\hat{X}_T = -4\hat{\xi}$, $\hat{\xi}_T = 2\bar{\alpha}\hat{X}$. If $\alpha = 0$, the soliton is trapped if $\bar{\alpha} > 0$ and repelled if $\bar{\alpha} < 0$.

If $\bar{\alpha}=0$, then $E=-2\alpha t$, $a=2\alpha t^2$ and the path of the soliton is described by $x \sim -2\alpha t^2$; namely if $\alpha>0$ (the case if the plasma density increases as x increases), all right going solution components decelerate, turn around and travel in the direction of the decreasing density gradient.

3b. Nonlinear Tunnelling. Consider the model provided by (1.14a) in the case where the parameter n, the plasma frequency (or the pinning frequency in a condensate) is a slowly varying function of X which increases from n_0^2 at some $X=X_0$ to ω^2 at $X=0$. Near $X=0$ $n^2 = \omega^2(1+2\sigma\varepsilon^2 X)$. Suppose a finite amplitude signal of frequency ω is initiated at some $X>X_0$. Because of the modulational (Benjamin-Feir) instability a signal of long but finite extent will break into a sequence of breather pulses (which for sufficiently small amplitudes are solitons of an appropriate NLS equation). Let us suppose that the frequency of the breather as seen by the observer at rest is ω. Then $\gamma\Omega=\omega$ and from (1.14d), $\frac{V^2}{c^2} = 1 - \frac{n^2}{\omega^2(1+A^2)}$. Now let us assume that this relation will continue to hold if $n^2(X)$ is slowly varying. If near $X=0$, $n^2 = \omega^2(1+2\sigma\varepsilon^2 X)$, then we might argue that the soliton will penetrate to a point X_p given by $2\sigma\varepsilon^2 X_p = A^2$. Furthermore if n^2 were to decrease before $X=X_p$, say $n^2 = \omega^2(1+2\sigma\varepsilon^2 X-\sigma_1(\varepsilon^2 X)^2)$, and $\varepsilon^2 X_p > \frac{\sigma}{\sigma_1}$, then we might suggest that the soliton penetrates the barrier in a lossless manner. We verified these conjectures [3] in a more precise way by examining the propagation of a NLS soliton in a situation where $n^2(X)$ is slowly varying. By setting $u(X,T) = \frac{2}{ck}\varepsilon B(\tau,y)\exp(i\int k dX - i\omega t) + (*)$, $k = \frac{1}{c}\sqrt{\omega^2-n^2(X)}$, $\tau=\varepsilon(T-\int k_1 dX)$, $k_1 = \frac{\partial\omega}{\partial k}$, $y = \frac{\varepsilon^2}{2}\int \frac{n^2}{c^4 k^3} dX$ we obtain a perturbed NLS equation for $B(\tau,y)$ which is

$$B_y - iB_{\tau\tau} - 2iB^2B^* = \frac{k_y}{2k} B - \frac{i\varepsilon\omega}{c^2k^2} B_{y\tau} + O(\varepsilon^2), \tag{2.10}$$

where the first term on the RHS reflects the change in n(X) and the second allows the envelope phase of the soliton to depend on amplitude. The solution obtained in this way is valid close to but not at X=0. Near X=0, we take $n^2 = \omega^2(1+2\sigma\varepsilon^2 X)$ and $u(X,T) = \varepsilon C(X,T)e^{-i\omega T}+(*)$, and obtain to leading order $-2i\omega C_T - C_{XX} = -2\sigma\omega^2\varepsilon^2 XC + \frac{1}{2} n\varepsilon^2 C^2 C^*$ which may be solved exactly. The soliton is matched to the incoming pulse of (2.10) and the penetration depth calculated. It agrees exactly with that predicted by the earlier argument. Therefore we might conjecture that in general nonlinear solitary pulses penetrate past the linear cut-off until they reach zero velocity at which point the local frequency of the pulse is equal to its initial external frequency.

3. Focussing and the Behavior of Unstable Waves.

There are many physical situations which are not simply conservative and, in addition to the balance of forces which give rise to wave motions, involve nonconservative effects such as diffusion and external influences which provide a reservoir of potential energy. The balance between the external forces and dissipation forces is usually characterized in the form of a parameter, such as the Reynolds, Rayleigh or Taylor numbers in fluid mechanics, the inversion number in lasers, the temperature in superconductors, the time step in a finite difference algorithm. At certain critical values of these parameters, a fundamental and nonanalytic change in the nature of the solution occurs; for example the change can be from a stationary state to a steady or wavelike, regular or irregular motion. When the resulting motion is ordered, like in a fluid heated from below (convection cells), or in a laser (synchronized light emission) or a superconductor (boson like behavior of superconducting electrons (Cooper pairs)), then we call the new state a cooperative phenomenon [15]. On the other hand, it is often the case (e.g., instability of the Blasius flow, certain plasma instabilities) that while at the onset of instability the flow appears regular, it quickly degenerates and exhibits local turbulent bursts. It is the goal of this section to suggest a possible explanation for the concentrated patches of irregular behavior.

Consider the model equation

$$L\left(\frac{\partial}{\partial t}, \frac{\partial}{\partial x_j}, R\right) u = N\left[\frac{\partial}{\partial t}, \frac{\partial}{\partial x_j}, R\right] (u^2, u^3, \ldots) \tag{3.1}$$

which describes the departure $u(x_j,t)$ of some quantity from the background state. L is a linear operator which depends on the time, the spatial derivatives and certain potentially critical parameters in an analytic way. For simplicity and in fact without loss of generality in the resulting equation we take L to be independent of x and t and the range of each x_j to be infinite. A linear stability analysis on the solution u=0 is carried out by setting $u(x_j,t) \propto \exp i(k_j x_j - i\sigma t)$, $\sigma = \omega + i\nu$ which leads to the generalized dispersion relation $L(-i\omega+\nu, ik_j, R) = 0$ which defines both the growth rate $\nu = \nu(k_j, R)$ and the dispersion $\omega = \omega(k_j, R)$. The neutral stability surface is $\nu(k_j, R) = 0$ and the critical wave-number k_{jc} is defined by $\frac{\partial \nu}{\partial k_j} = -\frac{\partial \nu}{\partial R}\frac{\partial R}{\partial k_j} = 0$. At $k_j = k_{jc}$, $R = R_c$ and we assume this value of R is lower than the one obtained by setting $\omega \equiv 0$, the principle of exchange of stabilities; namely the instability first sets in as a growing oscillation. Next suppose $R = R_c(1+\varepsilon^2\chi)$. Following NEWELL AND WHITEHEAD [16] we analyze the response of the system by writing down the behavior of the envelope $\varepsilon A(X_j, T)$, $X_j = \varepsilon(x_j - \frac{\partial \omega}{\partial k_j} t)$, $T = \varepsilon^2 t$ of the most unstable mode i.e., $u \simeq \varepsilon A(X_j, T) \exp(ik_{jc}x_j - i\omega(k_{jc}, R)t) + (*)$,

$$\frac{\partial A}{\partial T} - \frac{1}{2}\sum_{j,\ell}\left(i\frac{\partial^2\omega}{\partial k_j \partial k_\ell} + \frac{\partial\nu}{\partial R}\frac{\partial^2 R}{\partial k_j \partial k_\ell}\right)\frac{\partial^2 A}{\partial X_j \partial k_\ell} = \frac{\partial\nu}{\partial R} R_c \chi A - \beta A^2 A^*, \ \beta = \beta_r + i\beta_i \tag{3.2}$$

where we have assumed in this case that the only nonlinear response is a self interaction. One might expect that if $\chi, \beta_r > 0$, the amplitude A will grow to saturation $|A| = \sqrt{\frac{\chi}{\beta_r}}$ modified by a frequency adjustment $\exp -i\beta_i \frac{\chi}{\beta_r} T$. We have shown however, that this solution is stable only if the matrix

$$\beta_i \left(\frac{\partial^2\omega}{\partial k_j \partial k_\ell}\right)_{R \text{ fixed}} + \beta_r \frac{\partial\nu}{\partial R}\frac{\partial^2 R}{\partial k_j \partial k_\ell} \tag{3.3}$$

is positive definite. This criterion reflects the battle between the "cooperative" tendency of the system (namely, an initial power spectrum of the spatial correlation $\langle A(\vec{x})A(\vec{x}+\vec{r})\rangle - \langle A\rangle^2$ narrows around $\vec{k} = \vec{k}_c$ and the system develops an ordered structure, and the "dispersive" tendency in which case (3.3) is not positive definite and the spectrum broadens. This broadening in $\vec{k}$ space is simply a result of the modulational instability and is accompanied by a narrowing of the pulse in $\vec{x}$ space. In one dimension, and in the absence of the real terms in (3.2), (3.2) is the NLS equation and the pulses form solitons. In the presence of the real terms, we would expect the formation of some kind of solitary pulse.

However if the dimension of the system is greater than one, then the work of ZAKHAROV AND SYNAKH [17] suggests that a much more dramatic phenomenon can occur. If again we neglect the terms with real coefficients in (3.2), we obtain the higher dimensional NLS equation [19] of which the canonical form in two dimensions is $\phi_t - i(a\phi_{xx} + b\phi_{yy}) - 2ic\phi^2\phi^* = 0$. If a=b=c=1, the solution $\phi(x,y,t)$ collapses in a finite time in a self-similar manner for a sufficiently large value of the motion constant $\int_0^\infty r|\phi|^2 dr$, $r = \sqrt{x^2+y^2}$. If a=-b=c=1, the case of deep water gravity waves, then y independent solutions are unstable essentially as a result of the linear quartet resonance mechanism of the underlying carrier wave. In fact by setting $x = r\cosh\theta$, $y = r\sinh\theta$, $x^2 > y^2$ we see that the general solution has a tendency to collapse onto the resonance curves $x^2 = y^2$ (the portion of the PHILLIPS [19] figure of eight near the vertex). If a=b=-c=1, then the system disperses and its long time behavior is given by the two-dimensional version of the BENNEY-NEWELL [18] similarity solution

$$\phi(x,y,t) \simeq \frac{1}{t} B\left(\frac{r}{t}\right) e^{\frac{ir^2}{4t} + \frac{2i}{t} B^2\left(\frac{r}{t}\right) + \dots} \tag{3.4}$$

How are these results likely to apply to (3.2)? The first point we make is that if the dispersion tensor (3.3) is nonpositive, the long time behavior of the system is not cooperative in the usual sense. The second and strongest point is that if (3.3) is negative, then the solution begins to collapse. If $\beta_r > 0$, then the collapse is eventually stopped but the system begins to oscillate and can create local collapsing spots elsewhere and the process is repeated. Furthermore if β_r is very small (as in the case of BLASIUS flow [20]), then the large local amplitudes can give rise to secondary (e.g., inflexional point) instabilities. If $\beta_r < 0$, then even though $\chi < 0$, the subcritical case, the collapse can overcome the initial damping of the system and reach amplitudes at which the nonlinear instability occurs. Focussing provides a mechanism whereby the amplitude of linearly stable but nonlinearly unstable waves can reach the critical amplitude without the benefit of imperfections, end effects or large initial perturbations.

Focussing may also enhance the tunnelling capabilities of these nonlinear pulses in systems with dimension greater than one.

Finally we mention that we are not in agreement with arguments in [17] for the 2/3 collapse law. Let $\phi_t - i\nabla^2\phi - i\phi^2\phi^2 = 0$, $\phi = f^{-1}B\exp i\theta$, $x = rf^{-1}$, $z = t$ and $f(t) = o(1)$ as $t \to t_0$. We find

$$\frac{1}{2}xf^2(B^2)_z + (xB^2(\theta_x - xff'/2))_x = 0, \tag{3.5a}$$

$$\frac{1}{x}(xB)_x + B^3 = (f^2(\theta_z - x^2f'^2/4) + (\theta_x - xff'/2)^2)B. \tag{3.5b}$$

Note the exact solution $\theta = g + x^2ff'/4$, $B = B(x)$, and $\sigma = f^2g'$, $\alpha = f^3f''/4$ are constants. However, as pointed out in [17], this solution has infinite energy. Next take $\theta = g + x^2ff'/4 + \phi$, neglect the second term on the RHS of (3.5b), define $J_1 = \int_0^\infty xB^2dx$, $J_2 = \int_0^\infty xB_x^2dx$, $J_3 = \int_0^\infty xB^4dx$ and $J_4 = \int_0^\infty x^3B^2dx$ and show (by multiplying (3.5b) by xB and X^2B_x respectively) that

$$J_2 = J_3 - \gamma J_1 - \alpha J_4, \quad J_3 = 2\gamma J_1 + 4\alpha J_4, \tag{3.6}$$

where α, γ are as given above but may no longer be constant. Now the equation has the exact laws [17],

$$\frac{dI_1}{dt} = 0, \quad \frac{dI_2}{dt} = 0, \quad \frac{dI_4}{dt} = 8I_2 \tag{3.7}$$

where $I_1 = \int_0^\infty xB^2dx$, $I_2 = \frac{1}{f^2}\int_0^\infty (xB_x^2 + xB^2\theta_x^2 - \frac{1}{2}xB^4)dx$ and $I_4 = f^2\int_0^\infty x^3B^2dx$. But from (3.6)

$$I_2 = \frac{1}{8}(f^2)''J_4 \tag{3.8}$$

and thus $I_2 \to +\infty$ as $t \to t_0$ unless $f \propto (t_0 - t)^{1/2}$ or $(t_0 - t)^\lambda$, $\lambda \geq 1$. In [17], the authors set $B = B_0 + fB_1$ with B_1 constant in I_2, and balance the term

$f'^2/4\int_0^\infty x^3B_0^2dx$ with $2B_1/f\int_0^\infty xB_0^3dx$, an argument clearly inconsistent with the results stated above. Hence, to date, the theoretical arguments for the 2/3 law are not convincing. Nevertheless there are good arguments for some kind of quasi self-similar solution (exact solutions for $f\propto(t_0-t)^{1/2}$ have infinite energy). One argument is to note that the wavelength λ of maximum growth rate of the Benjamin-Feir instability of $\phi_t - i\nabla^2\phi - i|\phi|^{2a}\phi = 0$ is A^a or $A\propto\lambda^{-1/a}$. Thus a recurring Benjamin-Feir instability would suggest a solution of the form $f^{-1/a}\psi(\frac{r}{f})$.

References

1. Kaup, D.J. and Newell, A.C. (1978), Solitons as particles, oscillators and in slowly varying media: A singular perturbation theory. To appear, Proc. Roy. Soc. Lon. Ser. A.
2. Kaup, D.J. and Newell, A.C. (1978), A theory of nonlinear, oscillating, dipolar excitations in one-dimensional condensates. Preprint.
3. Newell, A.C. (1978), Nonlinear tunnelling, J. Math. Phys., 19, 1126.
4. Newell, A.C. (1978), Near integrable systems. Proc. Symposium on "Inverse Spectral Transform", Rome, June 1977. To be published 1978 Research Notes in Mathematics. Pitman, London, Ed. F. Calogero.
5. Zabusky, N.J. and Kruskal, M.D. (1965), Interaction of solitons in a collisionless plasma, Phys. Rev. Lett., 15, 240.
6. Gardner, C.S., Greene, J.M., Kruskal, M.D. and Miura, R.M. (1967), Method for solving the Korteweg-deVries equation, Phys. Rev. Lett., 19, 1095.
7. Gardner, C.S., Greene, J.M., Kruskal, M.D. and Miura, R.M. (1974), Korteweg-deVries equation and generalizations. VI. Methods for exact solution, Comm. Pure Appl. Math., 27, 97.
8. Zakharov, V.E. and Faddeev, L.D. (1972), Korteweg-deVries equation: a completely integrable Hamiltonian system, Functional Anal. Appl., 5, 280.
9. Flaschka, H. and Newell, A.C. (1975), Integrable systems of nonlinear evolution equations, Dynamical Systems, Theory and Applications, J. Moser, ed., Springer-Verlag, New York, 355.
10. Newell, A.C. (1978), The inverse scattering transform. A review article to appear 1978 in a Springer-Verlag volume on solitons. Editors: R. Bullough and P. Caudrey.
11. Keener, J. and McLaughlin, D.W. (1977), A Green's function method for perturbed solitons, Phys. Rev. A, 16, 777.
12. Benjamin, T.B. and Feir, J.E. (1966), The disintegration of wave trains on deep water. Pt. I, J. Fluid Mech., 27, 417.
13. Rice, M.J., Bishop, A.R., Krumhansel, J.A. and Trullinger, S.E. (1976), Weakly pinned Fröhlich charge-density-wave condensates: A new nonlinear current-carrying elementary excitation, Phys. Rev. Lett., 36, 432.
14. Currie, J.F. (1977), Ph.D. Thesis, Cornell University.
15. Haken, H. (1975), Cooperative phenomena in systems far from thermal equilibrium and in non-physical systems, Rev. Mod. Phys., 47, 67.
16. Newell, A.C. (1974), Envelope equations. Lectures in Applied Mathematics, 15, 157. Published by American Math. Society.
17. Zakharov, V.E. and Synakh, V.S. (1976), The nature of the self-focusing singularity, Soviet Physics JETP 41, 485.

18. Benney, D.J. and Newell, A.C. (1967), The propagation of nonlinear wave envelopes, J. of Math. and Phys. (Stud. in Appl. Math.), 46, 133.
19. Phillips, O.M. (1960), On the dynamics of unsteady gravity waves of finite amplitude, J. Fluid Mech., 9, 193.
20. Gaster, M. and Grant, I. (1975), An experimental investigation of the formation and development of a wave packet in a laminar boundary layer, Proc. Roy. Soc. Lon. A. 347, 253.

Novel Class of Nonlinear Evolution Equations Solvable by the Spectral Transform Technique, Including the So-Called Cylindrical KdV Equation

F. Calogero and A. Degasperis

Istituto di Fisica, Universita di Roma Istituto Nazionale di Fisica Nucleare Sezione di Roma, I-00185 Roma, Italy

1. Inverse Spectral Problem for the One-Dimensional Schroedinger Equation with an Additional Linear Potential

1.1 Spectral Problem

$$-\Psi_{xx}(x,z)+[x+u(x)]\Psi(x,z)=z\Psi(x,z), \qquad -\infty<x<+\infty, \quad u(\pm\infty)=0 , \tag{1}$$

$$\int_{-\infty}^{+\infty} dx\,|u(x)|\,/\,(1+|x|)^{1/2} < \infty \quad . \tag{2}$$

1.2 Definition of Jost Function f(z)

$$f(z)=\pi\left[F_x(x,z)\,\Phi(x,z)-F(x,z)\,\Phi_x(x,z)\right] \tag{3}$$

where F and Φ are the solutions of (1) characterized by the boundary conditions

$$\lim_{x\to+\infty}\left[\Phi(x,z)/Ai(x-z)\right]=1 \quad , \tag{4}$$

$$\lim_{x\to-\infty}\left[F(x,z)/Ei(x-z)\right]=1 \quad , \tag{5}$$

with

$$Ei(y)=Bi(y)-iAi(y) \tag{6}$$

and Ai(y), Bi(y) the standard Airy functions .

1.3 Solution of Inverse Problem

$$\rho(z)=|f(z)|^{-2}-1 \tag{7}$$

$$M(x,y)=\int_{-\infty}^{+\infty} dz\,\rho(z)Ai(x-z)Ai(y-z) \tag{8}$$

$$K(x,y)+M(x,y)+\int_{x}^{+\infty} dz K(x,z)M(z,y)=0, \qquad x\le y \tag{9}$$

$$u(x)=-2dK(x,x)/dx \quad . \tag{10}$$

1.4 Remark

There is a biunivocous correspondence between $u(x)$ and $f(z)$; and a constructive procedure is now available, involving only linear calculus, to construct f from u (direct problem; see eqs.(1-5)) and u from f (inverse problem; see eqs.(7-10)).

2. Solution by the Spectral Transform Technique of a Nonlinear Evolution Equation

2.1 Nonlinear Evolution Equation

$$u_t=\alpha_0(t)u_x+\alpha_1(t)\left[u_{xxx}-6u_xu-4xu_x-2u\right], \quad u\equiv u(x,t) \quad . \tag{11}$$

2.2 Corresponding Evolution of f(z,t)

$$f_t(z,t)=\left[\alpha_0(t)-4z\alpha_1(t)\right] f_z(z,t) \tag{12}$$

$$f(z,t)=f(z\ \exp\left[-4\int_{t_0}^{t} dt'\,\alpha_1(t')\right] + \\ + \int_{t_0}^{t} dt'\alpha_0(t')\exp\left[-4\int_{t_0}^{t'} dt''\alpha_1(t'')\right],\ t_0) \ . \tag{13}$$

2.3 Evolution Equations Related to Eq.(11) by Change of Variables

$$q(y,t)=\beta(t)u(x,t), \quad y=\gamma(t)x \quad , \tag{14}$$

$$q_t=aq_{yyy}+Aq_y+Bq_yq+(C_t/C)yq_y+(D_t/D)q \tag{15}$$

$$a(t)=-\frac{1}{6}B(t)D(t)E(t)/C^2(t) \tag{16}$$

$$E(t)=\int_0^t dt'B(t')C(t')D(t') \tag{17}$$

$$\beta=DE^{-1/3},\ \gamma=E^{2/3}/C,\ \alpha_0=ACE^{-2/3},\ \alpha_1=-\frac{1}{6}BCD/E \tag{18}$$

Remark: in Eq.(15), A,B,C and D are essentially arbitrary functions of t, while a is given by Eq.(16) and Eq.(17). A special case of Eq.(15) ($A=0$, $B=6$, $C=12$, $D=t^{-1/2}$, $a=-1$) is the "cylindrical KdV equation"

$$q_t+q_{yyy}-6q_yq+(2t)^{-1}\ q=0, \quad q\equiv q(y,t) \quad . \tag{19}$$

In this special case Eq.(13) simplifies to

$$f(z,t)=f(z(t/t_0)^{1/3},\ t_0). \tag{20}$$

3. Conservation Laws for the Evolution Equation (11)

An (infinite) sequence of conserved quantities (with explicitly time-dependent coefficients) is associated to the flow (11). It can be displayed by expressing the quantities $f_n(t)$ (being the coefficients of the asymptotic expansion as $z \xrightarrow{n} -\infty$ of $f(z,t)$) in terms of u. A constructive procedure to do this is available. The first few formulae read

$$f_1(t)=\int_{-\infty}^{+\infty}dx\,u(x,t) \quad , \tag{21a}$$

$$f_2(t)=\frac{1}{2}\left[f_1(t)\right]^2 \quad , \tag{21b}$$

$$f_3(t)=\frac{1}{6}\left[f_1(t)\right]^3-\int_{-\infty}^{+\infty}dx\,u(x,t)\left[2x+u(x,t)\right] \quad , \tag{21c}$$

$$f_4(t)=f_1(t)f_3(t)-\frac{1}{8}\left[f_1(t)\right]^4 \quad . \tag{21d}$$

On the other hand the explicit time-dependence of the f_n's is also known:

$$f_n(t)=\sum_{m=0}^{[[(n-1)/2]]} f_{n-2m}(t_\circ)\exp\left[2(n-2m)\int_{t_\circ}^{t}dt'\alpha_1(t')\right]\cdot \tag{22}$$

$$\cdot\left[A_\circ(t_\circ,t)\right]^m\cdot\frac{(n-2)!!}{2^m m!(n-2m-2)!!} \quad ,$$

$$A_\circ(t_\circ,t)=4\int_{t_\circ}^{t}dt'\alpha_\circ(t')\exp\left[4\int_{t'}^{t}dt''\alpha_1(t'')\right] \quad . \tag{23}$$

Note that Eq.(22) simplifies if in Eq.(11) $\alpha_\circ\equiv 0$, reading in such case simply

$$f_n(t)=f_n(t_\circ)\exp\left[2n\int_{t_\circ}^{t}dt'\,\alpha_1(t')\right] . \tag{24}$$

Exchanging in Eq.(22) t with $t_\circ$, and using Eq.(21), one gets explicit constants of the motion for the flow (11), for instance

$$C_1=\exp\left[-2\int_{t_\circ}^{t}dt'\,\alpha_1(t')\right]\int_{-\infty}^{+\infty}dx\,u(x,t) \quad , \tag{25a}$$

$$C_2=\frac{1}{6}\,C_1^3-2C_1\int_{t_\circ}^{t}dt'\,\alpha_\circ(t')\exp\left[4\int_{t'}^{t}dt''\,\alpha_1(t'')\right]$$

$$-\exp\left[-6\int_{t_\circ}^{t}dt'\,\alpha_1(t')\right]\int_{-\infty}^{+\infty}dx\;u(x,t)\left[2x+u(x,t)\right] . \tag{25b}$$

More details on all the results reported here will be published elsewhere.

The Complex Modified Korteweg-de Vries Equation, a Non-Integrable Evolution Equation

Charles F.F. Karney

Plasma Physics Laboratory, Princeton University, Princeton, New Jersey, USA

and Abhijit Sen

Plasma Fusion Center, Massachusetts Institute of Technology Cambridge, Mass., USA

and Flora Y.F. Chu

Electronic Systems Laboratory and Department of Electrical Engineering and Computer Science, Massachusetts Institute of Technology Cambridge, Mass., USA

1. Derivation of Equation

Some anisotropic media, e.g. a magnetized plasma, can support propagating electrostatic waves. We wish to study the dispersion and nonlinear self-modulation of such waves when the system is two-dimensional (x and y, x being the principal axis) and has reached a steady state [all field quantities $\sim \exp(-i\omega t)$, ω = constant]. We will take the medium to be homogeneous and non-dissipative, and will assume that the dielectric tensor depends on the electric field amplitude squared. The equation for the complex electric potential, ϕ, is

$$\nabla \cdot \overline{\overline{K}}(\nabla, |\nabla\phi|^2) \cdot \nabla\phi = 0 \quad . \tag{1}$$

We expand $\overline{\overline{K}}$ about $\nabla = 0$, $|\nabla\phi|^2 = 0$ (the long-wavelength, linear limit)

$$\overline{\overline{K}}(\nabla, |\nabla\phi|^2) = \overline{\overline{K}} + \varepsilon(1/2)(\partial^2\overline{\overline{K}}/\partial\nabla\partial\nabla):\nabla\nabla + \varepsilon(\partial\overline{\overline{K}}/\partial|\nabla\phi|^2)|\nabla\phi|^2 \quad , \tag{2}$$

where ε is a formal expansion parameter. Assuming $K_{yy} > 0 > K_{xx}$ then to order ε^0 (1) becomes the wave equation $-|K_{xx}|\partial^2\phi/\partial x^2 + K_{yy}\partial^2\phi/\partial y^2 = 0$. Expanding ϕ about the right-going solution we let $\phi = \phi(\tau, \xi)$, where $\tau \propto y$ and $\xi \propto x - (|K_{xx}|/K_{yy})^{1/2} y$ and $\partial\phi/\partial\xi = O(1)$, $\partial\phi/\partial\tau = O(\varepsilon)$. To order ε we obtain

$$v_\tau + v_{\xi\xi\xi} + (|v|^2 v)_\xi = 0 \quad , \tag{3}$$

where $v \propto \partial\phi/\partial\xi$, and subscripts denote differentiation. We call (3) the Complex Modified Korteweg-deVries equation [1].

2. Constants of the Motion

Only four constants of the motion are known. This is in contrast to equations soluble by the inverse scattering method which have an infinite number of constants of the motion. The four constants are:

$$I_1 = \int_{-\infty}^{\infty} v \, d\xi \quad , \tag{4}$$

$$I_2 = \int_{-\infty}^{\infty} |v|^2 \, d\xi \quad , \tag{5}$$

$$I_3 = \int_{-\infty}^{\infty} |v|^4/2 - |v_\xi|^2 \, d\xi \quad , \tag{6}$$

$$I_4 = \int_{-\infty}^{\infty} |v|^2/k \, dk \quad . \tag{7}$$

In (7) k is the Fourier-transform variable conjugate to ξ. Three of these constants have physical interpretations. I_1 = const. states that the electric field is derivable from a potential. I_2 = const. and I_4 = const. give the conservation of momentum and energy, i.e. the force and power balances.

3. Soluble Limits

Although (3) is not analytically soluble, it is closely related to the modified Korteweg-deVries equation,

$$v_\tau + v_{\xi\xi\xi} + \kappa|v|^2 v_\xi = 0 \ , \tag{8}$$

which is soluble by the inverse scattering method [2]. To see this relation we rewrite (3) in two ways,

$$v_\tau + v_{\xi\xi\xi} + 3|v|^2 v_\xi = 2i|v|^2 v\theta_\xi \ , \tag{9}$$

$$v_\tau + v_{\xi\xi\xi} + |v|^2 v_\xi = -v|v^2|_\xi \ , \tag{10}$$

where $\theta = \arg(v)$. In the limits of slow and rapid phase variation, the right hand sides of (9) and (10) respectively are negligible and in these limits (3) reduces to (8), although the strength of the nonlinear term, κ, is different. When neither limit applies, we must solve (3) numerically.

4. Numerical Solution

We choose initial conditions of the form,

$$v(\tau=0\,,\xi) = A\,\mathrm{sech}(\xi)\exp(ik_o\xi) \ . \tag{11}$$

Figs. 1 and 2 show two examples of the evolution. We see that there are two types of solitary pulses produced; one had a constant phase (Fig. 1), while the other is an envelope pulse (Fig. 2).

5. Constant Phase Pulses

The constant phase pulses are a special case of the solitons of (8). Their form is

$$v = \sqrt{2}\,a\,\mathrm{sech}[a(\xi - \xi_o - a^2\tau)]\exp(i\theta_o) \ . \tag{12}$$

The area of these pulses is $\sqrt{2}\pi$. However, these pulses do not behave as solitons in (3). Fig. 3 shows the collision of two of these pulses which have different phases, θ_o. We see that after the collision the phase and amplitude of the pulses have changed and some "radiation" is produced.

6. Envelope Pulses

The general form of the envelope pulses is $v(\tau\,,\xi) = V(\zeta)\exp(ik_o\xi - i\omega_o\tau)$, where V is complex, $\zeta = \xi - c\tau$, $c = a^2 - 3k_o^2$, and $\omega_o = k_o(3a^2 - k_o^2)$. Here $\underline{a}$ is the decay rate of V as $|\zeta| \to \infty$. V satisfies

$$(V_{\zeta\zeta} + |V|^2 V - a^2V)_\zeta + ik_o(3V_{\zeta\zeta} + |V|^2 V - 3a^2V) = 0 \ . \tag{13}$$

For $k_o = 0$, we recover the constant phase pulses. Numerically integrating (13), we find that for $0 < |k_o| \lesssim 0.5a$, V does not form a pulse, while for $|k_o| \gtrsim 0.5a$ we do obtain a pulse. The asymptotic form of the pulse for k_o large is

$$V = \sqrt{6}\, a \operatorname{sech}(a\zeta)\{1 + i\varepsilon \tanh(a\zeta) + \varepsilon^2[\tanh^2(a\zeta) - 1/2]/3 + O(\varepsilon^3)\} , \quad (14)$$

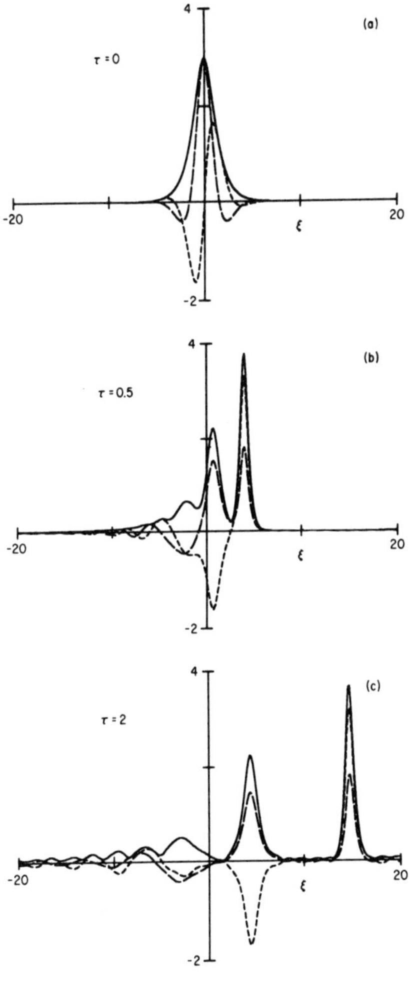

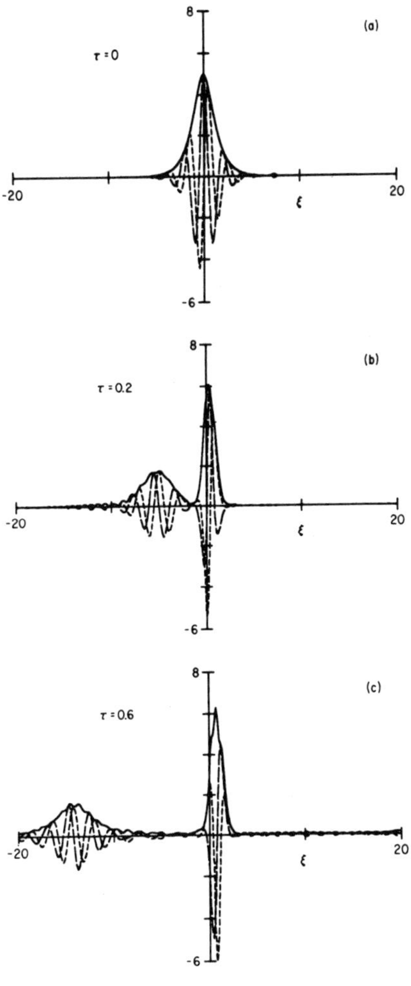

Fig. 1 (left) Evolution of (3) with initial conditions given by (11) with $A = 3$, $k_o = 1$. The solid line, long dashes, and short dashes denote $|v|$, Re(v), and Im(v) respectively

Fig. 2 (right) Same as Fig. 1, except $A = 5$, $k_o = 3$

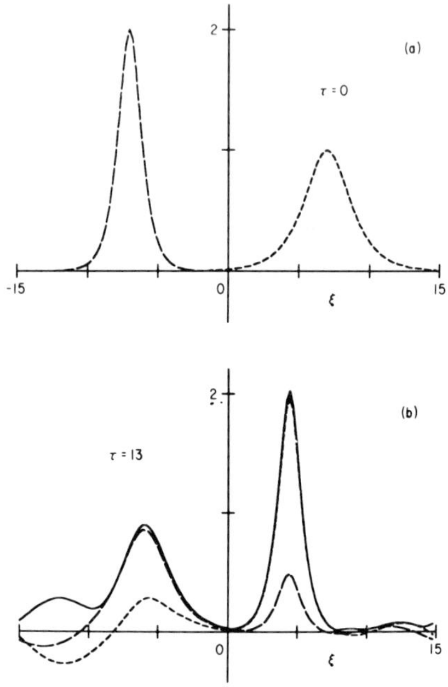

Fig. 3
Collision of two constant phase pulses

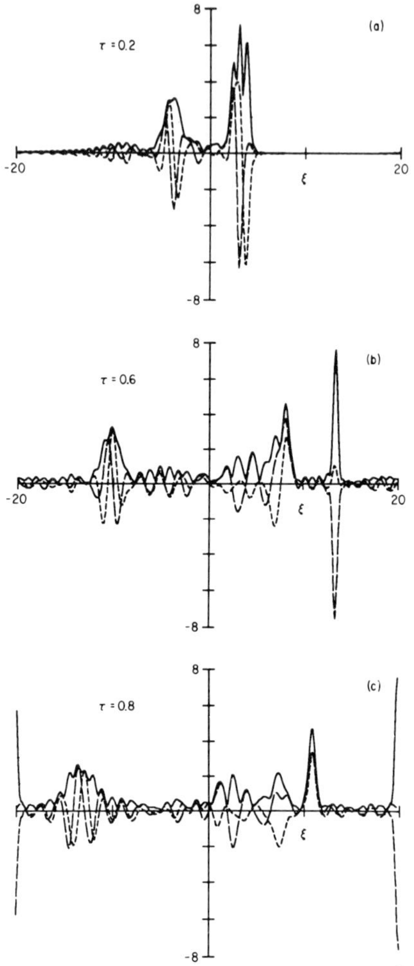

Fig. 4 (above)
Same as Fig. 2, except A = 6

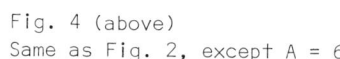

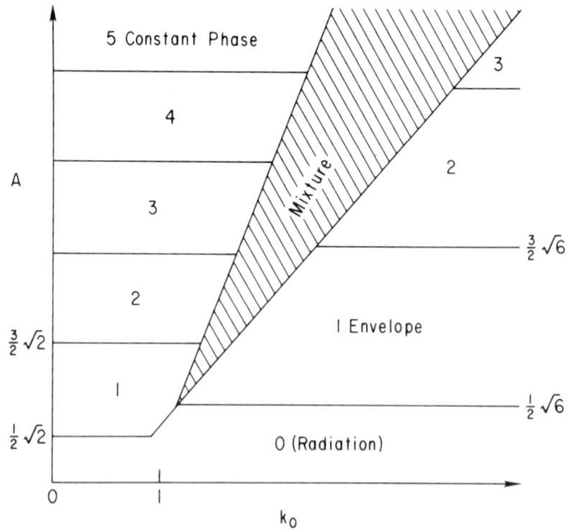

Fig. 5 (left)
Schematic showing numbers and types of pulses produced with initial conditions given by (11)

where $\varepsilon = a/k_o$. In the limit $\varepsilon \to 0$ we recover the soliton of (8) (with $\kappa = 3$). The "absolute" area of V is (cf. the area of the constant phase pulses)

$$\int_{-\infty}^{\infty} |V|\, d\zeta = \sqrt{6}\,\pi[1 + \varepsilon^2/4 + O(\varepsilon^4)] \qquad (15)$$

7. Transition From Envelope to Constant Phase Pulses

Since solitary pulses do not exist for the full range of k_o/a it is clear that there cannot be a continuous transition from envelope to constant phase pulses. If we take an envelope pulse and alter the initial conditions in such a way that $|k_o| < 0.5a$, we see from (15) that it has sufficient area to break up into about 3 constant phase pulses. For initial conditions of the form of (11) this happens when $A \simeq 2k_o$ (see Fig. 4). Fig. 5 shows schematically what pulses are produced for different initial conditions. The equivalent figure for (8) would consist of a single set of lines at $A(\kappa/6)^{1/2} = N - 1/2$.

Acknowledgments

We wish to thank J. D. Meiss and N. R. Pereira for making available their program for solving nonlinear evolution equations. One of us (FYFC) thanks the Theoretical Physics Group of Lawrence Berkeley Laboratory for its hospitality. This work was supported by USDoE (Contracts EY-76-C-02-3073 and EG-77-G-01-4107), NSF (Grant ENG-77-00340), and US Air Force (Grant AFOSR 77-3321).

References

1. C. F. F. Karney, A. Sen, F. Y. F. Chu: Plasma Physics Laboratory Report PPPL-1452, Princeton University (1978)
2. M. J. Ablowitz, D. J. Kaup, A. C. Newell, H. Segur: Stud. Appl. Math. 53, 249-315 (1974)

III. Statistical Mechanics and Solid-State Physics

Soliton-Bound States in the Magnetic Gap

A. Luther

Nordita, Blegdamsvej 17, DK-2100 Copenhagen, Denmark

Abstract

General spin correlation functions for the Ising model are reviewed and used to study the Ising model in a field. The quantum double sine-Gordon equation results. At the critical temperature this equation leads to six bound states within the magnetic gap, one at threshold, for a total of seven. It is conjectured that the number of bound states is a universal property of critical points.

Soliton States in the Ising Model

An interesting example of the quantum sine-Gordon equation appears in the statistical mechanics of the Ising model in a field. At first sight, there appears to be a little in common between these two problems. One is after all quantum mechanical in nature, while the other is a classical statistical mechanics problem. The relation between the two is provided by the transfer matrix method [1] , which leads to a Hamiltonian formulation of the Ising model. An analysis of the transfer Hamiltonian including the field provides some insight into the original Ising model. This analysis is not yet complete, for solutions of this Hamiltonian, the double sine-Gordon equation, are lacking, but the solutions which have been found are quite interesting, suggesting new physics.

There are many technical questions to deal with in deriving these relationships. Most have been treated before [2] , and it is only necessary to build the additional field into previous analysese. There is one caveat to this exercise however, because, to be precise, the double sine-Gordon equation corresponds to the eight vertex model [3] in the decoupling limit in an electric field. Now the decoupling limit is known to be trivially related to two non-interacting Ising models [4] , so solutions to this problem are solutions to Ising, once the algebra is untangled. However, in a field, the algebra is difficult and has not yet been carried through satisfactorily. Thus the information which is obtainable from the double sine-Gordon problem is of direct interest for the eight vertex problem in the electric field, and it is conjectured that many of the features survive the algebraic untanglement to the two sublattice Ising case.

The striking new feature of this problem, which make the exercise interesting, is the appearance of bound states in the transfer Hamiltonian in a field. It is not clear at present how to interpret these states, but it is intriguing to discuss their origins. It is well known on phenomenological grounds that an external field modifies the critical properties near a second order phase transition. The critical region is characterized by a length, the correlation length, which diverges as the critical temperature is approached. In the quantum mechanical formulation, this length corresponds to the reciprocal of a gap energy in the eigenvalue spectrum. Vanishing gap means infinite length, and the gap for the Ising case turns out to be proportional to the difference in temperature and the critical temperature, $T - T_c$, in the absence of a field.

In a similar manner, if the temperature is precisely equal to the critical temperature, but a field is present, the field induces a gap, the so-called magnetic gap. It is shown here that, in addition, there are bound states within this magnetic gap, states which in principle can modify the behavior of the critical fluctuations. The question of the appearance of these bound states in various correlation functions is not yet known, but their presence is established definately in the spectrum. For the model of interest here, there are six bound states within the gap, and a seventh at threshold.

The starting point for this discussion is the Ising model transfer Hamiltonian in zero field. Next the steps to the continuum limit field theory are traced to illustrate the relevance of the sine-Gordon equation for the critical behavior. It is emphasized that there are many operator equivalences in this limit, which permit the establishment of equivalent representations of a sine-Gordon problem, fermion problem, or the spin ½ one dimensional chain. This property suggests the phrase "operator democracy" reflecting the absence of preference among the representations. The field is added, and the double sine-Gordon, in that particular language, results.

The transfer Hamiltonian is given by the one dimensional Ising model in a transverse field [1]:

$$\mathcal{H}_I = \sum_i \sigma_i^z \sigma_{i+2}^z + \Delta \sigma_i^x \tag{1}$$

where the $\vec{\sigma}_i$ are Pauli matrices, $(\Delta - 1)$ is a measure of the temperature, and near the critical temperature, T_c, is proportional to $T - T_c$. The sum runs over the N sites of a chain. It should be recognized that this Hamiltonian contains two sublattices, since the even sites are not connected with the odd sites. The single sublattice problem, and correlation functions within a single sublattice, are in fact needed to solve the Ising model, and these will be extracted subsequently.

First a dual transformation on this Hamiltonian leads to another member of the democratic representation. Introduce additional spin ½ operators to be associated with bond variables by the relations $S_i^z = \sigma_i^z \sigma_{i+1}^z$. It then follows that the first part of Eq. (1), $\sigma_i^z \sigma_{i+2}^z$ becomes $S_i^z S_{i+1}^z$, and the transverse part σ_i^x has the same matrix elements in the set of states defined by the bond variables as the operator product $S_i^x S_{i+1}^x$. A rotation about the x-axis then leads to the x-y model:

$$\mathcal{H}_{xy} = \sum_i S_i^y S_{i+1}^y + \Delta S_i^x S_{i+1}^x \tag{2}$$

This spin ½ representation bears a close relation to the eight vertex model in the decoupling limit, that is, the limit of two non-interacting Ising sublattices (corresponding to the two sublattices in Eq. (1) obviously).

Through the magic of the Jordan-Wigner transformation, the spin problem can be transformed into a free fermion problem and solved [5]. The continuum limit of this problem is the charge density wave problem of a one-dimensional conductor, the non-interacting Luttinger model with a gap, the Dirac equation, or a special case of sine-Gordon problem. These steps have been discussed elsewhere, and a quick review of important results [2] is sufficient here.

The fermi problem at the critical temperature (it is important to emphasize that the temperature in the fermi system is always zero, as in Eq. (1), and the parameter Δ specifies the temperature in the statistical mechanics problem) is given by:

$$\mathcal{H}_F = \int \left[\frac{dx}{i} \psi_1^+ \partial_x \psi_1 - \psi_2^+ \partial_x \psi_2\right] \tag{3}$$

where the spinor fermi operators satisfy the canonical anti-commutation relations $[\psi_1, \psi_1^+]_+ = [\psi_2, \psi_2^+]_+ = \delta(x-x'), [\psi_1, \psi_2]_+ = 0$. Using the boson representation for these fermi fields, it is readily verified that the following bose problem results:

$$\mathcal{H}_B = \int \frac{dx}{2} [(\partial_x \varphi)^2 + \Pi^2] \tag{4}$$

Here the phase operator is related to the fermi operators through the relation

$$\varphi(x) = i\sqrt{\pi} \int^x dy\, [\rho_1(y) + \rho_2(y)] \tag{5}$$

The momentum $\Pi = \dot{\varphi}$, while $\sigma_1 = \psi_1^+ \psi_1$ and $\rho_2 = \psi_2^+ \psi_2$ are the fermion density operators.

It is now simple to follow these steps for the anisotropy, when $\Delta \neq 1$ (or $T \neq T_c$) in Eq. (2), and therefore have the fermionic equivalent to the temperature field. The results of these manipulations are [6]:

$$\mathcal{H} = \int \frac{dx}{2} \left[(\partial_x \varphi)^2 + \Pi^2 + (\Delta - 1)\alpha^{-2} \cos\sqrt{4\pi}\, \varphi \right] \tag{6}$$

where α is a short distance cutoff, corresponding to a lattice spacing in Eq. (2).

This Hamiltonian, called sine-Gordon, describes the critical region of the Ising model. The correlation functions are obtained from this directly, working the transformations backwards to Eq. (1). An Ising correlation function along a line is given by the equivalent object, for (say) even spacings,

$$\langle \sigma_{i1}^{z} \sigma_{i2}^{z} \ldots \sigma_{in}^{z} \rangle \tag{7}$$

It is now desired to obtain the equivalent operator in the sine-Gordon language. Following the dual transformation and subsequent rotation leads to the intermediate result:

$$\langle \sigma_{i1}^{z} \sigma_{i2}^{z} \ \ldots \ \sigma_{in}^{z} \rangle^{2} = \langle S_{i1}^{y} S_{i2}^{y} \ \ldots \ S_{in}^{y} \rangle \tag{8}$$

while the continuum limit of the Jordan-Wigner transformation [2] leads to the following identifications of the spin operators

$$S^{y}(x) = \alpha^{-\frac{1}{4}} \cos\sqrt{\pi}(\varphi(x) - \varphi(0)) \tag{9}$$

and this operator, at least along a line (corresponding to equal time), gives the correct Ising functions (squared).

There is an argument to find these functions correct at arbitrary positions. This, in essence, involves the use of Lorentz invariance together with the proper operator expression along a space direction (equal time). Lorentz invariance implies that a Wick rotation to the Euclidean region will give the proper correlation functions for the Ising model. It is easy to prove that the operator of Eq. (9) does not quite satisfy the Lorentz property, and therefore doesn't give the correct general functions. In the original construction of this operator, it was observed that an operator ambiguity was removed by the imposition of Lorentz symmetry. The ambiguity [2] arises from the $\varphi(0)$ operator, or in fact, any operator which has no spatial dependence. But if the equations of motion are to be Lorentz invariant, this operator must be dropped. Then from Eq. (8), the square of the proper Ising functions are given by the operator $\alpha^{-1/4}\cos\sqrt{\pi}\varphi(x,t) \equiv p(x,t)$, as:

$$\langle p(x_1,t_1)p(x_2,t_2) \; .... \; p(x_n,t_n)\rangle \tag{10}$$

and $t \to iy$ after the calculations are performed.

It is also possible to prove this operator to be unique (modulo the dual transform), for the $\cos\sqrt{\pi}\,\varphi$ for $\Delta = 1$ can be transformed to a current (or density operator) for a massless Thirring (Luttinger) model, at a coupling strength chosen to make the dimension of the current to be equal to the dimension of the $\cos\sqrt{\pi}\varphi$ in the free case (which is ½). Then the theorems about uniquess of currents [7] can be used to prove that there are only two operators which are Lorentz invariant and have a given anomalous dimension. One is the operator $\cos\sqrt{\pi}\varphi$, above, while the other is the $\sin\sqrt{\pi}\varphi$. Not surprising, since the dual transform implies that at T_c where the one dimensional models are self dual, that two equivalent representations must exist.

The result for the four spin correlation function [8] was confirmed using the construction of correlation functions on a lattice before taking the continuum limit. The present method for calculating the functions is actually trivial, for at T_c , correlation functions are Gaussian integrals, and the virtues of the "continuum limit first" espoused here, become obvious. It has been shown that for the special case of all spins on a line, the correlation functions calculated with this method [9] agree with previous results [10].

It is possible to identify the operator S_i^y with the electric field order parameter of the eight vertex model [2] , (sometimes called the arrow order parameter) and to add an electric field $\mathcal{E}\cdot P$ to the transfer Hamiltonian of Eq.(6). The result is the double sine-Gordon problem:

$$\mathcal{H} = \int\frac{dx}{2}\,[(\partial_x\varphi)^2+\Pi^2+(\Delta-1)\alpha^{-2}\cos\sqrt{4\pi}\varphi+\varepsilon\,\alpha^{-1/4}\cos\sqrt{\pi}\varphi] \tag{11}$$

where the electric field, $\mathcal{E}$, has been chosen in the y-direction. At the critical temperature $\Delta = 1$, it is possible to read off the spectrum of states, because the sine-Gordon equation results with the $\cos\sqrt{4\pi}\,\varphi$ replaced by $\cos\sqrt{\pi}\,\varphi$. The spectrum is given by [11] ,

$$\Delta_n = 2\Delta_0 \sin\frac{n\pi}{14} \tag{12}$$

for $n = 1,2 \,..\, 7$, and the gap Δ_0 is given by [3]

$$\Delta_0 \sim \varepsilon^{4/7} \tag{13}$$

Only the power law behaviour in Eq. (13) has been given; the numerical prefactor is non-universal and not of interest for the present discussion. It is remarkable that precisely seven bound states occur for the Ising case, and it is natural to ask if these states can appear in the spin pair correlation function.

However, the difficulty occurs in sorting out the algebra to check whether a relation analogous to Eq. (8) is valid. Eq. (8) has been derived only in the zero field limit. Since it has been used to calculate all spin correlation functions, it is possible to use these to derive the perturbation theory in the Ising model external field, as in fact done elsewhere [2] . In this way, there is a close relation between the Ising and eight-vertex models in their respective external fields, although it is clearly more complicated than Eq. (8). Further support for an equivalence between the two problems in an external field comes from the fact that the magnetic gap in the Ising case can be calculated from Eq. (8), giving the dependence $H^{8/15}$, the correct operator dimension. The same operator, P(x), determines the exponent 8/15 and 4/7 in both problems, the difference arising from the fact that the"square" of one is equal to the other.

The analytic structure of the Ising model in a field near T_c has been extensively analyzed, [12] although results precisely at T_c are lacking. The relationship between these two complementary viewpoints involves knowing the precise generalization of Eq. (8) including the external field.

In this language, the critical region is described by two gaps, one from the temperature field, $T - T_c$, the other from the external field, $\varepsilon^{4/7}$. The analytic structure of correlation functions near the critical point is obviously a complicated problem, particularly since the corresponding operators are composite operators, that is, they have matrix elements between multi-particle states. The arrow correlation function has the operator $\cos\sqrt{\pi}\varphi$ which can have such matrix elements. If this operator has matrix elements between the ground state and single bound state, the arrow correlation function will have simple poles at the appropriate binding energies. Matrix elements to pairs of these bound states will introduce a threshold, and the first such threshold occurs at twice the energy of the lowest state, $2\Delta_1$, which is still greater than Δ_2.

It is interesting to consider the possibility that these bound states are a universal feature of critical phenomena. In other solveable models, such as the gaussian or spherical models, there are no bound states within the gap generated by an external field. This is not surprising, since these models are completely free field models, that is, the order parameter is a free field and the Hamiltonian is bilinear in these fields. Depending on the representation, the Ising case is either a free particle Hamiltonian with complicated order parameter, or a free field order parameter with an interacting particle Hamiltonian. Certainly, the general expectations are that real three dimensional models are truly interacting systems, and it is then possible that there are bound states within the scale gap. If so, the number of these would be a universal quantity, dependent only upon the symmetry of the order parameter.

The importance of the number seven has often been suggested on theological and historical grounds, and it is with some satisfaction that this number is now singled out as special in the Ising model. Perhaps the converse is true as well.

References

1. T.D. Schultz, D.C. Mattis, and E.H. Lieb, Rev. Mod. Phys. 36, 856 (1964)

2. A. Luther and I. Peschel, Phys. Rev. B12, 3908 (1975); A. Luther, Phys. Rev. B14, 2153 (1976)

3. R.J. Baxter, Phys. Rev. Lett. 26, 834 (1971); Ann. Phys. (N.Y.) 70, 323 (1972); J.D. Johnson, S. Krinsky, and B. McCoy, Phys. Rev. A8, 2526 (1973)

4. L.P. Kadanoff and F. Wegner, Phys. Rev. B4, 3989 (1971)

5. B. McCoy, Phys. Rev. 173, 531 (1968)

6. The sine-Gordon equation with $\beta^2=4\pi$ is plagued with divergence problems in the continuum limit. The discussion here strictly refers to the interacting fermi problem on a lattice which does not, but does have the same infrared behavior as the sine-Gordon equation.

7. B. Klaiber, Lectures in Theoretical Physics, edited by A. Barut and W. Britten (Gordon and Breach, New York, 1968) Vol. XA

8. Helen Au-Yang, to be published

9. M. Bander and J. Richardson, Phys. Rev. B, to be published

10. L.P. Kadanoff and H. Ceva, Phys. Rev. B3, 3918 (1970)

11. L.D. Faddeev and L.A. Takhtajan, Theor. Mat. Fiz. 21, 160 (1974); R.F. Dashen, B. Hasslacher, and A. Neveu, Phys. Rev. D11, 3424 (1975)

12. B. McCoy and T.T. Wu, to be published

Statistical Mechanics of Nonlinear Dispersive Systems[1]

Alan Bishop

Physics Department, Queen Mary College, University of London
Mile End Road, London E1 4NS, Great Britain

and

IBM Zurich Research Laboratory, CH-8803 Rüschlikon, Switzerland[2]

1. Introduction

The primary recurrent theme in 'soliton physics' as it is now evolving [1] in statistical physics, critical phenomena and condensed-matter physics generally (as well as quantum-field theory and gravity, etc.), is the simple recognition that linearization schemes are frequently misleading and (as witnessed by the contents of this Symposium) sometimes unnecessary. It is increasingly appreciated that we need to go beyond (linear) normal modes and finite-order perturbation theories in a spectrum of intrinsically nonlinear problems for which distinctive nonlinear, e.g., soliton, sectors of solution space are possible. Since the physical signatures of nonlinear modes, particularly spatially-limited ones, are often so crucial, it is essential that they are fully represented. In this situation, there is a growing tendency to develop theories which *ab initio* include *all* fundamental modes (including nonlinear ones) even if this is only possible semi-quantitatively. Such an approach has well-established precedents, e.g., phonon-roton gas models in liquid-helium-4 or vortices in superconductors.

It is precisely this idea which will motivate our discussion of the classical statistical mechanisms for a class of model, nonlinear, dispersive Hamiltonian systems below. Essentially, we argue that a path-integral partition function (Z) representation for a Hamiltonian density $\mathcal{H}(\underline{x})$, i.e.,

$$Z = \int \mathcal{D}[\phi] \quad \exp[-\beta \int \mathcal{H}(\underline{x})\ d\underline{x}] \qquad (1)$$

[with field $\phi(\underline{x},t)$: $\underline{x}$ = space-variables; t = time; $\beta^{-1} = k_B T$], will be dominated by extremal paths minimizing $\int \mathcal{H}(\underline{x}) d\underline{x}$, i.e., by *all* solutions of the Euler-Lagrange equation of motion for $\phi(\underline{x},t)$. We then try to build a 'configurational phenomenology' based upon a mixture of these dominant trajectories with an appropriate thermal weighting. This is entirely analagous to the procedure adopted (when practicable [2]) in quantum-field or critical phenomena, where the dominant contributions are classical or mean-field approximations. To make this kind of approach rigorous it is important to have some knowledge of mode stability and of mode-mode interactions generally. In steepest-descent (or stationary-phase) theory these arise in first order as 'Gaussian corrections' from the interference effects of trajectories neighbouring the dominant ones, 'dressing' the bare mode with quantum [2],

---
[1] Supported in part by SRC (U.K.). [2] Guest Scientist, July 1978.

critical [3] or thermal (§3) fluctuations as appropriate.

In a system supporting *nonlinear* modes, superposition is, of course, *not* possible. Thus 'dressing' includes contributions from *interactions* between modes and we will emphasize that it can be vitally important to include these to ensure a physically satisfying phenomenology. Specifically, we will need to consider the effects of spatially-limited nonlinear modes (e.g., solitons) on the phase space available to extended ones (e.g., phonons) which are affected by boundary conditions. In fact, this problem is the natural corollary to the nonlinear conceptual revolution, namely, how to renormalize a nonlinear ground state, mean-field configuration, classical solution, etc., especially in view of symmetries (e.g., translation) broken by the nonlinear configurations. Apart from technical differences, exactly this same question is presently posed in statistical mechanics and Renormalization-Group approaches to critical phenomena (e.g., interfaces in mixtures [3]; discommensurations [4]; vortices in isotropic 2-D planar spin models [5]), quantization theories [2], etc. As yet *complete* understanding is limited to special models, particularly those in one-space dimension which we consider in §2. These examples have analytically tractable features which have led to their popularity in all the areas mentioned above, but it is important to stress that the basic physical features and nonlinear consequences are not limited to these special models. Certainly, numerical studies [6] and models supporting topological or non-topological defect structures [1,2,5] have strongly supported the broad concept of a configurational approach.

The parallel roles of critical, quantum and thermal fluctuations in statistical mechanics are widely appreciated (see [1] for a summary). Indeed, quantum fluctuations can sometimes actually destroy a classical phase transition. Although we will deal here only with classical statistical mechanics, it is possible to include quantum fluctuations [2,7] and conclude that at sufficiently low T the concept of 'ideal' gases of nonlinear excitations is still valid. It is also worthwhile noting the formal similarity between 'instantons' in quantum-field theory [2,8] and the classical statistical mechanics of §3, although instantons are quantum-mechanical tunneling phenomena between classical vacuum states.

The term 'soliton' is used with a variety of connotations in the natural sciences, and even in the context of this Symposium. We should be clear about the terminology now used in statistical mechanics. To the mathematician [9,10] the term is usually reserved for particle-like kink, pulse or envelope excitations appearing in totally integrable Hamiltonian systems. These examples are extremely rare and are associated with equally special properties (§2.3). Real physics is never concerned with such strict soliton systems. Fortunately, many of the *physically* important soliton characteristics (e.g., the existence of topologically stable particle-like excitations) do not demand the esoteric strict soliton environment. Consequently, the delightfully tractable strict soliton systems are ideal and quite representative models *if* realistic physical limitations are introduced [1,11]. (They can usefully be considered as the nonlinear generalization of a purely harmonic system.) This is particularly so for perturbation studies [10,11] and statistical mechanics where they provide natural models for a configurational approach if ergodicity-restoring interactions are introduced (§3). Again, their simplifying role in quantization schemes is well-documented [2].

In field theory and condensed matter it is now fashionable to use 'soliton' to describe a wide variety of stable, finite-energy, particle-like field patterns in various dimensions, which retain some central physical nonlinear properties but few of the precise mathematical ones. We will adopt this weaker nomenclature, referring to the mathematicians' by 'strict' (or 'aristocratic' [12]) soliton. The need to identify the more general nonlinear soliton modes has cultivated powerful generic tools which now form equally central foundations to 'soliton physics' - the concept of 'frustration' in the presence of competing interactions [1] (e.g., spin-glass, incommensurate structures), the use of classical techniques of algebraic *topology* (e.g., homotomy groups) to classify topologically stable 'order-parameter defects' allowed w.r.t. a given ordered medium [13] (e.g., liquid crystals, helium -3 and -4), etc. Topological arguments are usually possible if *degenerate* ground states are available, when homotopic classification reduces to considerations of boundary conditions. Alternatively, non-topological defects are still possible without degeneracy if non-trivivial conservations are imposed (of angular momentum, isospin, baryon number,etc.) [14]. Ultimately, specific minimization of a free energy, Lagrangian, etc. is necessary to locate metastable states or the detailed form of stable ones. Special models (§2) and numerical techniques are then valuable.

Although we will not be considering limiting critical effects in dimension $D > 1$ in this article, it is useful to place our nonlinear concerns in the wider framework of nonlinearity and critical or subcritical phenomena. Elsewhere [1] we have suggested four categories: (i) first-order phase transitions, including comparisons with droplet models and metastable states in field theory, or the thermodynamics of mixtures [8]; (ii) continuous phase transitions, and the emphasis on short-ranged, long-lived coherent cluster excitations in addition to conventional continuum modes and associated critical exponents and time scales [15]; (iii) defected mean fields where the mean-field solution (viewed as a, say, temperature-dependent ground state) is itself nonlinear or 'defected', requiring critical fluctuations to be chosen carefully (c.f. above). Examples include [4,16,17] incommensurate epitaxial, spin- or charge-density-wave structures; intermediate phases in ω-transforming alloys; interfacial structures, etc., and (iv) defect-induced transitions, where phase transitions may be intrinsically associated with the appearance of inhomogeneous (defect) states. Examples here are dislocations at ordinary melting transitions, discommensurations at commensurate-incommensurate transitions, transitions between various phases in liquid crystals [13], vortices in the isotropic 2-D XY model [5]. Note that inhomogeneous states need not always induce a transition, although they will seriously affect dynamics and correlations (e.g., the isotropic 2-D Heisenberg system [18]).

The examples we treat below fall into class (ii) so that nonlinearity enters as fluctuation structure. Nevertheless, the concept of a configurational approach and precursor effects is central to all the above classes. We will also confine ourselves to Hamiltonians with degeneracy in their bare form (e.g., through a phase or spin variable), rather than being generated by renormalizations as in a Ginzburg-Landau-Wilson approach. In fact, the relationship between these two sources of nonlinearity is not entirely clear, especially in isotropic systems [15]. Topological arguments are available in our cases (degenerate ground states). The models are actually quite directly relevant to some quasi-one-dimensional electronic [19] and magnetic [20] materials - low dimensionality frequently enhances nonlinear features, providing a substantial regime of low-dimensional fluctuations (with respect to an ordered condensate if below a long-range ordering temperature) outside any ultimate limiting 3-D critical region. Most generally, we are interested in nonlinear lattice dynamics both critically and subcritically, and our examples illustrate possible *intrinsic* mechanisms for soft modes and central peaks [15].

We reiterate that the primary emphasis is on configurational phenomenologies: both experimentalists and theorists should be at least as concerned with order-parameter *structure*, as with critical exponents. Specific configurations can imply very physical structural, response, energetic, transport properties. It should be emphasized that *finite* lifetime modes are a very relevant concern in statistical mechanics. Although we will concentrate on dispersive systems it will be appreciated that nonlinear *dissipative* equations are an equally challenging area of great physical importance - e.g., in diffusive critical phenomena, population dynamics, spinodal decomposition, chemical reaction fronts, interface dynamics, etc.

2. Model Hamiltonians and Mode Analysis

The models we will discuss below are chosen because of their analytical convenience but are also considered representative of very general nonlinear statistical mechanical systems (§1). In particular, we will emphasize the idea of a real space configurational approach and therefore need a rather detailed analysis of the nonlinear modes available and of how they interact. This is one reason why the models chosen are in one space dimension (D = 1), where the information is most accessible, but also in 1-D we have powerful alternative schemes for evaluating a partition function (§3). Thus the configurational phenomenology can be carefully monitored. The generality of the basic lessons are, however, becoming appreciated much more widely (see §1 and elsewhere in these Proceedings), particularly through the roles of 'inhomogeneous states'. An extremely popular example is provided by the 2-D isotropic planar spin model [5] (an example with two-component order-parameter dimension: n = 2). Here, topological vortex excitations are essential to the curious non-ordering phase transition which occurs. In fact, the partition function can be specifically decomposed into vortex and spin-wave configuration contributions [5]. Symmetry-breaking terms restore a conventional continuous phase transition [5] but then different nonlinear excitations are operative - in fact, 2-D generalizations of the 1-D, n = 2 'walls' we describe in §2.2. As we mentioned in §1, all of our models will have a degenerate local potential in the *bare* Hamiltonian. In real systems, this may appear through a complex order-parameter phase dependence, or from microscopic energy calculations or because we operate well below a mean-field scale temperature, so that broken symmetry has already developed in a mean-field effective Hamiltonian. We will omit any Hamiltonian dependence on thermodynamic fields (temperature, pressure, magnetic field, etc.), although this is easily restored and essential in some problems (e.g., multiphase equilibria [1], ω-phase-transforming alloys [16]; incommensurate phases in epitaxial, spin-phonon, electron-phonon systems [4,17], etc.).

2.1. Generalized Klein-Gordon Models (D = 1)

These have motivated much recent field-theory [2] and statistical-mechanics [21] literature. We consider the classical Hamiltonian density [22,23]

$$\mathscr{H}[\phi] = A\,\{\tfrac{1}{2}\,\dot{\phi}^2 + \tfrac{1}{2}\,c_0^2\,\phi_x^2 + \omega_0^2\;V(\phi)\}, \tag{2}$$

where $\phi(x,t)$ is a dimensionless, one-component field variable in one space (x) and time ($\phi_t \equiv \dot{\phi}$). 'A' sets the energy scale, ω_0 and c_0 are characteristic frequency and velocity, and the only constraint on the local (on-site) potential $V(\phi)$ is the topological one that it has at least two degenerate minima ($\phi = \phi_1$ and ϕ_2, say) and is bounded on (ϕ_1, ϕ_2). Topologically stable, single-soliton (kink, solitary wave, etc.) solutions will then follow from the equation of motion for ϕ

$$c_0^2\phi_{xx} - \ddot{\phi} = \omega_0^2\,\frac{dV}{d\phi}\,. \tag{3}$$

We simply look for a travelling-wave solution and apply boundary conditions [22] for a kink, ϕ_K^v with velocity v: using $s \equiv x - vt$, the solution describing a single kink is found as $[V(\phi_{1,2}) = 0]$

$$\frac{d\phi_K}{ds} = \pm\left[\frac{2\omega_0^2}{c_0^2}\left(1 - \frac{v^2}{c_0^2}\right)^{-1} V(\phi_K)\right]^{1/2} \tag{4}$$

or

$$s = \pm\left(1 - \frac{v^2}{c_0^2}\right)^{1/2} \frac{d}{\sqrt{2}} \int_{\phi_K(0)}^{\phi_K(s)} d\phi\,[V(\phi)]^{-1/2}\,. \tag{5}$$

Here, $|v| < c_0$ and the (static) kink 'width' is $2d \equiv 2c_0/\omega_0$. The 'relativistically invariant' form of (2) or (3) guarantees the covariant form of (5) and similarly assures the velocity dependence of the energy - direct calculation of the excess energy, E_K^v, associated with a bare kink gives

$$E_K^v = E_K^0\left(1 - \frac{v^2}{c_0^2}\right)^{-1/2} = \left(E_K^{0\,2} + p^2c_0^2\right)^{1/2}, \tag{6}$$

with 'relativistic momentum' $p \equiv M_Kv(1 - v^2/c_0^2)^{-1/2}$, where $E_K^0 = M_Kc_0^2$, and the 'rest mass' is given by

$$M_K = \frac{A\sqrt{2}}{d}\int_{\phi_1}^{2} d\phi\,|V(\phi)|^{1/2}\,. \tag{7}$$

Popular analytic examples of this class include the 'ϕ^4':$V=(\phi^2-1)^2/8$ [21]; sine-Gordon (SG): $V=1-\cos\phi$ [9]; multiple SG: e.g., $V=\pm(\cos\phi+\cos\phi/2)$ [24]; double quadratic: $V=1/2(\,|\phi|-1)^2$ [17,25]; rectangular wells and others. SG and ϕ^4 are representative of sub-classes in that they display bounded period and unbounded non-periodic local potentials, respectively, with some different physical consequences [22]. Explicit solutions follow readily from (5) and (7): e.g.,

$$\phi^4 : \quad \phi_K^v(x,t) = \tanh\left[+ \frac{(x-vt)}{2d(1-v^2/c_0^2)^{1/2}} \right] \quad ; \; E_K^0 = \frac{2}{3} Ac_0\omega_0 . \qquad (8)$$

$$SG : \quad \phi_K^v(x,t) = 4\tan^{-1}\left[\exp \pm \frac{(x-vt)}{d(1-v^2/c_0^2)^{1/2}} \right] ; \quad E_K^0 = 8Ac_0\omega_0 . \qquad (9)$$

Having found bare (classical) solutions the next question is invariably (§1) how these are 'dressed' with small fluctuations. Mathematically this coincides with familiar linear stability analysis [11]. SG and ϕ^4 enjoy some special simplifications in this regard (below) which is a major reason for their popularity. However, we emphasize again that these models are only singularly simple mathematically - the important *physical* conclusions almost all generalize completely, even though analytical convenience may be partially lost.

Linearization about any of the constant vacuum states in (2) (e.g., $\phi = \phi_1$ or ϕ_2) produces familiar *extended* periodic excitations (harmonic in the linear limit) - quite different from the particle-like, large-amplitude solitary waves. These extended modes will correspond to phonons, mesons, spin-waves, etc. and their unsuitability as a basis for expanding solitary waves is precisely the theme of §1. Specifically, if (3) is *linearized* about a potential minimum $\phi_{1,2}$ and we assume $V(\phi)$ is symmetric, the phonon modes are

$$\phi(x,t) - \phi_{1,2} \propto \exp\left[i(\kappa x - \omega_\kappa t)\right] \qquad (10)$$

with dispersion relationship

$$\omega_\kappa^2 = \omega_0^2 + c_0^2\,\kappa^2 , \qquad (11)$$

where $V(\phi)$ has been scaled so that

$$d^2V/d\phi^2\big|_{\phi_{1,2}} = 1.$$

A possibility always remains that linearized modes of this sort are modulationally unstable if higher-order effects are included. Thus, expanding to third order about a potential minima for ϕ^4 is known [1] to lead to the possibility of localized envelope solitons. Whilst such questions do not affect the validity of formal expansions and term-by-term comparisons in statistical mechanics (§3), they are important in interpreting precisely what field patterns are observed in numerical simulations (or experimentally if the pure model is directly relevant), or for establishing realistic configurational phenomenologies (§3). Thus, molecular-dynamics simulations [26] for a ϕ^4 chain exhibit evidence for pulse-soliton configurations as well as kinks (or domain walls). The pulse-soliton modes can be viewed [1] as (third-order) approximations to the finite-lifetime 'quasi-breather' modes found numerically in the ϕ^4 system. However, the infinite potential walls of this system preclude any total instability. On the other hand, the bounded potential in SG may mean that linear phonons are modulationally unstable to fully developed 'breathing' modes, known to be 'normal modes' of that very special system [9]. Such conclusions are subject to boundary conditions - e.g., natural normal modes of finite SG, Korteweg-de Vries, etc., systems with periodic boundary conditions appear as cnoidal waves [27] - although physical differences should be small in the large system limit.

Complete classes of *travelling-wave* solutions can generally be found as elliptic functions from which kinks or phonons appear as limits. For example, in ϕ^4 three classes are necessary [28] : the cn, dn and sn Jacobi elliptic functions, describing oscillations above the double-well structure, about the potential well minima, and about the potential hump, respectively. Naturally, they have different regimes of stability.

We have implicitly assumed a continuum limit in (2). Typical physical contexts will carry an intrinsic discreteness length scale, ℓ. The simplest generalization is the Hamiltonian

$$H = \sum_i A\,\ell\,\{\frac{1}{2}\dot{\phi}_i + \frac{1}{2}\frac{c_0^2}{\ell^2}(\phi_{i+1} - \phi_i)^2 + \omega_0^2 V(\phi_i)\}, \tag{12}$$

with the ensuing coupled differential-difference equations of motion replacing (3). Kinks remain stable excitations, although their energy, structure and dynamics are modified [29]. Similarly, phonons are allowed with modified dispersion

$$\omega_\kappa^2 = \omega_0^2 + 4(c_0/\ell)^2 \sin^2(\frac{1}{2}\kappa\,\ell). \tag{13}$$

We will need to use (13) in §3.2, rather than the Debye limit (11). The continuum or 'displacive' limit is valid for $d >> \ell$, i.e., when the coupling between neighbours is large compared to the on-site potential so that ϕ-variations from site to site are gradual.

Linear kink stability analysis proceeds [11] by substituting $\phi = \phi_K + \psi$ in the governing equation of motion (2) and linearizing in ψ, which then satisfies a 'scattering' equation with solutions of the form $\psi = f(x)\exp(i\omega t)$. In all cases there will be at least one bound state (B) with $_0\omega_B = 0$, representing (to linear order) rigid translation of the kink: $f_B(x) \propto d\phi_K/dx$. This may be termed a Goldstone translation mode. There may be further finite frequency bound states describing localized kink oscillations. In addition there will be a scattering continuum $f_\kappa(x)$ with $\omega = \omega_\kappa$. In fact for SG or ϕ^4 the scattering problem is purely reflectionless [30]. The phonon dispersion (11) is unchanged by the presence of a kink, whose only effects are to produce an *asymptotic phonon phase shift* and trap the finite number of modes described above. The number of modes trapped is related to the asymptotic phase shift in a Friedel sense, (c.f., impurities in an electron gas). In the SG case, for example [11], $f_B(x) = (2/d)\,\mathrm{sech}(x/d)$ and there are no other bound states. The continuum modes are

$$f_\kappa(x) = (2\pi)^{-1/2}(c_0/\omega_\kappa)\exp(i\kappa x)\,[\kappa + (i/d)\tanh(x/d)].$$

Examining the asymptotic forms of $f_\kappa(\pm\infty)$ we find that the phonons have suffered an asymptotic phase shift

$$\delta(\kappa) = \pi\kappa/|\kappa| - 2\tan^{-1}(\kappa d) \tag{14}$$

due to the (static) kink. The nonlinear modes satisfy orthogonality and completeness relations and the energy, E, is given to this linear order by

$$E = M_K c_0^2 + \frac{1}{2} M_K v^2 + \int d\kappa\, \rho(\kappa)\, \omega_\kappa . \tag{15}$$

Here v is the kink velocity and $\rho(\kappa)$ the density of phonon states. In these senses the 'kink-phonons' and kinks are 'nonlinear normal modes'.

The most important point is that the phonon density of states is *changed* by the phase shift, since these modes are extended. If we introduce periodic (Born-von Karman) boundary conditions on a chain of finite length L, then the allowed wave-vectors are

$$L\kappa_n + \delta(\kappa_n) = 2\pi n \quad (n = 0, \pm 1, \ldots) \tag{16}$$

and

$$\rho(\kappa) = \frac{L}{2\pi} + \frac{1}{2\pi}\frac{d\delta(\kappa)}{d\kappa} = \rho_0 + \Delta\rho(\kappa). \tag{17}$$

From (14) and (16) we see that the $\kappa_0 = 0$ mode is no longer allowed and all other modes are shifted towards $\kappa = 0$ slightly. (For a kink of velocity v, the mode at $\kappa_v d = -v/(c_0^2 - v^2)^{-1/2}$ is lost.) As required by degree of freedom conservation, the total number of phonon states is decreased by the number of bound states: $⨍ d\kappa\Delta\rho(\kappa) = -\pi^{-1}\delta(0+) = -1$, where $⨍$ signifies Cauchy Principal value. We see here the *essential* mechanism by which the kink's creation and translation are compensated for by two degrees of freedom in the phonon sea.

For ϕ^4 a second mode is removed from the continuum at frequency $\omega_1 = 1/2\sqrt{3}\,\omega_0$ and the maximum asymptotic phase shift, $\delta(0+)$, is 2π. The multiple SG and double quadratic systems are examples without the simplification of a reflectionless potential. In fact, the latter case presents a textbook δ - fn, scattering potential [25], so that again only one mode $(\omega_B = 0)$ is trapped. The asymptotic phase shift is actually the same as for SG, i.e., (14). [$f_\kappa(x)$ is a superposition of $\pm\kappa$ phonons.] These circumstances explain the popularity of ϕ^4 and SG, but again demonstrate that the central physical conclusions are much more general, even if potentials are reflecting or if numerical implementation of stability analysis is necessary. The results for class (2) are given in complete generality in [23].

2.2. Complex Scalar Fields with Phase Anisotropy

Multi-component order-parameter Hamiltonians $(n > 1)$ are also important in many physical problems, e.g., with coupled order parameters, complex order parameters $(n = 2)$, spin systems, etc. For $D > 1$ various defect types are possible [13,14,18] but even in $D = 1$ stable solitons are possible if we introduce anisotropy in n-space.

Several models have been studied but an $n = 2$ system with Ising (uniaxial) symmetry is especially tractable [31]: the Lagrangian density is

$$\mathscr{L} \propto \frac{1}{2}|\psi_t|^2 - \frac{1}{2}|\psi_x|^2 + \frac{1}{2}A|\psi|^2 - \frac{1}{4}B|\psi|^4 - D|\psi|^2(1 - \cos 2\phi) \tag{18}$$

with $\psi(x,t) = u(x,t)\exp[i\phi(x,t)]$ and A,B,D positive constants. The last three terms in (18) can be thought of as the local potential $V(\psi) \equiv -Au^2/2 + Bu^4/4 + Du^2(1 - \cos 2\phi)$ which corresponds to the shape of the 'bottom of a wine bottle' (c.f., the ϕ^4 model, §2.1) whose continuous rotational symmetry has been broken and replaced with a (2-fold) discrete one. There are thus two degenerate potential minima and solitary-wave profiles will now correspond to *coupled* variations in both u and ϕ as the field evolves between the two vacuum states. The amount of coupling should clearly depend on the relative magnitudes of the central hump and phase 'dimple' potentials, or equivalently the ratio D/A.

Linearization about a ground state readily produces decoupled amplitude-phonon and phase-phonon ('amplitudon' and 'phason') approximate, extended modes analogous to those for class (2). The complex kink solutions are best derived by transforming to the variables $\xi = \mathrm{Re}(\psi)$, $\eta = \mathrm{Im}(\psi)$ in which the only asymmetry derives from symmetry-breaking, i.e., $D \neq 0$. Two classes of solitary waves can be found [31] upon transforming to $s \equiv (x - vt)(1 - v^2)^{-1/2}$ ($|v| < 1$)

$$\begin{aligned} \xi_1(s) &= \pm u_0 \tanh[(\tfrac{1}{2}A)^{1/2}s] \\ \eta_1(s) &= 0 \end{aligned} \tag{19}$$

and

$$\begin{aligned} \xi_2(s) &= \pm u_0 \tanh(2D^{1/2}s) \\ \eta_2(s) &= \pm u_0(1 - \frac{8D}{A})^{1/2} \operatorname{sech}(2D^{1/2}s). \end{aligned} \tag{20}$$

Solution 2(20) is admissable only if $A > 8D$. Eqs. (19) and (20) coincide when $A = 8D$. We therefore have a classic example of ground state (or mean field) 'bifurcation' (or equivalent) 2D phase transition [1,31]. Although both solutions exist for $A > 8D$ only type 2(20) is stable, as can be proven [31] by the same linear stability analysis described above. The same conclusion is suggested by the relative magnitude of kink energies [31]

$$\frac{E_2^0}{E_1^0} = 3(\frac{2D}{A})^{1/2}(1 - \frac{8}{3}\frac{D}{A}) < 1, \quad \text{if} \quad 8D < A. \tag{21}$$

Notice that because of the Ising symmetry, type 1 (19) is no more than the ϕ^4 solution (§2.1) and D-independent. Similarly, in the limit $D \ll A$, amplitude variations are costly in energy and the solitary-wave evolves at essentially constant amplitude. This 'phase-dominated' regime [19] in fact corresponds to the SG equation, as we see from (20). Model (18) nicely demonstrates how the amount of phase-amplitude coupling is determined by the relative magnitudes of the phase and amplitude potential components.

2.3. Integrable Hamiltonian Systems

Clearly, travelling waves do not represent a complete solution to class (2). They are unable for example to describe a multi-kink configuration. In some cases more general special solutions are known, together with special techniques for generating them, e.g., Bäckland transformation, Hirota's technique, the inverse scattering transform, etc. [9,10]. In practice, all of these devices and special solutions apply (almost invariably) to a *very special* subset of nonlinear equations, namely those deriving from fully integrable Hamiltonian systems. It is well known [9] that examples of this type are extraordinarily generic physically but also highly singular mathematically. We will not describe examples in detail (see elsewhere in these Proceedings) except to point out that SG is the only member in class (2). Fully integrable systems make very beautiful idealized theoretical models (c.f. the role of the harmonic system in linear physics), primarily because a *complete* knowledge of solutions to the corresponding equations of motion is possible and (nonlinear) normal modes can be identified, lending themselves directly to the idea of a configurational basis. In these special cases, pulse, kink or envelope solutions correspond to our terminology 'strict soliton'. Such soliton-bearing equations are associated with a variety of related and quite remarkable special properties. For example, with appropriate boundary condition assumptions, they are *completely* soluble, via (for example) the 'inverse scattering transform' [9,10], a canonical transform to functions of generalized action-angle variables which serve to label allowed types of excitation ('nonlinear normal modes') in the system. The allowed excitations are infinitely long-lived, retaining their asymptotic identity exactly under 'collision' with other excitations. They simply suffer pair-wise additive asymptotic shifts. (c.f., *linear* order results of §2.1 for SG and ϕ^4.) Thus, general solutions can be labelled by their nonlinear normal mode components, and indeed can be expressed in the form of a generalized Fourier transform with respect to these [10]. A number of related properties also follow including recurrence phenomena, an infinite number of conservation laws, etc. For our statistical mechanics purposes, the most suggestive consequence [c.f., the linear order result (15)], is that the associated Hamiltonian (and momenta) *are exactly separable* [32] into contributions from the nonlinear normal modes. (See also parallel uses of these ideal models in perturbation studies [10] and quantum field theory [2].)

It should be noted that the most frequently cited integrable system forms apply strictly only to the continuum, infinite system limit with boundary condition $\phi_x \to 0$ sufficiently rapidly as $|x| \to \infty$. (e.g., the SG 'normal modes' are then [1,32] solitons (antisolitons), 'breathers' and continuum excitations.) This may be important since partition function or quantization calculations are regularly performed (§3) on a finite, discrete system and with a variety of boundary conditions; continuum and infinite system limits are taken last. In fact there is a good deal of interest from mathematicians in rigorous studies of finite versions of, e.g., SG with periodic boundary conditions [27]. Again, several discrete integrable models soluble by inverse scattering are available, although it must be noted that these do not coincide with the local discretizations usually posed in condensed matter [i.e., (12)].

3. Equilibrium Statistical Mechanics: Exact Results and Configuration Phenomenology

The ideas described in §1 have prompted attempts [21,23] to construct configurational phenomenologies emphasizing elementary nonlinear excitations. We describe here how this approach can be made rigorous for the low-T limit of class (2) *if* mechanisms for degrees of freedom conservation and free energy sharing are incorporated, arising from the non-superpositional interactions of §2.1. An extended discussion of these results will be given elsewhere [23], but the principal ideas are summarized below. We first describe an exact calculation of the classical partition function Z in the original variables $(\phi,\dot{\phi},\phi_x)$, with which we then compare phenomenology.

3.1. Exact Partition Function Calculation

For the *discrete* system governed by (12) we can use a transfer integral technique to calculate Z, details of which can be found elsewhere [21]. We ultimately compare results with a phenomenology based on §2.1, so that we will primarily consider the continuum limit $\ell/d \to 0$. Classically, Z factors as $Z = Z_{\dot{\phi}} Z_{\phi}$ with [23] (L = chain length)

$$Z_{\dot{\phi}} = (2\pi A \ell/\beta h^2)^{N/2} \tag{22}$$

$$Z_{\phi} = \sum_n \exp(-\beta A \omega_0^2 L \epsilon_n), \tag{23}$$

where $\beta = (k_B T)^{-1}$, h is Planck's constant and $L = N\ell$. Eq. (25) assumes periodic boundary conditions $\phi_{N+1} = \phi_1$, but others are readily possible [23]. $\{\epsilon_n\}$ in (23) are the eigenvalues of the transfer integral operator [21], which can be solved by various techniques. However, in the displacive limit, much more intuition is possible from a differential approximation valid to $O(\ell/d)$ [21,23]

$$H(\phi)\, \Psi_n(\phi) = \epsilon_n \Psi_n(\phi) \tag{24}$$

with

$$H(\phi) = -\frac{1}{2m^*}\frac{d^2}{d\phi^2} + V(\phi) + V_0 \tag{25}$$

$$m^* = A^2 \omega_0^2 c_0^2 \beta^2 \propto (\beta E_K^0)^2 \tag{26}$$

$$V_0 = (2\beta\omega_0^2 \ell A)^{-1} \ln (A c_0^2\beta/2\pi\ell). \tag{27}$$

We are thus faced with a pseudo-Schrödinger equation for a single particle of dimensionless 'mass' m^*, in 1-D, moving in the nonlinear potential $V(\phi)$ [33].

Physical interpretations are possible at both high and low T. Here we concentrate on $m^* >> 1$. Note first from (23) that in the thermodynamic limit $(L,N \to \infty\ ;\ \ell = \text{const.})$, Z_{ϕ} is dominated by the lowest eigenvalue ϵ_0. Thus the configurational (i.e., potential energy) free energy density contribution is

$$F_\phi/L \equiv -(k_BT/L)\ \ln Z_\phi \overset{L\to\infty}{\to} A\,\omega_0^2\,\epsilon_0. \tag{28}$$

We therefore need to determine ϵ_0. At low T, i.e., $\beta E_K^0 >> 1$, the low-lying eigenspectrum will be 'tunnel-split'. Specifically, if E_0 is the lowest level in a single isolated well of $V(\phi)$, then

$$\epsilon_0 = E_0 - t_0 \tag{29}$$

where t_0 is the tunneling component. For large m^*, E_0 is approximated by the lowest harmonic oscillator level (suppressing V_0): $E_0 = 1/2\ m^{*-1/2} + O(T^2)$. The splitting is given reasonably accurately by the WKB tunneling formula [34]

$$t_0 \cong 2\pi^{-1}\ E_0(T)\ \exp[-I(T)] \tag{30}$$

$$I(T) = (2m^*)^{1/2} \int_{\phi_1+\delta\phi_1(T)}^{\phi_2-\delta\phi_2(T)} d\phi \left| E_0 - V(\phi) \right|^{1/2}. \tag{31}$$

Here $\delta\phi_{1,2}(T)$ determine turning points for the WKB integral.

Periodic potentials (e.g., SG) differ from unbounded ones in several ways. For example, the pseudo-Schrödinger equation (24) poses a 1-D 'band structure' problem (Hill equation). Thus the eigenfunctions have Bloch form and the eigenvalues lie in a sequence of continuous bands - despite the absence of a discrete spectrum, result (28) still holds [23]. For the SG case, the eigenproblem (24) is actually equivalent to the Mathieu equation, which is extensively documented [35]: asymptotic expansions in the low T ('tight-binding') limit give the centre of each band as $(n = 0,1,2,\ldots)$

$$\bar{\epsilon}_n = 4(2n + 1)(\beta E_K^0)^{-1} - [(2n + 1)^2 + 1]\ (\beta E_K^0)^{-2} + \ldots \tag{32}$$

The linear T term is the harmonic oscillator level E_0. Eigenvalues in the n-th band differ only by tunneling. An improved WKB calculation [34,35] [c.f. (30)] gives the bandwidth

$$2\ t_n(T) \cong 2^{4(n+1)}(n!)^{-1}(2/\pi)^{1/2}(\tfrac{1}{4}\beta E_K^0)^{n-1/2}\ \exp(-\beta E_K^0). \tag{33}$$

Contrast this with the band gap T-dependence

$$\bar{\epsilon}_{n+1} - \bar{\epsilon}_n \cong 8(\beta E_K^0)^{-1} + O(k_BT/E_K^0)^2. \tag{34}$$

3.2. Configurational Phenomonology

We now ask whether the kink (5) and phonon (13) excitations appear as natural elementary modes in the formal statistical mechanics of §3.1.

Following the qualitative suggestions of KRUMHANSL and SCHRIEFFER [21] in the ϕ^4 example, it is not difficult to demonstrate that the component E_0 of ϵ_0, taken together with *all* of $F_{\dot\phi}$ [from (22)] *and* F_{V_0}, corresponds *exactly* to the free energy, F_0, of a set of one-dimensional classical ($k_BT >> \hbar\omega_0$) *harmonic* phonons (13) - calculated to $O(\ell/d)$ to be consistent with (26). It is essential that the discrete dispersion relation (13) is used in order to be consistent with the transfer operator technique. We find

$$F_0/L = \frac{k_BT}{2\pi} \int_{-\pi/\ell}^{+\pi/\ell} d\kappa \; \ln(\beta\hbar\omega_\kappa)$$

$$\overset{\ell\to 0}{\to} \; k_BT \{\ell^{-1} \ln(\hbar\omega_0\beta d/\ell) + (2d)^{-1}\}$$

$$= A\omega_0^2 E_0 - k_B T L^{-1} \ln Z_{\dot\phi} + A\omega_0^2 V_0. \tag{35}$$

This identification of E_0 leads naturally to the speculation [21] that t_0 corresponds to the free energy of a gas of independent kinks. It is easy to show in general [22] that as $T \to 0$, $\delta\phi_{1,2} \to 0$, $E_0 \to 0$, and so $I(T) \to I(0) = \beta E_K$, from (7) and (31). This result certainly suggests [21] an exponentially low kink density but includes no thermal kink dressing (§1) and consequent kink-phonon non-superpositional interactions. Indeed in the strict limit $T \to 0$ no modes survive at all! Note similarly that *all* of the dynamical free energy $F_{\dot\phi}$ is apparently taken up by the phonon modes. Mode interactions are essential to account for kink dynamics (§2.1), but formally (§3.1) we must calculate $\delta\phi$ to the *same* order as E_0 to be consistent. Explicit calculations [23] to linear T (harmonic order) show that periodic members of class (2) have the 'universal' [22] form

$$F_t/L \propto - A\omega_0^2 (\beta E_K^0)^{-1/2} \exp(-\beta E_K^0). \tag{36}$$

The numerical proportionality constant depends on the particular $V(\phi)$ and the accuracy of the WKB formula (30) [34].

To understand these formal results phenomenologically, we must include kink-phonon interactions explicitly, as found in §2.1. The simple lesson was that the density of states of extended modes (e.g., phonons) are *changed* by the presence of local defects (e.g., solitons). This results in a change in the phonon free energy which we view below as a 'kink self-energy'.

At low T only low velocity ($v << c_0$) kinks need be considered. The *change*, ΔF, in phonon free energy is thus [c.f. (35)]

$$\Delta F = k_BT \fint_{-\pi/\ell}^{+\pi/\ell} d\kappa \; \ln(\beta\hbar\omega_\kappa) \; \delta\rho(\kappa) \tag{37}$$

$$\overset{\ell\to 0}{\to} - k_BT \, N_B \ln(\beta\hbar\omega_0) \\ + k_BT (2\pi)^{-1} \int_{0+}^{\infty} d\kappa \left(\frac{d\delta}{d\kappa}\right) \ln(1 + d^2\kappa^2), \tag{38}$$

where we have used (17) and N_B is the number of bound states for the particular $V(\phi)$ (see §2.1). From (38), ΔF is generally negative [c.f.(14)] - the phonon free energy is *reduced* by the kink's presence, as expected from the reduction in the number of allowed phonon states. We interpret this as a contribution to the kink self-(free) energy, which is similar in spirit to that adopted [2] in the quantum renormalization of the kink mass from its effect on the vacuum zero-point energy [36]. Similarly, another contribution to the kink self-free energy will come from any localized internal harmonic oscillations corresponding to additional 'bound' states (§2.1). The total kink self-(free) energy is thus

$$\textstyle\sum_K(T) = \Delta F(T) + k_B T \sum_{n=2}^{N_B} \ln(\beta\hbar\omega_n), \tag{39}$$

where $\{\omega_n\}$ are the internal oscillation frequencies. (If $N_B = 1$ there is no second term.) For $k_BT \ll E_K^0$, the kink density is exponentially small and we can associate the full $\sum_K$ with each kink (or antikink) as well as neglecting kink-kink interactions or ordering restrictions. For periodic boundary conditions the number of kinks (N_K) and antikinks ($N_{\bar K}$) are equal, but [particularly for periodic $V(\phi)$] the net 'winding number' density, $\equiv L^{-1}(N_K - N_{\bar K})$, can itself be physically relevant and serve as an additional thermodynamic variable [37]. We restrict ourselves here to periodic boundary conditions for brevity.

We can now construct a grand canonical partition function Ξ in mode representation. Simple factorization (as in a linear system) is *not* possible because of the (phase-shifting) interactions. However, ΔF (total) is proportional to the number of kinks (at low T), which leads us to write [38]

$$\Xi = \Xi_p^{(0)}\,\Xi_K\,\Xi_{\bar K} \tag{40a}$$

$$\Xi_p^0 = \exp(-\beta F_0) \tag{40b}$$

$$\Xi_K = \Xi_{\bar K} = \sum_{N_K=0}^{\infty} e^{\beta\mu_K N_K} Z_K(N_K) \tag{40c}$$

$$Z_K(N_K) = (B^{N_K} h^{N_K} N_K!)^{-1}\left\{\int_0^L dq_K \int dp_K \exp\left[-\beta\{(p_K^2c_0^2 + E_K^{0\,2})^{1/2} + \textstyle\sum_K\}\right]\right\}^{N_K}. \tag{40d}$$

Here Ξ_p^0 is the free phonon result and $Z_K(N_K)$ is the classical partition function for an ideal gas of N_K indistinguishable 'relativistic' particles with self-energy $\sum_K$. The dimensionless constant B is a T-independent phase space factor (see later). Since there are no external constraints (in the thermodynamic limit) we eventually set $\mu_K = \mu_{\bar K} = 0$. Similarly, there are no constraints on 'phonon' numbers ($L \to \infty$).

In the limit $\beta E_K^0 \gg 1$, (40d) gives [23]

$$\Xi_K = \exp\left\{e^{\beta\mu_K}\left(\frac{LE_K^0}{Bhc_0}\right)\left(\frac{2\pi}{\beta E_K^0}\right)^{1/2} e^{-\beta[E_K^0 + \sum_K]}\right\}. \tag{41}$$

The grand canonical potential density is $\Omega \equiv -k_BT L^{-1} \ln \Xi$. The average total kink number density is then

$$n_K^{TOT} = L^{-1}(N_K + N_{\bar{K}}) = -(\partial\Omega/\partial\mu_K)_{T,L}$$

(evaluated at $\mu_K = 0$), and the free energy density is $F/L = F_0/L - k_B T\, n_K^{TOT}$. We find explicitly from (35), (40) and (41) that

$$n_K^{TOT}(T) = (2E_K^0/Bhc_0)(2\pi/\beta E_K^0)^{1/2} \exp[-\beta(E_K^0 + \textstyle\sum_K)] \tag{42}$$

$$\textstyle\sum_K(T) = -k_B T \ln(\beta\hbar\omega_0) - k_B T \sum_{n=2}^{N_B} \ln\left(\frac{\omega_0}{\omega_n}\right) \tag{43}$$

$$+ k_B T\,(2\pi)^{-1} \int_{0+}^{\infty} d\kappa \left(\frac{d\delta}{d\kappa}\right) \ln(1 + d^2\kappa^2).$$

From (43) the T-dependence of $\sum_K$ is quite simple and yields a 'universal' [22] dependence of n_K on (βE_K^0):

$$n_K^{TOT}(T) = (2\pi)^{-1/2}\,(2e^{\sigma}/Bd)(\beta E_K^0)^{1/2} \exp(-\beta E_K^0), \tag{44}$$

where $\sigma = -\beta \sum_K(T) - \ln(\beta\hbar\omega_0) \neq \sigma(T)$. The numerical factor B has to be obtained by comparing the phenomenological and exact free energies [23] - $B(SG) = 1$; $B(\phi^4) = 4$; B (double quadratic) = $2\sqrt{2}$; $\sigma(SG) = \sigma$ (double quadratic) = In2; $\sigma(\phi^4) = \ln(4\sqrt{3})$. The exact agreement for SG is presumably a reflection of its strict soliton interactions. Other thermodynamic functions now follow. For example, the internal energy density $u \equiv L^{-1}\partial(\beta F)/\partial\beta$ can be written in the suggestive form [38]

$$u = (\ell^{-1} - N_B\, n_K^{TOT})\, k_B T + n_K^{TOT}\,[E_K^0 + \tfrac{1}{2} k_B T + (N_B - 1)\, k_B T]. \tag{45}$$

This is simply the internal energy of a system with $(L/\ell - N_B L\, n_K^{TOT})$ classical harmonic modes, and $L n_K^{TOT}$ 'Newtonian particles' of rest energy E_K^0 each having $k_B T/2$ translational energy and thermal energy $k_B T$ for each of the $(N_B - 1)$ internal oscillation modes.

At higher temperatures the complete phenomenology described above is not available (except in principle for integrable systems). Higher-order terms in E_0 [c.f.(32)] are purely anharmonic phonon contributions and include any 'breather' contributions. Mode-mode interactions (virial, etc.) are increasingly complex and have only been included approximately, although with some success even near the incipient first-order transition [22] $\beta E_K^0 \sim 2$.

It will be clear from §3.2 and the above discussion that *fully integrable* systems are in principle *ideal* for configurational phenomenologies at all temperatures - the energy is exactly separable into 'normal modes' whose interactions are purely asymptotically phase-shifting and pair-wise additive (not requiring low T). Thus, the formulation (40) immediately generalizes to the complete basis. The familiar SG example (with decaying boundary conditions) has been investigated in detail [23]. However, the 'breather' contribution for that example produces some anomalies [23] when compared with the transfer integral results, which have yet to be fully understood. They may arise from the discrete-lattice-finite L limits or the different boundary conditions - an ideal comparison should be with a separable, finite, discrete model with, e.g., periodic boundary conditions. Intensive thermodynamic quantities are insensitive to these questions but appropriate modes for a configurational phenomenology may be more so (see also §2.3).

We have already mentioned (§1) the separable Hamiltonian form for the 2-D XY model partition function in terms of spin waves and vortices, which emphasizes the conceptual similarity with the 1-D models above. This is again a *nonlinear* separation as is shown by Renormalization-Group studies [5] which self-consistently generate mode interactions.

4. Correlation Functions

Appropriate static correlation functions (see below) are sensitive tests of kinks in particular, and dynamic correlations can be even more so, leading to central-peak and soft-mode descriptions [15]. The latter are best discussed elsewhere in these Proceedings but we note that again mode interactions are crucial and should be totally manageable in integrable systems. In practice, this can be complex and is frequently omitted [39], although approximate phenomenology can be very satisfactory.

Thermodynamic functions such as F or u are dominated by phonon contributions at low T, since the kink density is exponentially small. However, kinks will dominate those properties insensitive to phonons - e.g., DC transport in periodic $V(\phi)$ models [40] or appropriate correlations. Static correlations follow formally [21] from the transfer integral technique (§3.1):

$$\begin{aligned} C(x) &\equiv \langle g[\phi(x)]g[\phi(0)]\rangle \\ &= \sum_n |\langle n|g(\phi)|0\rangle|^2 \exp[-\beta A\omega_0^2 (\epsilon_n - \epsilon_0)x] \end{aligned} \tag{46}$$

Typical correlations consider $g(\phi) = \delta\phi, \delta\phi^2, \exp(i\phi)$, etc., where $\delta\phi(x) \equiv \phi(x) - \langle\phi\rangle$, etc., and $\exp(i\phi)$ is most relevant if $V(\phi)$ is periodic.

At large distances $C(x)$ will be dominated by the state with the smallest eigenvalue which is coupled to the $n = 0$ level [i.e., has non-zero matrix element in (46)]. For periodic $V(\phi)$ these are determined from a 'Bragg condition' if $g = \exp(i\phi)$. The ϕ^4-type example is very familiar [21,23]. Here the order-parameter correlation length ($g = \delta\phi$) is dominated by the lowest tunnel split levels and is thus proportional to an 'activation' term $\exp(\beta E_a)$, where E_a is the kink creation energy: $E_a = E_K^0$ ($T \to 0$). This is expected physically since ϕ is changed by the presence of a kink (from $\sim \pm 1$ to ∓ 1). In fact the correlation length is then essentially the inverse kink density [21]. On the other hand $\delta\phi^2$ is insensitive except in the immediate kink vicinity and shows no long-range correlations, which we see formally since the lowest coupled levels in $C(x)$ then differ by a term $\propto T$, as $T \to 0+$.

The same ideas apply [31] to the complex order-parameter system §2.2. There the 'Schrödinger' problem (24) has a 'band-structure' distribution of eigenvalues (as for SG). There are no long-range correlations as in $\delta|\psi|^2$ (i.e., correlation length $\not\to \infty$ as $T \to 0+$) because the lowest coupled levels are in different bands and differ by $\sim T$ [c.f. (34)]. Physically, $|\psi|^2$ is unchanged by the kinks and again the correlation length is characterized by the defect width rather than average separation. However, complex kink energies *will* characterize activated behaviour for correlations in the field itself [$<\delta\psi(x)\delta\psi^*(0)>$]: ψ changes locally from $\pm u_0$ (§2.2) in passing through a kink and correspondingly the lowest coupled levels are the top and bottom of the first band, whose width is exponentially small at low T. In view of the crossover from SG to ϕ^4 as $D/A \to 1/8$ described in §2.2, it is interesting to isolate the activation energy E_a as a function of D/A and compare it with the complex kink creation energy. A numerical determination of E_a as $T \to 0$ has been found [31] to agree excellently with the kink energy (21).

Finally, we emphasize the general importance of order-parameter defects for correlation behaviour. To choose one example, the non-topological inhomogeneous states in the 2-D isotropic Heisenberg model [18] do not result in a transition (as for 2D XY [5]) but do change temperature dependence from power law to exponential in the order-parameter correlation length.

5. Summary

The roles of nonlinearity in lattice dynamics, critical phenomena and statistical mechanics are numerous and at a point of rapid development as will be seen from other contributions to these Proceedings. I have tried in this article to focus on a few simple models which can be handled in some analytic detail, but nevertheless illustrate very general nonlinear lessons - see, e.g., the molecular-dynamics results reported by SCHNEIDER and STOLL. [In addition the models of §2 are directly useful for some physically accessible quasi-1-D systems - see, e.g., articles of M.J. RICE and M. STEINER.] I have emphasized especially the usefulness of a configurational phenomenology as a conceptual and practical scheme. As a corollary, it was necessary to consider (nonlinear) mode-mode interactions very carefully to properly account for conservation of degrees of freedom and essential mechanisms of free-energy sharing (§3).

It has not been possible to describe recent activity [40] in *non-equilibrium* statistical mechanics for the nonlinear models in class (2) of SG type, although these properly complement the equilibrium motivations above. One approach has been to employ a Fokker-Planck formulation and then the analogue of steepest descent trajectories in the partition function are the most probable evolution paths in phase space. The transfer integral technique is useful here [40] as for the equilibrium problem (§3.1). The driven, damped SG system in the presence of thermal noise is generically interesting as a model for nonlinear transport or phase slippage [1]. Once again, identifying *configurations* leads to an appealing physical interpretation of formal results and phenomenological nucleation descriptions have been suggested [41]. In particular, at low constant driving fields and temperatures, solitons are found to be the dominant transport mechanism, as confirmed by molecular-dynamics simulations [42].

I am grateful to many colleagues, especially J.A. Krumhansl for his foresight, and S.E. Trullinger for most enjoyable collaborations.

References

1. A.R. Bishop: Chalmers Soliton Symposium (Göteborg, June 1978, Sweden) (to appear in Physica Scripta)
2. R.F. Dashen, *et al.*: Phys. Rev. D 11, 3424 (1975); S. Deser, *et al.*: Nucl. Phys. B 114, 29 (1976); R. Jackiw: Rev. Mod. Phys. 49, 681 (1977)
3. J. Rudnick, D. Jasnow: Phys. Rev. B 17, 1351 (1978)
4. See, e.g., P. Bak: these Proceedings
5. J.V. Jose, *et al.*: Phys. Rev. B 16, 1217 (1977); see also T. Schneider, E. Stoll: these Proceedings
6. See, e.g., T. Schneider, E. Stoll: these Proceedings
7. S.E. Trullinger: Preprint
8. See, e.g., D.J. Wallace: these Proceedings
9. A.C. Scott, *et al.*: Proc. IEEE 61, 1443 (1973); R. Bullough: these Proceedings
10. See A.C. Newell: these Proceedings
11. A.R. Bishop: In *Solitons in Action*, ed. by K. Longgren, A.C. Scott, (Academic Press, New York 1978)
12. I am indebted to T.D. Lee for the imaginative terminology!
13. M. Kleman, L. Michel: J. de Phys. Lett. 39, L29 (1978); see also K. Maki: these Proceedings
14. R. Friedberg, *et al.*: Phys. Rev. D 13, 2739 (1976); G. Toulouse, M. Kleman: J. de Phys. 37, L149 (1976)
15. See, e.g., A.D. Bruce: these Proceedings
16. See B. Horovitz: these Proceedings
17. See S. Aubry: these Proceedings
18. A.A. Belavin, A.M. Polyakov: JETP Lett. 22, 245 (1975)
19. See, e.g., M.J. Rice: these Proceedings
20. See M. Steiner: these Proceedings
21. J.A. Krumhansl, J.R. Schrieffer: Phys. Rev. B 11, 3535 (1975)
22. A.R. Bishop: Physica (in press)
23. J.F. Currie, *et al.*: in preparation
24. See, e.g., P.W. Kitchenside *et al.*: these Proceedings
25. S.E. Trullinger, R.M. DeLeonardis: Preprint
26. T.R. Koehler: (unpublished results 1975)
27. S.P. Novikov: JETP 40, 1058 (1975); G. Costabile, *et al.*: Appl. Phys. Lett. 32, 587 (1978)
28. S. Aubry: J. Chem. Phys. 64, 3392 (1976)
29. J.F. Currie, *et al.*: Phys. Rev. B 15, 5567 (1977)
30. L.D. Landau, E.M. Lifshitz: In *Quantum Mechanics - Non-relativistic Theory* (Pergamon Press, Oxford 1965), pp.78-80
31. S. Sarker, *et al.*: Phys. Lett. 59A, 255 (1976); J.F. Currie, *et al.*: Preprint; see also J. Lajzerowicz: these Proceedings
32. V.E. Korepin, L.D. Fadeev: Theor. Mat. Phys. 25, 1039 (1975)
33. The formal analogy with a zero-dimensional field theory and *instantons* is striking here (see Refs. [2] and [8])
34. Formula (30) should be corrected by the universal factor $(\pi/e)^{1/2}$ for symmetric members of class (2) (R.M. De Leonardis: Private communication) [see also (33)]
35. M. Abramowitz, I.A. Stegun: In *Handbook of Mathematical Functions* (U.S. Department of Commerce)
36. Although the rigorous procedure is to identify 'collective coordinates' and orthogonal modes (Ref. [2]), Fadeev has shown that, to each order in

quantum perturbation theory, free continuum modes can be used (no phase shifts) if the zero frequency modes are excluded to avoid double-counting
37. J.F. Currie, *et al.*: Phys. Rev. A 16, 796 (1977)
38. These decompositions are *not* unique but aesthetically satisfying in view of our configurational approach or comparisons with quantum field-theory treatments (Ref. [2])
39. K. Kawasaki: Prog. Theor. Phys. 55, 2029 (1976); Y. Matsuno: Phys. Lett. 64A, 14 (1977); H.J. Mikeska: J. Phys. C 11, L29 (1978)
40. S.E. Trullinger, *et al.*: Phys. Rev. Lett. 40, 206 (1978)
41. R. Landauer: Private communication
42. T. Schneider, E. Stoll: these Proceedings

Some Applications of Instantons in Statistical Mechanics

D.J. Wallace

Physics Department, The University, Southampton SO9 5NH, Great Britain

Abstract

Instantons are solutions of non-linear equations associated with tunnelling phenomena. We discuss how instantons characterise the behaviour of divergent perturbation series and how this information can be used to obtain convergent results from these series. Applications to ε-expansions and phase transitions are reviewed. We indicate how the same instanton techniques apply to the problem of the electron in a random potential and to metastability.

1. Introduction

This talk is concerned with handling problems in statistical mechanics where the Hamiltonian looks unbounded below. The basic idea is to define the partition function in this case by analytic continuation from a well-defined problem with a bounded Hamiltonian. An example which we will consider in some detail is the Landau-Ginzburg-Wilson Hamiltonian of a fluctuating field $\phi(x)$

$$\mathcal{H} = \int d^dx\{\tfrac{1}{2}(\vec{\nabla}\phi)^2 + \tfrac{1}{2}m^2\phi^2 + \tfrac{1}{4}g\phi^4\}, \tag{1}$$

in the case in which the coupling g is negative; this is defined by continuation from the usual theory for positive g. (Here $\mathcal{H}$ means the reduced Hamiltonian, containing $1/k_BT$.) As we shall see, a typical result of the continuation is that for negative g the partition function, which is real for positive g, picks up an imaginary part. This talk is about techniques for calculating this imaginary part.

The main justification for including this topic in this symposium is that non-trivial, extended solutions of the classical field equation

$$\nabla^2\phi = m^2\phi + g\phi^3 \tag{2}$$

play a central role in these techniques. These solutions are the instantons referred to in the title. These calculations are to be contrasted with the standard perturbation expansion in the anharmonic coupling g by Feynman graph expansion; at any order in this perturbation theory, the partition function is real for real coupling g (regardless of its sign). The result of the instanton calculations is an imaginary part (for $g<0$) which is exponentially small in g ($\sim \exp(\text{const}/g)$).

Despite one's anticipation that the partition function should be real in any physical problem, the imaginary part provides extremely interesting information for a number of problems in physics. In this talk I shall indicate its importance in three areas: (i) to obtain the asymptotic form of the perturbation in g i.e. the behaviour of the high order terms; (ii) to obtain the density of states in the tail of the band for an electron in a random potential; (iii) to elucidate metastability and the nature of the singularity on the coexistence curve ($H\to 0$, $T<T_c$ in magnetic notation) of Ising - like systems.

Some remarks about the historical development of this subject are probably worth-while. The use of non-trivial solutions of field equations in problems (ii) and (iii) above was recognised over ten years ago [1, 2 and references therein]. Identical techniques were developed independently by quantum field theorists to discuss the tunnelling phenomena which occur in Non-Abelian Gauge Theories of elementary particle interactions [3,4,5,6] ; the word instanton was coined in this work. The particle physicists also revisited [7] the problem of condensation considered by Langer [2] . Following some pioneering work by Bender and Wu [8 ; see also 9] it was recognised that the same techniques could be used to obtain the asymptotic behaviour of perturbation expansions in quantum mechanics, quantum field theory and statistical mechanics [10,11].

The form of instanton solutions will be clarified later. At this point it is sufficient to say that they have some features in common with kinks and solitons; here I am characterising instantons as objects associated with tunnelling in quantum physics, or analogous phenomena in statistical physics.

2. An Example in Complex Variable Theory

In order to illustrate the basic ideas behind instanton techniques, let us consider the one-dimensional integral

$$Z(g) = \int d\phi \exp - (\tfrac{1}{2}\phi^2 + \tfrac{1}{4}g\phi^4) \; ; \qquad (3)$$

it may be considered as the partition function of a zero-dimensional field theory (with real field ϕ defined at a single point - the integral on ϕ represents the statistical average). We want to show that Z has an imaginary part for $g<0$, and to calculate it.

For the "physical" case $g > 0$, it is understood that the integral runs from $-\infty$ to ∞ and $Z(g)$ is real. By expansion in g within the integral it is then straightforward to obtain the perturbation expansion

$$\begin{aligned} Z(g) &\sim \sum_{K=0}^{\infty} (-g)^K (\tfrac{1}{4})^K \frac{1}{K!} \int_{-\infty}^{\infty} d\phi \; \phi^{4K} \exp - \tfrac{1}{2}\phi^2 \\ &= \sum_{K=0}^{\infty} (-g)^K . \sqrt{2} . \frac{\Gamma(2K+\frac{1}{2})}{\Gamma(K+1)} \end{aligned} \qquad (4)$$

by the substitution $t = -\frac{1}{2}\phi^2$ and the definition of Euler's gamma-function. This is the analogue of the Feynman graph expansion. It is clear that the perturbation series (4) has zero radius of convergence and the function $Z(g)$ therefore has a singularity at $g = 0$.

In order to elucidate the nature of this singularity it is useful to consider Z for complex values of g. Clearly the integral $\int_{-\infty}^{\infty}$ in (3) converges whenever $\mathrm{Re}\,g > 0$ and the function Z is defined as an analytic function of g in the right hand plane. By the standard method of rotating the integration contour in the ϕ plane, we can extend this definition of Z to all complex values of g. In particular for $g < 0$ the two cases $\arg g = \pm\pi$ are defined by contour integrals rotated by $\mp\pi/4$ respectively (so that $g\phi^4$ remains positive and the integral always converges). These two contours of ϕ integration are the complex conjugate of one-another and thus $Z(g)$ has equal and opposite imaginary parts for the cases $\arg g = \pm\pi$, indicating that it is analytic in the g plane cut from $-\infty$ to 0.

We can now proceed to evaluate the imaginary parts for small (negative) g by the method of steepest descents [See e.g.12]. This consists of finding the saddle points with respect to ϕ of the function

$$\mathcal{H} = \tfrac{1}{2}\phi^2 + \tfrac{1}{4}g\phi^4 \tag{5}$$

and the lines in the complex ϕ plane of constant (in this case zero) imaginary part of $\mathcal{H}$. The contours of integration can then be distorted to follow the path of steepest descent along these lines. For the case $g < 0$, the saddle points are:

$$\frac{\partial\mathcal{H}}{\partial\phi} = 0 \qquad \phi = 0 \text{ or } \pm\phi_c : \phi_c = (-g)^{-\frac{1}{2}} \tag{6}$$

They are shown together with lines of zero imaginary part in Fig.1

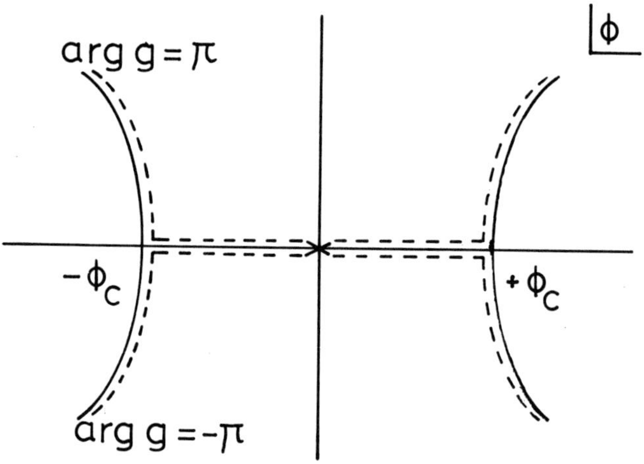

Fig.1 Saddle points and lines of constant imaginary part of the function $\mathcal{H}$ (Eq.5) for $g < 0$. The integration contours of steepest descent are shown as dotted lines

Since the integrand is entire, we can distort the original contours at angles $\pm\pi/4$ into the steepest descent contours shown in Fig.1. Clearly the part of the integral between the saddle points $\pm\phi_c$ is purely real; the largest contribution occurs near $\phi=0$, and this part yields the usual perturbation expansion of Z as in (4). The imaginary part has its largest contribution from the saddle points $\pm\phi_c$. After the change of variables $\phi = \phi_c + i\tau$, Taylor expansion in τ yields

$$\operatorname*{Im}_{\arg g = \pi} Z(g) = -\exp -\mathcal{H}(\phi_c) \cdot \int_{-\infty}^{\infty} d\tau \exp - \tau^2 [1+O(g)]$$

$$= -\sqrt{\pi} \exp \frac{1}{4g} \; [1 + O(g)] \tag{7}$$

(Note that since $g < 0$ this is indeed exponentially small for small g).

We discuss the generalisation of this kind of calculation to field theories in the next section and complete this one by outlining how the result (7) can be used to obtain the behaviour of late terms in the series (4). The function Z(g) is analytic in the g plane cut from $-\infty$ to 0. We can therefore write a dispersion relation

$$Z(g) = \frac{1}{2\pi i} \int_C \frac{Z(g')}{g'-g} \, dg'$$

where C is the contour shown in Fig.2.

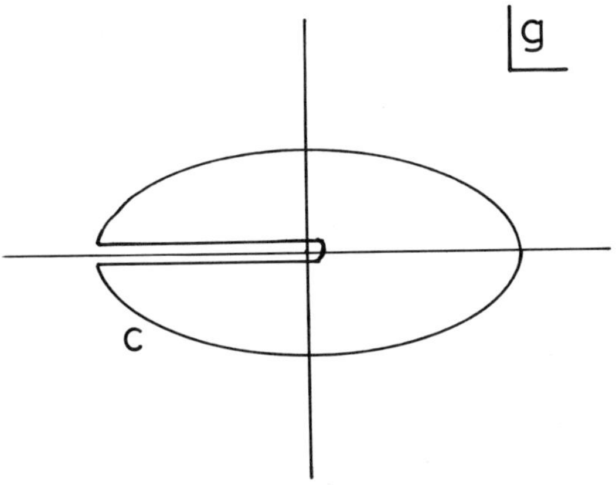

Fig.2 Dispersion relation contour C used to obtain high order estimates

The contour at infinity may be discarded and the real parts of Z cancel between the contours $-\infty + i0$ to 0 and 0 to $-\infty - i0$. Therefore we have

$$Z(g) = \frac{1}{\pi} \int_{-\infty+i0}^{0} \frac{\operatorname{Im} Z(g')}{g' - g} \, dg' \tag{8}$$

Expanding this integral in g and using the result (7) yields

$$Z(g) \sim \sum_{K=0}^{\infty} g^K \left(\frac{-1}{\sqrt{\pi}}\right) \int_{-\infty}^{0} dg' \, (g')^{-K-1} \exp 1/4g' \; [1+O(g')]$$

$$= \sum_{K=0}^{\infty} (-g)^K \frac{1}{\sqrt{\pi}} \; \Gamma(K) . 4^K \; [1 + O(K^{-1})] \tag{9}$$

We have used the change of variables $t = -1/4g$ to obtain again the Euler gamma function. Note particularly that the (uncalculated) corrections of order g' produce a $\Gamma(K-1)$ instead of a $\Gamma(K)$ and are hence negligible for large K. The reader may verify by Stirling's formula that expression (9) agrees with the form for large K of the exact series (4).

To summarise, these calculations show that Z(g) has an imaginary part for $g < 0$. For small g, the value of ImZ(g) is determined by a non-trivial saddle point, ϕ_c, of the function $\mathcal{H}$(Eq.5). The asymptotic behaviour of the perturbation series in g can then be obtained by means of a dispersion relation in g.

3. Instantons and Field Theory Calculations

It is clearly desirable to be able to extend the above calculations to the partition function of a field theory in d space dimensions,

$$Z(g) = \frac{\int D\phi \exp - \mathcal{H}(\phi)}{\int D\phi \exp - \mathcal{H}_o(\phi)} \tag{10}$$

where $\int D\phi$ denotes the functional integral over all field configurations. The Hamiltonian $\mathcal{H}$ is as in (1), for example, and $\mathcal{H}_o$ is the free Hamiltonian ($g = 0$), so that Z is normalised by $Z(0) = 1$. Unfortunately, we do not have sufficient control of steepest descent methods in function space to follow completely the calculation of section 2. However, we can extend sufficient key components of that calculation to enable us to obtain the analogue of (7) for many field theories.

The first key component is the analogue of the saddle point, ϕ_c, (Eq.6) from which the imaginary part receives its dominant contribution. The equivalent in field theory is a solution of $\delta\mathcal{H}/\delta\phi = 0$ i.e. a solution $\phi_c(x)$ of the classical field equation (2). From the complex variable analysis, it is clear that the trivial solution $\phi(x) = 0$ will not do. We need a non-trivial solution, the instanton, $\phi_c(x)$. We can then formally follow Eq.(7). Writing $\phi(x) = \phi_c(x) + \hat{\phi}(x)$ in the numerator and expanding $\mathcal{H}(\phi) = \mathcal{H}(\phi_c) + \frac{1}{2}\hat{\phi}\mathcal{M}\hat{\phi} + O(\hat{\phi}^3)$, where $\mathcal{M} = \delta^2\mathcal{H}/\delta\phi^2(\phi_c)$, we obtain

$$\underset{(g<0)}{\mathrm{Im}Z(g)} \sim \exp - \mathcal{H}(\phi_c) . \frac{\int D\hat{\phi} \exp - \int \frac{1}{2}\hat{\phi}\mathcal{M}\hat{\phi}}{\int D\phi \exp - \int \frac{1}{2}\phi\mathcal{M}_o\phi} \left[1 + O(g)\right]$$

$$= \exp - \mathcal{H}(\phi_c) . \left[\frac{\mathrm{Det}\mathcal{M}}{\mathrm{Det}\mathcal{M}_o}\right]^{-\frac{1}{2}} . \left[1 + O(g)\right] \tag{11}$$

For the Hamiltonian (1), $\mathcal{H}(\phi_c)$ is always of the form $-1/(ag)$ (a = constant) and the determinants of the differential operators $\mathcal{M}$ and $\mathcal{M}_o$ are obtained by expanding the fields $\hat{\phi}$ and ϕ in terms of their respective eigenfunctions. The corrections of order g come from the neglected anharmonic terms $O(\hat{\phi}^3)$ in the numerator, as in (7).

Of course the non-linear equation (2) has many non-trivial solutions and we must select the appropriate solution $\phi_c(x)$ for expression (11). Since Im Z(g) is always exponentially small in g, we obtain the leading behaviour of ImZ for small g by taking the non-trivial solution with minimum $\mathcal{H}(\phi_c)$ (and we thereby assume it is the first saddle point we meet in a proper steepest descent calculation). In particular, $\mathcal{H}(\phi_c)$ must be finite, so that $\phi_c(x) \to 0$ as $|x| \to \infty$ in order for the integral (1) to converge, i.e. for large $|x|$ the field ϕ is constrained to the "metastable" value $\phi = 0$.

One may further show [13] that for large classes of Hamiltonians, including (1) for $d < 4$, the solutions with minimum $\mathcal{H}$ are radially symmetric. In this case the field equation (2) becomes an ordinary differential equation

$$\frac{d^2\phi}{dr^2} + \frac{(d-1)}{r}\frac{d\phi}{dr} = -\frac{d\tilde{V}}{d\phi} \tag{12}$$

where

$$\tilde{V} = -\tfrac{1}{2}m^2\phi^2 - \tfrac{1}{4}g\phi^4 \quad . \tag{13}$$

The form of $\tilde{V}$ for $m^2 > 0$, $g < 0$ is shown in Fig.3 .

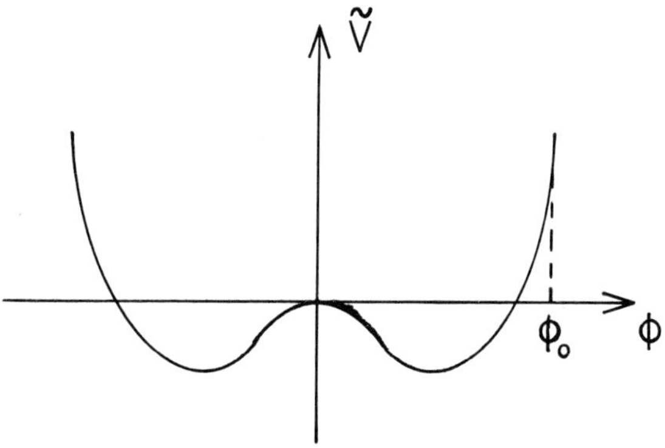

Fig.3 Potential $\tilde{V}$ for the motion in one dimension (ϕ) described by equations (12) and (13)

Clearly (12) can be interpreted as an equation of motion in one dimension ϕ with "time" variable r, damping depending on $d\phi/dr$, and in potential $\tilde{V}$. One can anticipate a solution starting at some value $\phi(0) = \phi_o$ with sufficient potential energy to overcome the damping and climb to $\phi = 0$ at $r = \infty$. This is the instanton solution required because other solutions starting at a larger value of ϕ and performing two oscillations before coming to rest at $\phi = 0$ will have a greater value of $\mathcal{H}$. (This argument clearly breaks down if the damping is too great and in fact the case $d = 4$ requires special treatment). In many cases the instanton solution cannot be obtained analytically in terms of simple functions, but it is clear from the above discussion that it can always be constructed numerically.

Having identified the relevant classical solution, we turn to some technical problems in evaluating the determinants in (11). For the Hamiltonian (1), the differential operator $\mathcal{M}$ takes the form

$$\mathcal{M} = -\nabla^2 + m^2 + 3g\phi_c^2(x) \quad (g < 0) \tag{14}$$

Since $\phi_c(x)$ is localised about $x = 0$ and vanishes at infinity, the calculation of the determinant of $\mathcal{M}$ is equivalent to finding the eigenvalues (whose product gives the determinant) in a quantum mechanical problem with potential $V = m^2 + 3g\phi_c^2(x)$ whose qualitative form is shown in Fig.4.

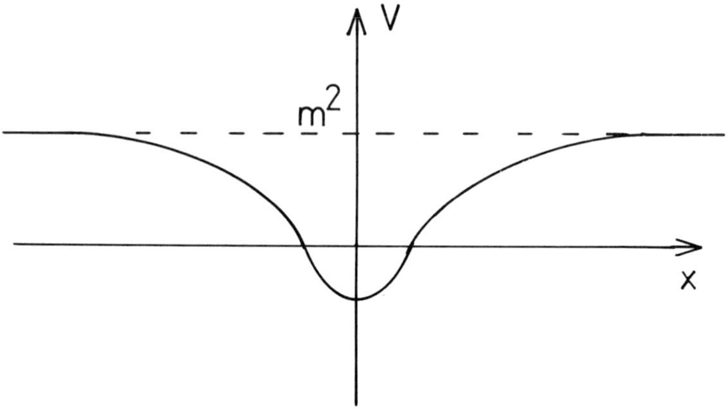

Fig.4 Qualitative form of the potential whose eigenvalues give det$\mathcal{M}$ in (14)

Clearly V is an attractive potential and one must pay particular attention to the possible existence of bound states which have eigenvalue < 0. The following analysis resolves this problem in a very simple way. We have remarked that $\phi_c(x)$ is radially symmetric about x = 0. Clearly, by the translation invariance of the original Hamiltonian, the function $\phi_c(x - x_o)$ is also a solution of the field equations with the same value of $\mathcal{H}^c$ for any fixed x_o; it just corresponds to an instanton centred at x_o. Now if we take $\partial/\partial x_o^\mu$ ($\mu = 1,2,\ldots,d$) of eq.(2), obeyed by $\phi_c(x - x_o)$, we obtain

$$(-\nabla^2 + m^2 + 3g\phi_c^2)\ \frac{\partial\phi_c}{\partial x_o^\mu} = 0$$

Therefore the differential operator $\mathcal{M}$ has a d-fold degenerate eigenfunction $\partial\phi_c/\partial x_o^\mu$ with zero eigenvalue. This eigenfunction transforms under the $\ell = 1$ representation of the rotation group O(d); one must therefore expect that for the attractive potential V in Fig.4 there is at least one $\ell = 0$ eigenfunction with negative eigenvalue. Now, for negative eigenvalues of $\mathcal{M}$ the steepest descent contours of integration in (11) for each of the corresponding eigenmodes must branch into the complex plane, in presumably the way indicated in Fig.1. It is precisely this imaginary "Dϕ" which produces the imaginary part we wish to calculate. Thus there is a further constraint on ϕ_c in these calculations; it should produce a differential operator $\mathcal{M}$ which has one and only one negative-energy bound state. In fact this condition is probably automatically satisfied in most cases by the solution with minimum $\mathcal{H}$ because the solutions with higher $\mathcal{H}$ correspond to larger values of ϕ and hence more and deeper potential wells in V.

Thus the existence of one bound state of $\mathcal{M}$ is precisely what ensures we are calculating the imaginary part of the partition function Z.

Other problems remain for the evaluation of the determinant of $\mathcal{M}$. We have recognised the existence of zero modes in $\mathcal{M}$. These exist for every continuous symmetry of $\mathcal{M}$ broken by ϕ_c. (They are Goldstone modes of a spontaneously broken continuous symmetry). For the corresponding mode integration in (11), the Gaussian approximation there is inadequate because the $O(\phi^2)$ terms vanish, and in fact we have to integrate exactly over these modes. This can be done by the method of collective coordinates [2,14] The basic idea is to take the centre x_o of the instanton as an integration variable, writing

$$\phi(x) = \phi_c(x-x_o) + \sum_n a_n \phi_n(x-x_o) \quad ,$$

where ϕ_n are (normalised) eigenfunctions of $\mathcal{M}$, so that

$$d\phi(x) = dx_o^\mu \cdot \frac{\partial\phi}{\partial x_o^\mu}\left[1+O(\sqrt{g})\right] + \sum da_n \phi_n(x-x_o) \quad .$$

Since $\partial\phi_c / \partial x_o^\mu$ is the zero mode, we can exclude it from the sum on eigenmodes ϕ_n and still span the space of all modes by varying x_o. If the eigenfunctions ϕ_n are normalised one picks up simply a Jacobian factor:

$$D\phi = dx_o^\mu \prod da_n \cdot \left[\int(\partial\phi_c/\partial x_o)^2\right]^{d/2}\left[1+O(g)\right] \quad . \tag{15}$$

The integrals on a_n now give the determinant with the zero modes removed and the integral on x_o restores the original translation invariance of the theory. In particular for the partition function Z, the integrand is independent of x_o and one obtains simply an extensive factor proportional to the volume of the system. (In fact the effect of a dilute gas of instantons is to produce exponentiation of the result (11), so that this extensive factor does indeed appear in $\ln Z$ as one should anticipate; c.f. the appearance of the free energy density in (22).)

Collective coordinates must also be introduced if there is a continuous internal symmetry (such as in the Heisenberg model) or a dilatation invariance (such as in four dimensions).

There is a further problem in evaluating $\det\mathcal{M}$ in that the contribution to it from modes with high angular momentum ℓ contains the usual ultraviolet divergences of the Hamiltonian (1), which can be removed by conventional renormalisation. We refer the reader to the literature for further details on these technicalities [4,10,11,15,16,17].

Let us complete this rather technical section by noting that the general form for the imaginary part of any N-point correlation function for the Hamiltonian (1) is

$$\text{Im}\, G^{(N)} = \exp {}^{1}/ag \cdot (-g)^{-b} \cdot c\left[1 + O(g)\right] \quad . \qquad (g < 0) \tag{16}$$

Since $\phi_c \propto (-g)^{-\frac{1}{2}}$, the constant b is determined by the number s of spontaneously broken continuous symmetries (from the Jacobian factor such as (15)) and the number N of external fields according to

$$b = \tfrac{1}{2}(s + N) \quad . \tag{17}$$

The constant a comes from the classical calculation of $\mathcal{H}(\phi_c)$ and is positive; the hard work of calculating the determinant is required to evaluate c. We remark finally that in four dimensions the form (16) is complicated by the dilatation invariance which is broken by renormalisation.

4. Some Applications of High Order Estimates

The result (16) for the imaginary part can be used directly to obtain high order estimates via the dispersion technique in (8). Following equation (9) one obtains

$$G^{(N)} \sim \sum g^K (-a)^K \, \Gamma(K+b) . c/\pi \, [1+O(K^{-1})] \tag{18}$$

This indicates the asymptotic nature of the Feynman graph expansion anticipated by Dyson [18].

The question naturally arises of how best to use the information (18) to reconstruct convergent approximations for $G^{(N)}$ from the first few terms of the Feynman graph expansion which have been calculated explicitly for any given problem. The natural starting point is to use the Borel transform technique based on the usual integral representation of the gamma function. If one formally interchanges the sum and integral representation in (18), one has

$$G^{(N)} \sim \frac{c}{\pi} \int_0^\infty dt . t^{b-1} e^{-t} \sum (-agt)^K \, [1+O(K^{-1})] \quad . \tag{19}$$

The sum on the right hand side has now finite radius of convergence $[gt=1/a]$. Provided the analytic continuation of the sum has no singularities on the positive t axis, Padé methods or conformal mapping techniques may be used to obtain a unique convergent expression for $G^{(N)}$. It is instructive to note that for $g > 0$ the oscillatory nature of the series (18) implies that the singularity closest to the origin is indeed on the negative t axis (at $-1/ag$). Indeed it has been proved [19] that in the anharmonic oscillator in quantum mechanics, Padé-Borel resummation does converge to the exact answer. Conversely when $g < 0$, the series (18) is not oscillatory and the Padé-Borel method fails because there is a singularity in the region of t integration (at $-1/ag$); this is just a reflection of the fact that the perturbation theory is incomplete when exponentially small terms such as (16) are present. The requirement that the perturbation series should oscillate is very important for models of physics.

The coefficients a, b and c in (16) have been calculated for many models in quantum and statistical physics [8,10,11,20]. So far the most useful application in statistical physics has been to the problem of phase transitions in three dimensions. The coefficients a, b and c have been obtained numerically [21] for various quantities arising in the field-theoretic renormalisation group approach to phase transitions. Resummation techniques based on the Borel transform and using the information in (18) and (19), can then be applied to the explicit sixth order calculations of Baker et al [22]. The results [21] for various critical exponents in Ising-like systems are shown in Table 1; similar results exist for n-component Heisenberg-like models [21].

High order estimates have also been made for ε-expansions of critical exponents in $d = 4 - \varepsilon$ dimensions [11, Houghton et al 20]. Because of the dilatation invariance of the Hamiltonian (1) for $m^2 = 0$ and $d = 4$, much of the calculations may be done analytically - for example the instanton solution in this case is the function

$$\phi_c(x) = \sqrt{\frac{8}{-g}} \cdot \frac{1}{1 + x^2}$$

Technical problems have so far prohibited the calculation of the coefficient c. The results for the ε-expansions of critical exponents of n-component Heisenberg models, in the form $\Sigma \varepsilon^K (-a)^K \Gamma(K+b)$ are

$$a = 3/(n+8) \qquad b = \begin{cases} 4 + n/2 & \text{for } \eta \\ 5 + n/2 & \text{for } 1/\nu \end{cases} \tag{20}$$

With the appearance of explicit fourth order calculations [24] one should be able to obtain another set of independent results with errors similar to those in Table 1.

Table 1 Comparison of field theory results [Le Guillou and Zinn-Justin 21] (first column), series expansion results [23] (second column) and experiment [24] (third column) for Ising-like critical exponents in three dimensions.

γ: $\chi \propto (T-T_c)^{-\gamma}$	1.2402 ± 0.0009	$1.250^{+0.003}_{-0.007}$	1.240 ± 0.002
η: $G^{(2)} \propto q^{-2+\eta}$	0.0315 ± 0.0025		0.018 ± 0.015
ν: $\xi \propto (T-T_c)^{-\nu}$	0.6300 ± 0.0008	$0.638^{+0.002}_{-0.008}$	0.625 ± 0.006

5. Other Applications

One of the earliest problems to which these techniques were applied was that of the electron in a random potential [1]. The problem is to find the average density of states and conductivity for electrons in a random potential $V(x)$ with a probability distribution $P\{V\}$. For a Gaussian distribution $P\{V\} \propto \exp -\gamma^{-1}\int d^d x V^2(x)$, one may show that the average density of states $\overline{\rho(E)}$ for electrons of energy E is given in terms of the imaginary part of the two point correlation function of an n-component LGW Hamiltonian (1), in the limit $n \to 0$. The parameter m^2 in (1) is given by the energy E, and the coupling g by the width γ of the potential distribution $P\{V\}$: in units in which $\hbar^2/(\text{electron mass}) = 2$ one has

$$m^2 \equiv -E \quad : \quad g \equiv -\tfrac{1}{4}\gamma \tag{21}$$

Thus for a sensible Gaussian distribution with $\gamma > 0$, the coupling g is negative. For low lying states with negative E it is trivial to read off the γ dependence of the imaginary part of $G^{(2)}$ from equations (16) and (17). We can identify $N = 2$ and $s = d-1$ Jacobian factors (for the translation modes and $n-1 \to -1$ Goldstone modes of the O(n) symmetry) and hence write down the average density of states in the form

$$\overline{\rho(E)} = \exp -\frac{1}{a\gamma} \cdot \gamma^{-(d-1)/2} \cdot c\left[1 + O(\gamma)\right] \tag{22}$$

The E dependence which we really want can now be obtained by careful dimensional analysis from equations (1) and (2) and the identification (21). For example if L is a length dimension, we have $[E] = [L]^{-2}, [\gamma] = [L]^{(d-4)}$ so that $\gamma/(-E)^{(4-d)/2}$ is dimensionless. The final result is

$$\overline{\rho(E)} = \text{const.}(-E)^{d(5-d)/4}\gamma^{-(d+1)/2}\exp -\left[\text{const}(-E)^{(4-d)/2}/\gamma\right] \cdot \left[1 + O(\gamma/(-E)^{(4-d)/2})\right] \tag{23}$$

This is the result quoted in a recent paper by Cardy [26]. The powers of E differ apparently from those in Zittartz and Langer [1]; I believe that (21) is correct (For example the reader may check that $\overline{\rho(E)}$ has dimensions number/unit volume/unit energy). Note finally that for $d < 4$ (21) is valid for any distribution width γ provided we are deep enough in the band. Correspondingly, the approximation breaks down if we increase E towards the mobility edge; this is an extremely interesting strong-coupling problem to which the renormalisation group cannot be immediately applied because there is no available infra-red stable fixed point [27].

Finally I remark that the same technique was also applied by Langer [2] to the problem of the singularity at a first order transition. In magnetic language this involves the Hamiltonian (1) with an external field term $H\int d^d x \phi(x)$ added. For $T < T_c$ ($m^2 < 0$ in the mean-field approximation) the problem is to find the nature of the singularity of the free energy as a function of H as H goes through zero (The coupling g is now held fixed). The idea is that for $H > 0$ say, the configurations of the field ϕ are constrained to have the positive expectation value $+M$ at infinity. If we attempt to continue analytically this free energy as a function of H to negative values of H, the continuation of the functional integral must be performed for fixed boundary conditions on the field ϕ. Thus for $H < 0$, the fluctuations are still constrained to tend to the positive expectation value at infinity. In this circumstance there is an instanton, which is conventionally called the critical droplet, which gives rise to an imaginary part of this free energy analytically continued to negative H. The general form of the result can be obtained from a simple droplet-model picture, with a critical droplet of radius $r_o \sim \sqrt{-m^2}\, M/|H|$. The small oscillations have indeed a single negative energy state. The final result for the imaginary part of the free energy density is

$$\begin{aligned} \underset{(H\to 0^-)}{\mathrm{Im}\,\mathfrak{F}} \sim\ & \text{const}\ |H|^{-(2-d+d^2)/2}(g)^{-(2+d+d^2)/4}(-m^2)^{(6+3d+2d^2)/4} \\ & .\exp -\left[\text{const}(-m^2)^{(2d+1)/2}|H|^{-(d-1)}(g)^{-(d+1)/2}\right] \\ & .\ (1 + O(g/(-m^2)^{(4-d)/2})\ ) \end{aligned} \tag{24}$$

The factors in the first line are from the Jacobian and the eigenvalue of the negative energy bound state of $O(H^2)$. A discussion of the long wavelength singularities occurring in $\det\mathfrak{M}$ can be found in reference [2].

The form of the singularity in $\mathfrak{F}$ as $H \to 0^+$ (i.e. in the stable state) is obtained by a dispersion relation in H as in equation (8). Since a range of H contribute in this dispersion relation, "virtual" droplets of many scale sizes contribute to this singularity, and this is a problem which one may attempt to tackle by lattice renormalisation group methods [28]. The strength of the instanton calculation of the imaginary part (24) is that the single length scale r_o of the critical droplet dominates the problem.

Thus the form of the essential singularity on the coexistence curve can be elucidated; the interpretation of the imaginary part as a decay rate of a metastable state requires further clarification.

ACKNOWLEDGEMENTS

I am very grateful to M.Stone and to all colleagues at Southampton for illuminating discussions.

REFERENCES

1. J.Zittartz and J.S.Langer: Phys. Rev. 148, 741 (1966)
2. J.S.Langer: Ann.Phys. 41, 108 (1967)
3. A.A.Belavin et al: Phys. Lett. 59B, 85 (1975)
4. G. 't Hooft: Phys. Rev. Lett. 37, 8 (1976) ; Phys.Rev. D14, 3432 (1976)
5. A.A.Belavin and A.M.Polyakov: Nordita preprint 77/1
6. C.G.Callan, R.Dashen and D.J.Gross: Princeton preprint (1977); S.Coleman: Harvard preprint and references therein (1978)
7. M.B.Voloshin, I.Yu Kobzarev and L.B.Okun: Sov.J. Nucl.Phys. 20, 644 (1975); S.Coleman: Phys. Rev. D15, 2929 (1977); C.Callan and S.Coleman: Phys. Rev. D16, 1762 (1977); M.Stone: Phys Lett. 67B, 186 (1977); H.J.Katz: Phys Rev. D17, 1056 (1978)
8. C.M.Bender and T.T.Wu: Phys. Rev. 184, 1231 (1976); Phys. Rev. Lett. 27, 461 (1971); Phys. Rev. D7, 1620 (1973)
9. C.S.Lam: Nuovo Cim. 55A, 258 (1968)
10. L.N.Lipatov: Sov. Phys. JETP 44, 1055 (1976) and 45, 216 (1977)
11. E.Brézin, J.C. Le Guillou and J.Zinn-Justin: Phys. Rev. D15, 1544 and 1558 (1977); E. Brézin: Proc. European Conference on Particle Physics, Bucharest, and references therein (1977)
12. E.T.Copson: Theory of Functions of a Complex Variable (Oxford University Press) p.330
13. S.Coleman, V.Glaser and A.Martin: Comm. Math. Phys. 58, 211 (1978)
14. J-L.Gervais and B.Sakita: Phys. Rev. D11, 2943 (1975); N.H.Christ and T.D.Lee: Phys Rev. D12, 1606 (1975)
15. J.Zinn-Justin: Cargèse Lectures 1977; G.Parisi: Cargèse Lectures 1977; R.Jackiw: Rev. Mod.Phys. 49, 681 (1977)
16. S.Chadha et al: Phys. Lett. 72B, 103 (1977)
17. A.J.McKane and D.J.Wallace: Southampton preprint THEP 77/78-18
18. F.J.Dyson: Phys. Rev. 85, 631 (1952)
19. S.Graffi, V.Grecchi and B.Simon: Phys.Lett. 32B, 631 (1970)
20. G.Parisi: Phys. Lett. 66B, 167 (1977); E.Brézin, G.Parisi and J.Zinn-Justin: Phys. Rev. D16, 408 (1977); C.Itzykson, G.Parisi and J-B. Zuber: Phys. Rev. D16, 996 (1977); A. Houghton, J.S.Reeve and D.J. Wallace: Phys. Rev. B18, 2956 (1978); J.C.Collins and D.E. Soper: Ann. Phys. (NY) 112, 209 (1978)
21. J.C.Le Guillou and J.Zinn-Justin: Phys. Rev. Lett. 39, 95 (1977); G. Parisi and E. Brézin, to be published
22. G.A.Baker et al: Phys. Rev. Lett. 36, 1351 (1976); B.G.Nickel, to be published
23. W.J.Camp et al: Phys. Rev. B14, 3990 (1976)
24. R.F.Chang et al: Phys. Rev. Lett. 37, 1481 (1976); R.F. Chang, H.Burstyn and J.V.Sengers: Maryland preprint
25. F.M.Dittes, Yu.A.Kubyushin and O.V.Tarasov: Dubna preprint 1977
26. J.L.Cardy: Santa Barbara preprint UCSA TH-1 1978
27. D.J.Thouless: J.Phys. C: Solid State Phys. 8, 1803 (1975); M.V. Sadovskii: Sov. Phys. JETP 43, 1008 (1976)
28. W.Klein, D.J.Wallace and R.K.P.Zia: Phys.Rev. Lett. 37, 639 (1976)

The Theory of Structural Phase Transitions: Cluster Walls and Phonons

A.D. Bruce

Department of Physics, University of Edinburgh
Edinburgh EH14 4AS, Scotland

Abstract

The essential problems posed by systems undergoing a structural phase transition are discussed in the context of a simple double-well potential model. The traditional conceptions of the contrasting behaviour at the model's order-disorder and displacive limits are reviewed, and set alongside a contemporary view of the displacive system, which assigns a central role to the formation of precursor-clusters of short range order, displaying a two-time-scale, order-disorder behaviour. An assessment is made of the extent to which this view is substantiated, directly or indirectly, by dynamic and static theory (analytic, Renormalisation Group and molecular dynamics methods), by experiment ("central peaks") and speculative analogy (precursor effects at lock-in phase transitions).

1. Introduction

Systems undergoing structural phase transitions are essentially and inescapably anharmonic. This principle is established by the most rudimentary phenomenology of phase transitions, and is expressed in the essential ingredients of any microscopic model. Nevertheless, until relatively recently, the full subtlety of its consequences has tended to be lost in the inadequacies of Landau phenomenology, and the inconsistencies of the classical approximation schemes traditionally invoked in the solution of microscopic models.

Much of the recent activity is attributable to the motivating impact of two key experiments [1,2], which revealed the need for techniques specifically designed to accommodate the effects of strong non-linearity [3,4].

Firstly, the observation [2] of non-classical order-parameter behaviour near the 105 K structural phase transition (s.p.t.) in $SrTiO_3$ initiated interest in the distinctive problems of the critical regime, and motivated extensive application of Renormalisation Group (R.G.) techniques. Though employing a conventional (harmonic) basis, these methods provide a prescription for the systematic treatment of the non-linear coupling among critical fluctuations: their explicit predictions for the universal quantities (chiefly exponents) characterising (principally) static [4,5] and (less extensively) dynamic [6] behaviour, near the critical point, represent the most clearly-identifiable successes of recent years.

These techniques, however, have (as yet) offered little to illuminate the underlying character of the spectrum of excitations promoting the (calculable but opaque) deviations from classical exponent values. In revealing an unexpected double-peaked structure in the soft mode spectral function, above the critical temperature in $SrTiO_3$, the second key experiment [1] focussed attention on this issue, thereby motivating a series of studies, probing the dynamical consequences of strong non-linearity - either by direct computer simulation [7,8], or by analytic techniques shunning the customary assumptions of perturbation theory about a phonon basis [3,8-12]. Partly because the productive questions are less evident in this area, and partly because those questions that are evident are also hard to answer, the success of these efforts has, perhaps, been more muted than that of exponent-generating R.G. calculations. Nevertheless, there has emerged a qualitatively-compelling (if, as yet, largely unsubstantiated) picture of the grass-roots "physics" of systems undergoing s.p.t.'s - a picture which I shall develop here, in the context of the ubiquitous double-well potential model.

2. Model and Background

In keeping with the bulk of the literature I will discuss the properties of the simple d-dimensional model (Fig.1) defined by the classical Hamiltonian

$$H = T + V$$

$$T = \frac{1}{2m} \sum_{\ell} p^2(\ell) \; ; \quad V = \frac{C}{2} \sum_{\ell,\ell'}^{n.n.} (u(\ell) - u(\ell'))^2 + \sum_{\ell} V_s(u(\ell)) \tag{1}$$

describing the behaviour of a set of atoms of mass m, each located in its own double-well potential

$$V_s(u) = \frac{\overline{A}}{2} u^2 + \frac{B}{4} u^4 \; ; \quad \overline{A} < 0 \tag{2}$$

and coupled to its 2d immediate neighbours by harmonic springs. It is evident that the model's impressive array of simplifying features (the supposed perfection of the crystal lattice, the rigidity of the sublattice setting up the local potential V_s, the short range forces, and the scalar character of the coordinates, reflecting a uniaxial anisotropy) limit its relevance to real (dirty, compressible, frequently ferroelectric and multi-component-order-parameter) systems. Nevertheless the model embodies the one key feature common (in some form) to all systems undergoing s.p.t's. - the local double-well potential which (for $d > 1$: c.f. Section 6) may be shown [13] to constitute a sufficient condition for a phase transition, at a non-zero T_c, to a homogeneously-ordered phase ($\overline{u} \equiv \langle u(\ell) \rangle \neq 0$)

The model offers, moreover, a unified context in which to examine the behaviour of the two limiting ("order-disorder" and "displacive") classes of systems undergoing s.p.t.'s, which traditionally, and perhaps unfortunately, have tended to be treated in isolation from one another. This classification may be made according to the ratio

$$h = E_W/E_B = |\overline{A}|/4dC \tag{3}$$

of the two energies characteristic of (1) - the depth of the local potential well $E_W = |V_S(\overline{u}_0)| = |\overline{A}|^2/4B$, and a representative bond energy

$E_B = dC\bar{u}_0{}^2 = C|\bar{A}|/B$. Since, in the d=3 case (to be understood in the absence of any contrary indication) the ratio k_BT_c/E_B is of order unity, and only mildly model-dependent, the ratio h closely reflects the parameter-dependence of the conceptually more transparent quantity

$$g = E_W/k_BT_c = |\bar{A}|^2/4Bk_BT_c \tag{4}$$

In the order-disorder limit h >> 1 it is then clear that, except at temperatures $T >> T_c$, each atom will be predominantly localised in the neighbourhood of one or other of the well minima ($\pm\,\bar{u}_0 = \pm\,\sqrt{|\bar{A}|/B}$), as envisaged in the Ising model; within the harmonic approximation one finds that all modes of vibration about the high-temperature (h.t.) phase equilibrium positions (h.t. phonons) are unstable and the inescapably non-linear character of the dynamics is immediately apparent.

In contrast, in the displacive limit, h << 1, the critical thermal energy is large in comparison with the local barrier height; within the harmonic approximation only a small group of long-wavelength h.t. phonons are unstable and a quasi-harmonic view of the dynamics seems tenable.

Remarkably, these fundamental differences are not reflected in the majority of exactly-established properties of (1). Specifically it is well-known that there are ("universal") features of the model's static critical behaviour - principally the critical exponents - which are quite independent of the system's location on the order-disorder/displacive spectrum, and which therefore necessarily coincide with those of the Ising model of ferromagnetism (now known with great accuracy). Similarly, it has been shown [14] that there are universal features of the dynamic behaviour, which are those of a model with a scalar non-conserved order parameter, and conserved energy ("model C" in the nomenclature of [6,15]). In particular the relaxation time [6]

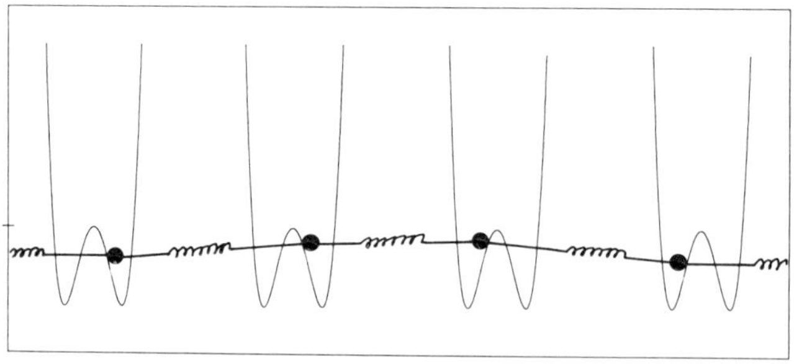

Fig.1 Schematic representation of the d=1 form of the model (1)

$$\tau_u \equiv \chi(0) \left. \frac{\partial\chi^{-1}(\Omega)}{\partial(-i\Omega)} \right|_{\Omega=0} \tag{5}$$

characterising the critical slowing down displayed by the frequency-dependent order parameter response function $\chi(\Omega)$ satisfies [14]

$$\tau_u^{-1} \approx D_T \kappa^2 \sim t^{2\nu+\alpha} \tag{6}$$

where t is the reduced temperature, D_T is the thermal diffusivity whose inverse has the critical behaviour of the specific heat ($c_v \sim t^{-\alpha}$), and κ is the inverse of the correlation length ($\xi \sim t^{-\nu}$).

This result is explicit, but conceptually opaque, giving no hint as to the physical character of the collective mode in which the critical slowing down occurs. Traditionally, the critical slowing down in order-disorder and displacive systems has been envisaged in quite different terms. The traditional picture of the order-disorder system (Section 3) emerges immediately from a recognition of the strongly non-linear character of the deep-well limit; qualitatively this picture is not in doubt. The traditional picture of the displacive system has emerged from classical approximation schemes (Section 4): its basis is therefore less secure - the classical schemes are known to yield incorrect critical exponents - and, indeed, it has come to be replaced (Section 5) by a picture (that alluded to in Section 1) which has much in common with the traditional conception of the order-disorder system, and which illuminates the origins of universality.

3. Order-disorder Systems: A Qualitative Discussion

In the order-disorder (deep-well) limit it is clear that the fundamental dynamical processes (i.e. those that maintain/express the thermodynamic equilibrium of the high-temperature phase) are the collective analogues of single-particle (thermally activated) hopping across or (in a quantum theory) tunnelling through the large potential barrier separating the two wells. It seems little more than a truism to assert that these processes represent the motion of domain walls, whereupon it becomes natural to interpret the relaxation time (6) for the order parameter degrees of freedom as the time taken for a domain wall, with "diffusion constant" $D_{D.W.} \approx D_T$ [16] to diffuse the distance $\xi = 1/\kappa$.

The established treatments of the order-disorder regime - the (classical) Glauber model [17] and the (quantum-mechanical) pseudo-spin model [18,19] focus exclusively on these "inter-well" degrees of freedom, which must evidently dominate the critical dynamics in general and, in particular, the spectral response $S(\Omega) = kT \operatorname{Im} \chi(\Omega)/\Omega$ in which they will manifest themselves as a quasi-elastic peak of width $1/\tau_u$. However, these treatments do not, in the first instance, accommodate the additional "intra-well" degrees of freedom, possessed by the Hamiltonian (1), which will be reflected in additional collective modes - the phonons of a quasi-statically-disordered lattice. (A direct "independent site approximation" (i.s.a.) to the $h >> 1$ limit of (1) does some justice to <u>these</u> degrees of freedom, but the non-ergodic character of the isolated <u>classical</u> double-well, to whose properties the problem is reduced in i.s.a., leads to an artificially narrow-δ-function! - representation of the "inter-well" contribution to $S(\Omega)$ [20]). The spectral function $S(\Omega)$ should therefore exhibit additional (resonant) response at the frequency of this collective bottom-of-the-well motion, which should itself remain largely unaffected by the approach to criticality.

In effect, then, the order parameter degrees of freedom have not one but two distinct time scales - one long (and critical) time scale associated with collective inter-well motion ("domain wall diffusion") and the other generally much shorter time scale associated with collective intra-well motion (low temperature, ℓ.t., phonons). Indeed, the existence of both low-frequency (diffusive) and high frequency (resonant) components in the spectral response of a "single mode" (c.f. the schematic Fig.2) is an experimentally well-established (though not invariably appreciated) feature of order-disorder systems, a prime example being $NaNO_2$ where dielectric measurements reveal a (critical) central response at frequencies lower than 10^9 Hz. [21] in addition to a non-critical response at the frequencies (10^{12} Hz.) of infra-red phonons [22].

4. Displacive Systems : Classical Approximation Schemes and the Soft Phonon

The history of the dynamical properties of displacive systems is dominated on both theoretical and experimental fronts by the concept of the soft phonon mode [23]. Thus, revelations of the idea's limitations have, if anything, come as more of a cultural shock to the s.p.t. community than the discovery of non-classical behaviour in equilibrium properties. Both the idea itself, and some pointers to its limitations, emerge from the two classical decoupling schemes which I shall now briefly outline, beginning with the independent site approximation [20,5].

The essential feature of the i.s.a. is the neglect of inter-site correlation - and thence of short range order (s.r.o). Explicitly, one makes

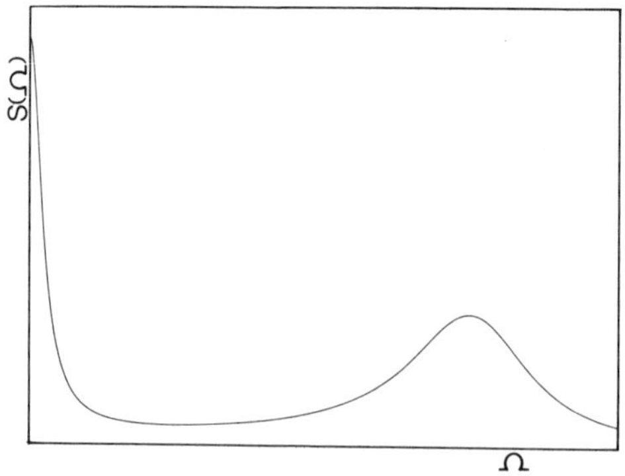

Fig.2 Schematic representation of a spectral function $S(\Omega)$ with both low-frequency (diffusive) and high-frequency (resonant) components

the replacement

$$u(\ell)u(\ell') \rightarrow u(\ell)\langle u(\ell')\rangle^{i.s.} = u(\ell)u^{i.s.} \tag{7}$$

in the Hamiltonian (1) when the $T > T_c^{\,i.s.}$ ($u^{i.s.} = 0$) response function may be written as

$$\chi(\Omega) = \chi^{i.s.}(\Omega) \equiv \frac{\chi_0(\Omega)}{1 - 4dC\chi_0(\Omega)} \tag{8}$$

where $\chi_0(\Omega)$ is the analytically-tractable [20] response function for an isolated atom in the effective potential

$$V^{i.s.}(u) \equiv V_s(u) + 2dCu^2 \tag{9}$$

In the displacive limit $C/|\bar{A}| \gg 1$, so that the effective potential $V^{i.s.}$ has only a single minimum, and the strongly non-linear character of the underlying system is suppressed. The dynamic response function $\chi^{i.s.}(\Omega)$ is then found to have a single pole (and the spectral function $S(\Omega)$ a single peak) located by the ("soft mode") frequency [20]

$$(\omega^{i.s.})^2 \approx \frac{|\bar{A}|}{m}\left(T/T_c^{\,i.s.} - 1\right) \tag{10}$$

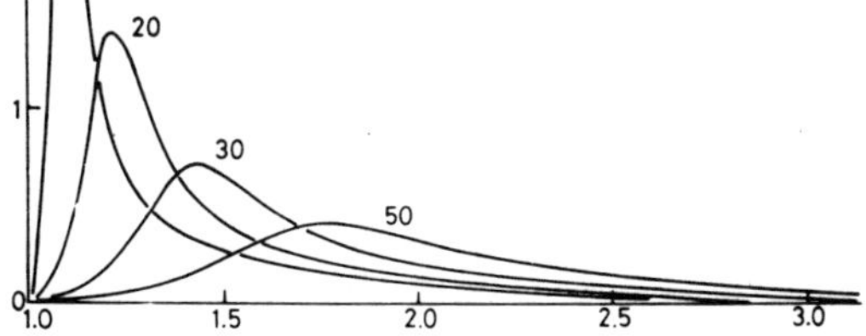

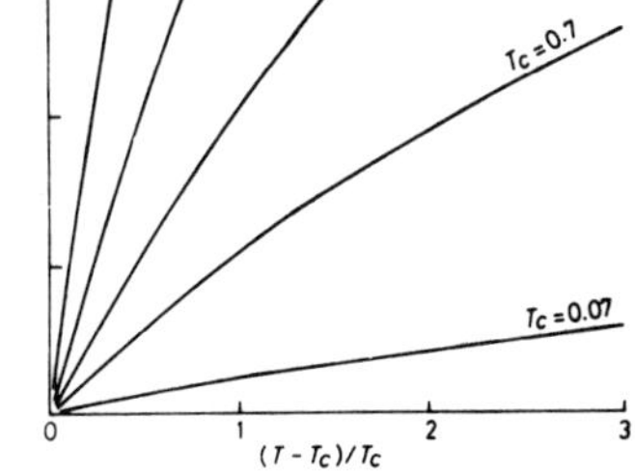

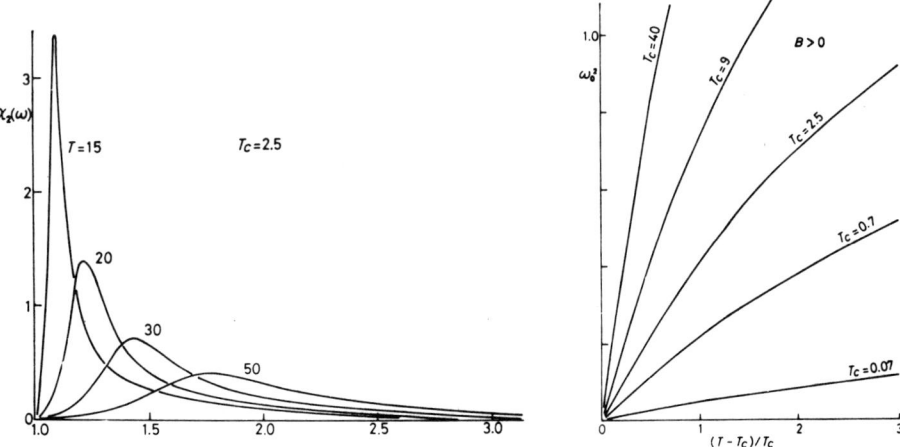

Fig.3 The imaginary part of $\chi(\Omega)$ as a function of frequency (left) and the squares of soft mode frequencies as a function of reduced temperature, in the i.s.a. for the model (1), in the displacive limit [20].

whose approach to zero as $T \to T_c^{i.s} \approx 4C|\overline{A}|/B$ (c.f. Fig.3) is the dynamic signal of the divergence in $\chi(\Omega=0)$, and is the explicit realisation of a critical slowing down with relaxation time $\left(\gamma \equiv \frac{d\chi_0''(\Omega)}{d\Omega}\Big|_{\Omega=0}\right)$

$$\tau_u^{i.s.} \approx \gamma(4dC)^2/m(\omega^{i.s.})^2 \tag{11}$$

Below the phase transition (not shown in Fig.3) the frequency of this mode (then a ℓ.t. phonon) hardens from zero, under the stabilising influence of order parameter growth.

A qualitatively similar picture emerges from the "self-consistent phonon", alias "independent mode", approximation (i.m.a.) in which variational techniques (see, e.g. [24]) are used to establish the optimum harmonic representation for the equilibrium properties of (1), thus eliminating explicitly non-linear effects by fiat. In effect (in characterising the h.t. phase) one makes the replacement

$$u^4(\ell) \to 3u^2(\ell)\langle u^2(\ell)\rangle^{i.m.} \tag{12}$$

in (1); a self-consistent treatment of the resulting harmonic equations of motion then again leads to a spectral response with a soft phonon resonance at a frequency $\omega^{i.m.}$, tending to zero (Fig.4) as $T \to T_c^{i.m.} \approx 2.6\, C|\overline{A}|/B$. (In i.m.a., however, $(\omega^{i.m.})^2$ vanishes as $(T-T_c^{i.m.})^{\frac{1}{2}}$, and the soft phonon is undamped in the simplest version of the theory; more importantly, see also Section 5.2.1.)

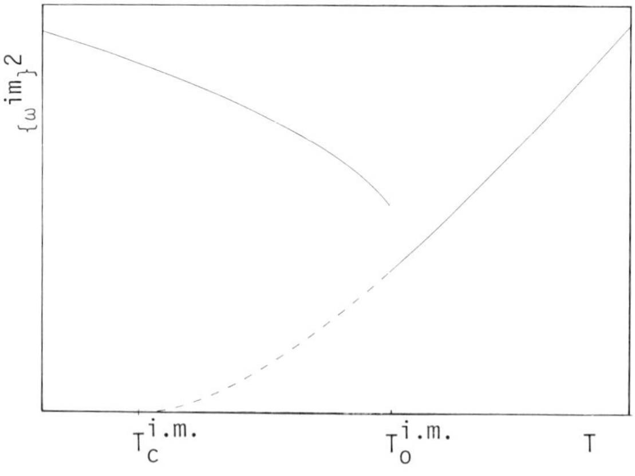

Fig.4 The qualitative temperature-dependence of the square of the soft mode frequency in the i.m.a. for the model (1); $T_0^{i.m.}$ locates the (first order) phase transition

Both approximation schemes thus imply (or assume) the same underlying view of the $T > T_c$ dynamics: each atom exhibits an essentially harmonic motion about its high-temperature phase equilibrium position (the top of its well), while the collective mode in which the critical slowing down resides is a (generally) damped long wavelength "well-to-well", h.t. phonon.

There can be no doubt that this picture gives a qualitatively correct account of the gross features of the behaviour of displacive systems, of which $SrTiO_3$ is the accepted prototype: the experimental evidence supporting the soft phonon concept is compellingly extensive (c.f. Fig.5, and [25]) altogether dwarfing the tenuous evidence for its deficiencies. It is, nevertheless, to the latter that the remainder of this paper is devoted.

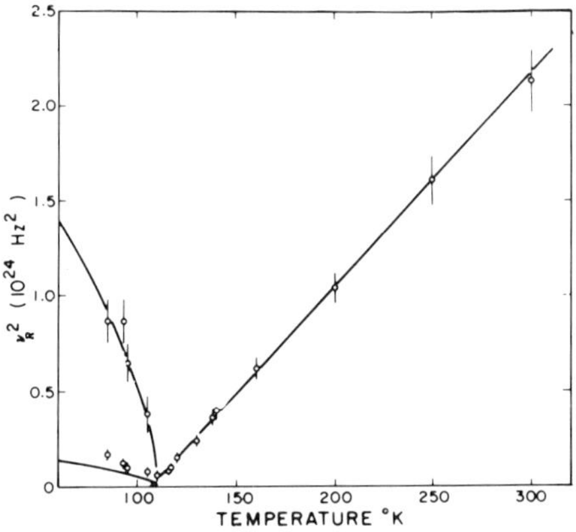

Fig.5 The squares of soft mode frequencies in $SrTiO_3$, as determined by neutron scattering techniques [25]

5. Displacive Systems: Beyond Classical Theory

5.1 A Contemporary View

In the last few years the soft phonon picture sketched above has been called in question, not merely in detail ("With what exponent does the soft phonon frequency vanish as $T \to T_c$?") but in its very essence ("Does a soft phonon mode frequency vanish at T_c?"). From these questionings there has emerged a new view - I believe it to be the consensus of those whose names appear in the references - of the essential physics of displacive systems near their critical points: this view I shall now describe in outline, before assembling the evidence in which it originated, and through which it may be refined and assessed.

The essential idea is that, with the onset of criticality, the growth in short range order promotes a crossover from a regime in which the collective behaviour has the classical displacive form, to a regime in which the collective behaviour displays features that are better described in the language traditionally reserved for order disorder systems. This "displacive-order-disorder" (or "phonon-Ising" [26]) crossover, occurring at some temperature T_I, is signalled by the appearance of "microdomains" or "clusters" - groups of atoms whose motion is strongly correlated - within which the average atomic displacement ("cluster order parameter") is non-zero: each atom then oscillates about the equilibrium position set by the value of the "order parameter" of the cluster of which it is a member, this position of local equilibrium itself changing sign as a cluster wall diffuses by.

Thus, while the collective behaviour within the displacive regime has but a single time scale, set by the soft high-temperature phonon (giving a single region of possibly overdamped response in $S(\Omega)$) the collective behaviour in the regime $T_c < T \lesssim T_I$ has the two time scales characteristic of an order-disorder system. The critical time scale is set by the diffusion of the walls of the clusters, giving rise to a central response in $S(\Omega)$ - the s.r.o.-induced dynamic precursor of the long-range-order (ℓ.r.o)-induced static (Bragg) peak to appear below T_c. The second, non critical, time scale is set by the collective oscillations about the positions of local equilibrium, giving rise to a (possibly overdamped) phonon side band in $S(\Omega)$ - the residue of the high temperature phonon, and the natural precursor of the low temperature phonon, into which it evolves without ever exhibiting a complete softening.

Evidently this picture constitutes a graphic and aesthetically appealing expression of universality - a perspective I shall develop, in discussing its origins and assessing its validity.

5.2 The Theoretical Background

A balanced assessment of the theoretical basis for the picture I have described must discriminate between the evidence that is purely circumstantial, and the evidence that is explicit: the following discussion is loosely subdivided along these lines.

5.2.1 Circumstantial Evidence

The failure of the soft mode concept is illuminated by the inadequacies of the classical theories from which it emerges. The formal inconsistencies of both the i.s.a. and the i.m.a., close enough to T_c and in systems with dimensionality $d < 4$, are well known (see e.g. [5]).

The i.s.a. fails because, though predicting a divergent correlation length, it takes no account of correlations: its failure emphasizes the need to do justice to the behaviour's growingly coherent - and therefore growingly non-linear-character. (It will be recalled that it is in the assumption of incoherent, uncorrelated motion that the double-well character of the local potential is lost.)

The i.m.a. fails because, though allowing for the growing coherence of the motion, it requires that the increasing coherence be expressed simply through a growing population of soft harmonic phonons: its failure emphasizes that the behaviour of the (long wavelength) collective degrees of freedom is non-linear in an essential way.

In fact, the i.m.a. is more informative than the discussion of Section 4 would suggest: when extended to the ordered phase, the i.m.a. is found to imply that the complete softening of the high-temperature phonon mode is preempted by a first order phase transition (see e.g. [24], [5]) at a temperature $T_0^{i.m.} > T_c^{i.m.}$ (Fig.4). Though it is clear that this prediction is wrong - there is no doubt that (1) exhibits a continuous phase transition - it is not so clear that it is devoid of physical significance. In effect the i.m.a. analysis demonstrates that there exists a temperature $T_0^{i.m.}$ below which it becomes better (in a variational sense) to describe the behaviour in terms of a static homogeneous ordering, together with the corresponding ℓ.t. phonons, than in terms of h.t. phonons (whose softening is, at that stage, incomplete). It is then tempting [27,28] to regard $T_0^{i.m.}$ as a measure of the displacive-order-disorder crossover temperature T_c, below which (in the picture sketched earlier) the (softening) high-temperature phonon is indeed superseded by a (hardening) low-temperature phonon. Within this interpretation the i.m.a.'s principal failure is that it identifies the crossover temperature $T_I \approx T_0^{i.m.}$ with the macroscopic ordering temperature T_c, when in fact $T_c < T_0^{i.m.}$. This failure it is natural to attribute to the i.m.a.'s inability to accommodate the other feature of the order-disorder regime - the strongly non-linear order parameter configurations associated with cluster walls, against whose formation (one might surmise) a homogeneously ordered phase is actually unstable in the regime $T_c < T < T_0^{i.m.}$ [12], the consequent break up into microdomains destroying the ℓ.r.o. but leaving the s.r.o. necessary to sustain the low temperature phonons.

A formal demonstration that the true transition temperature T_c does indeed lie below $T_0^{i.m.}$ would lend further plausibility to these conjectures. (It is known that the stability limit $T_c^{i.m.}$ represents a lower bound for T_c [29,13], the ratio $T_c/T_c^{i.m.}$ tending to unity in the displacive limit [30,5].)

A second, substantially richer, vein of circumstantial support for the cluster picture is to be found in the properties of the analytically-tractable one-dimensional form of the Hamiltonian (1), which has been the subject of a number of extensive studies [3,8,9]. Since the explicit results of these studies will no doubt figure in the presentations of other speakers at this Symposium I shall simply discuss their implications, briefly and qualitatively.

Perhaps the most immediately-instructive consequence of the analytic tractability of the d=1 model, in the displacive limit, is that exact solutions to the non-linear equations of motion may be identified, and shown to constitute explicit realisations of the three excitations prominent in the picture promoted in Section 5.1 - low and high temperature phonons, and cluster walls bounding microdomains. These solutions have been shown to constitute a successful phenomenology both of the low-temperature equilibrium properties (exactly calculable by transfer-integral techniques [3]), and of the spectral response $S(\Omega)$ (itself not analytically tractable, but calculable by molecular dynamics methods [8]), which displays the expected features: a softening h.t. phonon response evolves gradually (with reducing temperature) into the response of a hardening ℓ.t. phonon, and a critically-narrowing central component that can be traced to microdomain-induced s.r.o.

That such a crossover should occur at low enough temperature is, of course, evident from the nature of the ground state of (1): thus, in a sense, the principal open question is whether (or to what extent) the crossover occurs before (at temperatures higher than) the onset of ℓ.r.o., at T_c - a question which is interesting (non-trivial!) only in systems which (unlike the d=1 model; c.f. also [11]) possess ℓ.r.o. at a finite temperature.

Nevertheless, the temperature T_I locating the d=1 crossover does deserve attention. It has been characterised according to a number of conceptually different but (in d=1) practically equivalent ways - as the temperature of the i.m.a.'s first order phase transition (c.f. above), as the temperature at which the specific heat exhibits its maximum [8], kink-excitations (domain walls) become thermally unstable [10], and the single-particle distribution function begins to exhibit two peaks [28]. Though additional criteria may seem less than essential, the following argument is, I think, illuminating.

As the temperature is lowered the correlation length of the d=1 displacive system grows (within the displacive regime) as

$$\xi \sim \frac{C^{2/3}}{(Bk_B T)^{1/3}} \cdot a \tag{13}$$

where a is the lattice spacing. While the growth in ξ will be expressed in the appearance of clusters of atoms moving coherently, the motion of the clusters will not reveal a distinctively Ising behaviour (and (12) will remain valid) until the potential barrier $E_{w,\xi}$ seen by the <u>cluster</u> (during the <u>collective</u> inter-well motion of its member atoms) becomes comparable with the thermal energy $kT = kT_I$. Evidently $E_{w,\xi} \sim \xi E_w/a$, whence, using (13),

$$kT_I \sim \frac{C^{\frac{1}{2}} |A|^{3/2}}{B} \tag{14}$$

in accord with the results based upon the other criteria, cited above, and with the crossover temperature located by d=1 molecular dynamics [8].

Since the form of this result is actually directly implied by the homogeneity properties of the displacive limit [8] it would be dangerous to accord great significance to the apparent "success" of the criteria from which it may be derived. Nevertheless, the argument presented above may be rigorously grounded in the methods of the R.G., which provide a framework within which, for any dimensionality d, a precise meaning may be attached to the idea of a "cluster coordinate" and the "potential" it sees.

Cluster coordinates describe the collective behaviour of atoms within a volume of linear dimension ξ, and are thus plausibly represented by the local degrees of freedom left after application of a R.G. transformation $\mathbb{R}_b$ [4] with scale factor $b = \xi/a$; the potential seen by a cluster (of the physical system) may thus be identified with the potential experienced by a local coordinate in the transformed system, with configurational energy $V_b = \mathbb{R}_b V$. In the limit in which criticality is approached the potential seen by a cluster tends to a fixed point form, V_ξ^*, with universal features: Figure 6 displays the form of V_ξ^* for various dimensionalities. (In d=1 the fixed point potential presumably consists of two infinitely sharp and deep Ising-like wells.) A natural measure of their order-disorder/displacive character is afforded by the fixed point value g^* (c.f. (4)) of the ratio of well to critical thermal energy. In d=3, Wilson's approximate recursion formula [4] gives $g^* \approx 0.55$; for d=4-ε, the result is $g^* = \varepsilon/4 + O(\varepsilon^2)$. (The fixed point value h^* of the ratio of well to bond energies displays similar behaviour - though the R.G.'s use of a spherical Brillouin Zone makes it difficult to determine h^* in a way that is entirely consistent with (3).)

These results should be viewed with a modicum of caution: it is not clear

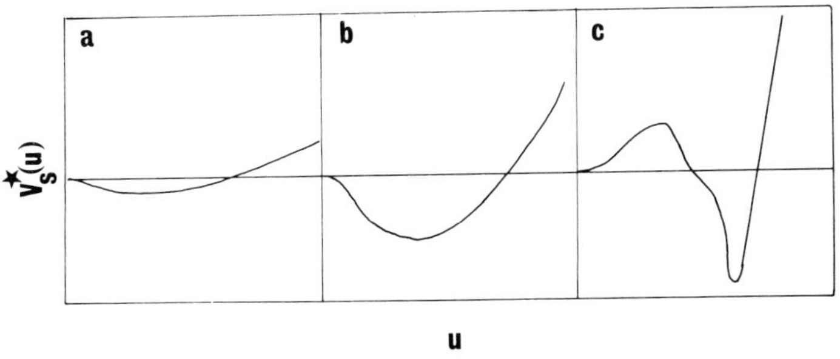

Fig.6 The fixed point form of the single particle potential [4]
(a) for d = 3.9 (b) for d = 3 (c) for d = 2

that the quantity g^* will be found to be strictly universal (cut-off-independent) in calculations going beyond the levels of approximation inherent in Wilson's recursion formula and the leading order ε-expansion; neither is it clear to what extent the universality of features of the fixed point configurational energy will be reflected in universal features of the critical dynamics.

Nevertheless these results do illuminate the nature and, potentially, the extent of the displacive-order-disorder crossover. The behaviour - at the very least, the equilibrium properties - of the long wavelength, "cluster", degrees of freedom in the near critical system is determined by a fixed-point potential which itself has some definite location on the order-disorder displacive spectrum - a location that is independent of the system's microscopic parameters (e.g. h or g) but which reflects, principally, those essential features (here, simply the dimensionality) that are instrumental in determining universal quantities. Thus, while the parameters g and h form a crude basis for a displacive/order-disorder subdivision in the non-critical regime of weak correlations (where, e.g., h determines the character of the potential seen by an atom: c.f. (9)), close to criticality it seems more plausible to base this subdivision on the fixed-point quantities g^* and h^*, whose values incorporate the all-important effects of correlations.

The quantitative inferences to be drawn from these arguments are not yet apparent; qualitatively, however, it seems safe to observe that the order-disorder "content" of the near critical behaviour (as measured, say, by g^*) decreases from a large value in d=2 (an infinite value in d=1) to zero in d=4, and is by no means insignificant in d=3.

This trend mirrors the dimensionality-dependence of the deviations of critical exponents from their mean field (i.s.a.) values, making it tempting to correlate the occurrence and extent of non-classical behaviour, with the formation and order-disorder content of precursor clusters, and to visualise the universality of the critical region as the universality of cluster behaviour.

5.2.2 Direct Evidence

As witnessed by the brevity of this section, the direct theoretical evidence for two-time-scale, order-disorder features in the near-critical behaviour of displacive systems, is not extensive. Such evidence as there is falls into two categories.

Firstly there are the results (arguably more appropriate to the "experimental" Section below) of a number of illuminating molecular dynamics investigations of the Hamiltonian (1) [7]. These studies will be reviewed elsewhere at this Symposium [31] and so I shall remark only that they provide considerable support for the cluster picture - having, indeed, played a major role in its development: specifically, calculations on a 2-dimensional "displacive" system ($h \approx 0.13$; $g = 0.19$) display, in the near-critical wavevector-dependent spectral response (Fig.7) both an (incompletely-softened?) overdamped h.t. phonon, and a critical central component (shifted to finite energy at finite wavevector?) which has been traced, quite explicitly, to the dynamics of Ising-like clusters.

There have, secondly, been a number of studies of (3-dimensional) displacive systems which, while employing a phonon basis, have attempted to treat non-linearity with greater care than classical (or low-order perturbation) theory. Infinite order perturbation theories (ad-hoc, but physically-motivated in the calculations of [26]; systematically-controlled by the dictates of the R.G. and the ε-expansion in [14]) have vindicated an earlier conjecture, expressed in a number of different guises [32], that in systems with the essential symmetry envisaged in the model (1), the critical slowing down (and thus the central response in $S(\Omega)$) will be dominated by a contribution to the soft h.t. phonon self-energy that arises from non-linear coupling to a thermal diffusion mode (c.f. Eq.6) built from some "ladder" series of phonon diagrams. The R.G. studies described in [14] seemingly capture

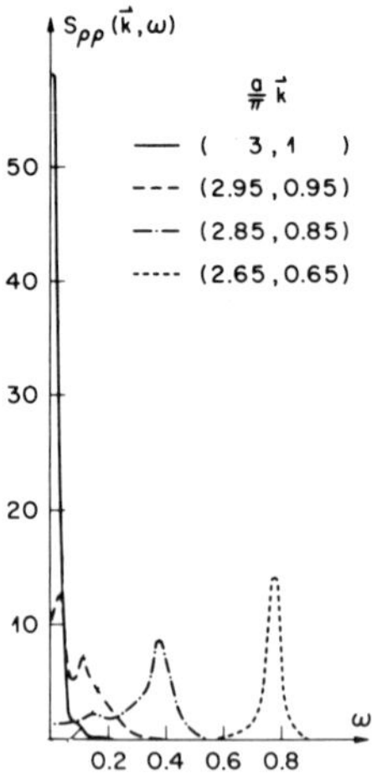

Fig.7 The frequency- and wave-vector-dependence of the spectral response of a d=2 displacive model at $t \approx 0.06$, as determined by molecular dynamics [7]

only this central response; however the calculations reported in [26] reveal both central and h.t. phonon responses, the former, at its first appearance, being broader than the latter, but narrowing and growing in intensity with the approach to T_c.

These calculations provide the most immediate analytic evidence for a two-time-scale dynamics in displacive systems; however, they vindicate the cluster picture itself, only to the extent that microdomain walls may be regarded as the principal microscopic realisations of phonon-density (temperature) fluctuations, and the diffusion of "thermal fluctuations" effectively synonymous with the diffusion of domain walls (as suggested, implicitly, in Section 3).

5.3 The Experimental Background

The experimental support for the cluster picture of displacive systems is sparse, resting largely, and somewhat insecurely, upon the key inelastic neutron-scattering study of $SrTiO_3$ [1] referred to in Section 1. This experiment revealed that the softening h.t. phonon observed in earlier, less-detailed, studies (Fig.5) in fact saturates at a finite frequency ($\approx 0.13\times10^{12}$ Hz) at T_c; a central component in the spectral response appears some 50 K above the phase transition, its intensity diverging with the approach to criticality (Fig.8).

While these observations are in qualitative accord with the cluster picture (and, indeed, constituted the major motivation for the theoretical efforts reviewed above) it now seems unlikely that they can be understood entirely within this framework. The principal problem lies in the frequency-width of the central component - less than the instrumental resolution, both in the original neutron-scattering experiments [1], and in subsequent investigations employing a high-resolution back-scattering spectrometer [33]: even on its first appearance the width is no greater than 2×10^7Hz - a Mössbauer γ-ray scattering experiment establishes an even smaller upper bound of 3×10^6Hz. [34].

It seems highly unlikely [14,16] that such a narrow response does, in fact, originate in the cluster dynamics of an ideal crystal: molecular dynamics computations [7] and the calculations reported in [26] both suggest a central component with a width not much different from that of the soft phonon - except, of course, for reduced temperatures $t \ll 1$, where critical slowing down produces an arbitrarily narrow response. It is possible that a response much narrower than that of the scalar (n=1), spatially-isotropic, models investigated so far may arise in systems which, like $SrTiO_3$, have an n=3-component order parameter (c.f. the pathological features of the critical dynamics of such systems alluded to in [35]), or, again like $SrTiO_3$, have a highly-anisotropic soft-phonon dispersion (c.f. the conjectures in [36]). Nevertheless, it seems likely that a large proportion of the central component observed in $SrTiO_3$ reflects non-ideal crystal behaviour: perhaps the influence of a strain-induced ordering of a surface layer, setting in well above the bulk T_c [34]; perhaps the influence of mobile impurities [16] acting as nucleation centres for clusters, and extending their life-time. The latter seems the more attractive possibility (since it salvages the central ideas propagated here!) and the more plausible, since the former mechanism would not (in itself) account for the saturation of the phonon side band, observed in the neutron scattering experiments [1].

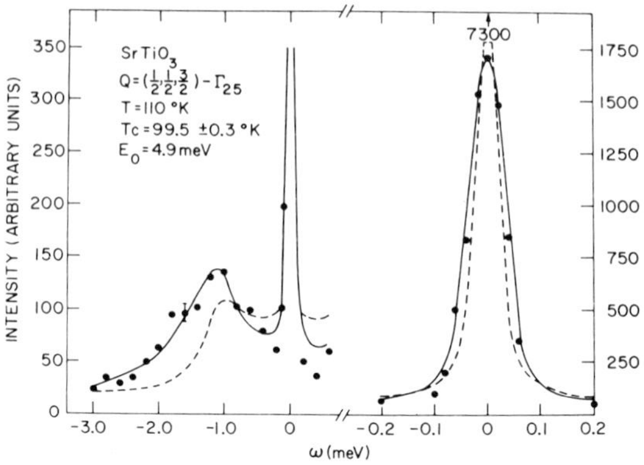

Fig.8 The soft mode spectral response in $SrTiO_3$, as determined by neutron scattering techniques [1]

It is arguable that this feature deserves as much attention as the central peak itself, since it is indisputably a bulk phenomenon, and (unlike the central component!) is apparently sample-independent. In this context it is worth observing that, if the cluster mechanism is, at least partially, responsible for the observed behaviour in $SrTiO_3$ (- note that the experiments do not rule out the existence of an additional central component, with a width compatible with the cluster picture -) the inelastic response at T_c might reasonably be expected to display precursor traces of the two (A and E) soft ℓ.t. phonons that evolve from the 3-fold-degenerate h.t. soft mode. (c.f. Fig.5)

5.4 A Suggestive Analogy: the Lock-in Phase Transition

Stepping briefly outside the framework of the Hamiltonian (1), I wish, finally, to make reference to a class of s.p.t. at which the elements of the cluster picture are demonstrably present, even within Landau theory [37,38].

Consider the double-well potential model, modified by the inclusion of additional, longer-range, interactions between atoms, such that the minimum in the h.t. phonon dispersion occurs at some wave-vector $\vec{q}_s$, lying close to a wave-vector $\vec{q}_c$ that is commensurate with the h.t. lattice (e.g. $\vec{q}_c = 2\pi/a(1/3,0,0)$) but shifted from it by some small amount $\vec{\delta}$, so that $\vec{q}_s = 2\pi/a(1/3+\delta_0,0,0)$ is effectively incommensurate [39] with the h.t. lattice.

On cooling, this system will undergo a phase transition to an incommensurate phase (Fig.9), characterised by the spontaneous ordering of the fourier coordinates of wavevectors $\pm\vec{q}_s$: the physics of this s.p.t. is not very different from that of the Hamiltonian (1).

On further cooling, however, such a system typically undergoes a further transformation (Fig.9) to a phase where the ordering is characterised by non-zero expectation values for the fourier coordinates of wavevectors $\pm\vec{q}_c$,

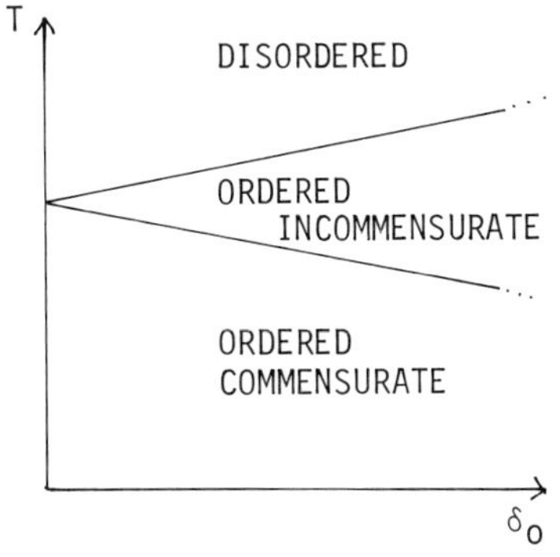

Fig.9 Schematic phase diagram for a system exhibiting an incommensurate phase and a lock-in s.p.t.

and thus is commensurate with the h.t. lattice: this "lock-in" s.p.t. has some remarkable features, the most notable of which is that precursor effects are inherent in strictly equilibrium properties, and hence are evident within Landau theory [37,38].

The Landau theory itself (though effectively one-dimensional) is not trivial, however, because the nature of the incommensurate phase structure, close to the lock-in temperature, T_L, is not simple - at least, it does not appear to be simple when viewed within the usual framework of an h.t. phonon-basis: non-linearities ensure that, once there exists a distortion with an incommensurate wavevector $\vec{q}_c + \vec{\delta}$, a whole series of harmonics (with wavevectors $\vec{q}_c - 2\vec{\delta}$, $\vec{q}_c + 3\vec{\delta}$ ....) are induced (so that $\vec{\delta}$ need <u>not</u> coincide with $\vec{\delta}_o$), the series being only slowly convergent near T_L.

The Landau theory becomes tractable when the problem is viewed in direct space [37,38]. Close to T_L it transpires that the incommensurate structure can best be described as consisting of a series of regions that already possess a commensurate order (within these the equilibrium displacement field $\langle u(x)\rangle$ varies as $\cos(\vec{q}_s.\vec{x}+\phi)$) separated by "domain walls" ("discommensurations" [37], "phase solitons" [38]), which are perpendicular to the [100]-direction, and at which there is a mismatch of the phases of the commensurate ordering in the adjoining regions (the phase angle ϕ evolves through $2\pi/3$ within the domain wall): c.f. Fig.10. The concentration of domain walls diminishes with the temperature, vanishing continuously at T_L, where true long-range (commensurate) ordering sets in.

The analogy with the situation envisaged in the cluster picture of (1) is apparent, and is reflected also in the scattering properties [40]. The increase, with the approach to T_L, of the mean separation between the domain walls is expressed as a decrease in the value of the parameter δ controlling the location of the Bragg-satellites (associated with the series of fourier components with wavevectors $\vec{q}_c+\vec{\delta},\vec{q}_c-2\vec{\delta}$.....), which thus converge upon the commensurate-phase Bragg point $\vec{q}_c$. Thus, just as in the cluster picture of (1), the growth of the scattering in the vicinity of the incipient Bragg point (as the s.p.t. is approached from above) reflects a decreasing population of domain walls, and a consequent growth in the coherence of the ordering of the precursor microdomains which they separate. In this case,

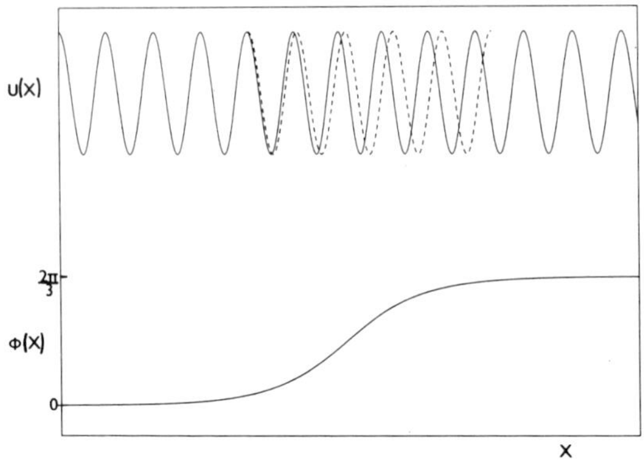

Fig.10 The spatial variation of the displacement field <u(x)>, and the phase field $\phi(x)$, in an incommensurate phase near a lock-in s.p.t.: the dashed line shows the displacement field in the absence of the phase soliton [37,38]

however, the domain walls are essential ingredients of the equilibrium configuration in the h.t. phase, and the scattering to which they give rise is elastic; in the cluster picture of (1), the domain walls form part of the excitation spectrum of the h.t. phase, and hence issue in scattering that is inelastic.

Lastly it is illuminating to notice that the lock-in s.p.t. occurs when the ℓ.t., commensurate, phase becomes unstable against the creation of domain walls, which destroy ℓ.r.o: in a real sense a domain wall is the "soft mode" for this phase transition.

6. Concluding Comments

The last observation prompts one final question which embodies, picturesquely, if imprecisely, the central issues raised here: is a displacive structural phase transition more properly regarded as an instability of the low-temperature phase against the creation of a ℓ.r.o.-destroying domain wall, than as an instability of the high-temperature phase against the condensation of a ℓ.r.o.-creating phonon? The answer to this question exhibits the dimensionality-dependence which we have already seen in other properties of the Hamiltonian (1), and which, therefore, merits a final resume.

In one dimension the R.G. fixed-point potential lies at the order-disorder end of the spectrum; two-time scales will be clearly identifiable in the near-critical dynamic behaviour; deviations from classical theory are at their largest; the transition temperature T_c=0 does indeed reflect the fact that (for any $T>T_c$=0) the ℓ.t. (ordered) phase is dynamically unstable against domain-wall formation [9].

In two dimensions the fixed-point potential is still quite strongly order-disorder; two time scales should be evident in the dynamic behaviour in a range of reduced temperature reflecting the size of some characteristic parameter such as h; critical exponents are strongly non-classical; it seems possible that the idea of the s.p.t. as an instability against a domain wall could be made the basis of as successful a phenomenology of the transition temperature in the displacive case as it has in the order-disorder limit (see, e.g. [41]).

In four dimensions the fixed point Hamiltonian is displacive ("gaussian"); a quasi-harmonic, single-time scale dynamics is expected; critical exponents have their classical values; the s.p.t. is best viewed as an instability against the soft h.t. phonon.

The stark differences between the low (1 and 2) and high (4) dimensional cases presumably reflect, at heart, the dimensionality-dependence of cluster topology. In d=1,2 clusters are "mutually exclusive": at any one time a given region of the crystal can be occupied by only a single cluster (though in d=2 pockets of "anti-ordering" can be embedded in larger clusters of "ordering"); the concept of a cluster wall seems viable here. In d=4 (indeed in $d > 2$) spongelike [42] clusters of ordering and anti-ordering can interpenetrate one another and the idea of a cluster wall seems less potent.

Appropriate use of the Renormalisation Group, together with further experiments (laboratory or computer) on displacive systems offer the best hope of establishing the picture appropriate for three-dimensional systems (doubtless a compromise between the d=2 and d=4 situations) - in particular, the relative importance of the soft phonon and the soft domain wall.

Acknowledgements

I am happy to acknowledge numerous illuminating conversations with, among others, S. Aubry, H. Beck, A.R. Bishop, R.A. Cowley, J.A. Krumhansl, T. Schneider, E. Stoll and H. Thomas.

References

1. T. Riste, E.J. Samuelsen, K. Otnes, J. Feder: Solid St. Commun. 9, 1455, 1971; S.M. Shapiro, J.D. Axe, G. Shirane, T. Riste: Phys. Rev. B6, 4332, 1972
2. K.A. Müller, W. Berlinger: Phys. Rev. Lett. 26, 13, 1971
3. J.A. Krumhansl, J.R. Schrieffer: Phys. Rev. B11, 3535, 1975
4. K.G. Wilson, J. Kogut: Phys. Rep. C12, 77, 1974
5. For a review see A.D. Bruce: Ferroelectrics 12, 21 1976; and to be published
6. P.C. Hohenberg, B.I. Halperin: Revs. Mod. Phys. 49, 523, 1977
7. T. Schneider, E. Stoll: Phys. Rev. Lett. 31, 1254, 1973; Phys. Rev. Lett. 35, 296, 1975; Phys. Rev. B13, 1216, 1976; Phys. Rev. B17, 1302, 1978
8. S. Aubry: J. Chem. Phys, 62, 3217, 1975; 64, 3392, 1976
9. C.M. Varma: Phys. Rev. B14, 244, 1976

10. A.R. Bishop, J.A. Krumhansl: Phys. Rev. B12, 2824, 1975
11. H. Beck: J. Phys. C. 9, 33, 1976
12. A.R. Bishop: Proc. Int. Conf. on Lattice Dynamics (1977), 144 (Flammarion, Paris, 1978)
13. S. Sarbach: Phys. Rev. B15, 2694, 1977
14. R. Bausch and B.I. Halperin: 1978 preprint
15. B.I. Halperin, P.C. Hohenberg, S. Ma: Phys. Rev. B10, 139, 1974; Phys. Rev. B13, 4119, 1976
16. B.I. Halperin, C.M. Varma: Phys. Rev. B14, 4030, 1976
17. R.J. Glauber: J. Math. Phys. 4, 294, 1963
18. R. Brout, K.A. Müller, H. Thomas: Solid St. Commun. 4, 507, 1966
19. R. Blinc, B. Zeks: Adv. in Phys. 91, 693, 1972
20. Y. Onodera: Prog. Theor. Phys. 44, 1477, 1970
21. I. Hatta: J. Phys. Soc. Japan 24, 1043, 1968
22. M.K. Barnoski, J.M. Ballantyne: Phys. Rev. 174, 946, 1968
23. W. Cochran: Phys. Rev. Lett. 3, 412, 1959; Adv. in Phys. 9, 387, 1961
24. E. Eisenriegler: Phys. Rev. B9, 1029, 1974
25. R.A. Cowley, W.J.L. Buyers, G. Dolling: Solid St. Commun. 7, 181, 1969. For a review of the evidence for soft phonons see J.F. Scott: Revs. Mod. Phys. 46, 83, 1974
26. K.K. Murata: Phys. Rev. B11, 462, 1975
27. S. Aubry, private communication
28. A.D. Bruce, T. Schneider: Phys. Rev. B16, 3991, 1977
29. J. Fröhlich, B. Simon, T. Spencer, Commun. Math. Phys. 50, 79, 1976
30. R. Oppermann, H. Thomas: Z. Phys. B22, 387, 1975
31. T. Schneider: this Symposium
32. R. Silberglitt: Solid St. Commun. 9, 2021, 1971; J. Feder: Solid St. Commun., 9, 2021, 1971; R.A. Cowley, G.J. Coombs: J. Phys. C6, 143, 1973
33. J. Töpler, B. Alefeld, A. Heidmann: J. Phys. C10, 635, 1977
34. C.N.W. Darlington, D.A. O'Connor: J. Phys. C9, 3561, 1976
35. K.K. Murata: Phys. Rev. B13 2028, 1976
36. S. Aubry: Ferroelectrics 16, 313, 1977
37. W.L. McMillan: Phys. Rev. B14, 1496, 1976
38. P. Bak, V.J. Emery: Phys. Rev. Lett. 36. 978, 1976; P. Bak: this Symposium
39. See, e.g., A.D. Bruce, R.A. Cowley: J. Phys. C (in press)
40. A.D. Bruce, R.A. Cowley, A.F. Murray: J. Phys. C. (in press)
41. M.E. Fisher: Essays in Physics (Academic Press) 4, 43, 1972
42. E. Stoll: Phys. Lett. 58A, 121, 1976

Nonlinear Lattice Dynamics: Molecular Dynamics Studies

T. Schneider and E. Stoll

IBM Zurich Research Laboratory, CH-8803 Rüschlikon, Switzerland

1. Introduction

Nonlinearity combined with dispersion were essential aspects in various contexts for many decades but have enjoyed remarkable revival of interest in the last few years [1,2]. The topic we wish to discuss here is structurally-commensurate lattice systems reducing in the continuum limit to the Φ^4 Hamiltonian and exhibiting a distortive phase transition. This transition takes the system from an ordered low-temperature displacement configuration to a different high-temperature pattern, where the mean displacements, representing the order parameter, vanish. The occurrence of a phase transition brings about novel aspects which were absent in previous studies of nonlinear systems such as the FERMI-PASTA-ULAM problem [3] or the TODA [4] lattice. In fact, a molecular-dynamics study [5] of a two-dimensional (2-d) model revealed, close to the transition temperature T_c, the formation and motion of clusters separated by walls, representing connected particles whose displacements have a sign opposite to that expected from the zero-temperature arrangement. The existence of these clusters and their dynamics has important consequences. It reveals that a given particle not only vibrates around a mean position, but also that the particles perform large amplitude motions which cannot be treated within the framework of conventional lattice dynamics. Subsequently, these numerical results have been confirmed [6] extended to 1-d [7-9] and 3-d systems [10] including models where the displacement vector is no longer a scalar but has two components ($n = 2$) [11]. The 1-d systems are special in the sense that for any finite temperature T no ordered phase occurs. Nevertheless, because $T = 0$ can be considered as the critical temperature and considerable analytic progress was possible, the 1-d models are also of considerable interest. The work of KRUMHANSL and SCHRIEFFER [12] on the 1-d Φ^4 problem, demonstrated that at sufficiently low temperatures the free energy is, to a good approximation, that of a gas of noninteracting quasiparticles of two types, namely phonons and kink excitations. The presence of kinks with a statistical weight is gratifying since it puts the numerical evidence for cluster walls (kinks) onto a firm physical ground. These analytic procedures are substantially limited to 1-d and low T. Nevertheless, numerical studies revealed that close to the transition temperature, clusters and cluster walls also occur in 2- and 3-d Φ^4 systems [5,6,10,11], and that the static and dynamic critical exponents are consistent with those of the corresponding Ising model.

It is also interesting to note that the partition function of the Φ^4 system reduces in a certain limit [13] to that of the Ising model. Moreover, according to the universality hypothesis for static critical phenomena [14], the Φ^4 system belongs to the Ising universality class with critical exponents which depend on d only, becoming classical for $d \geq 4$ and identical to the Ising model exponents. This expectation is consistent with the numerical results for $d = 2$ [5,6] and $d = 3$ [10]. Recognizing that in the Ising limit, where the scalar displacement can assume two values only, let us say +1 and -1, cluster and cluster walls represent the only possible displacement pattern, one might question whether or not cluster and cluster walls dominate the static and dynamic properties close to the phase transition. If so, we would have a pictorial explanation of universality, according to which the Φ^4 system belongs to the Ising universality class, in terms of the dominating displacement configurations, the clusters being separated by walls.

It is the main purpose of this paper to shed some light onto this question. In Sec. 2 we define the models, a conventional 3-d Φ^4 system and an extension of this model, which also exhibits an acoustic mode. Section 3 is devoted to a sketch of our molecular-dynamics technique. Some numerical results, demonstrating the dominating role of clusters and cluster walls, and giving the temperature dependence of the zero wave-vector excitation spectrum are presented in Sec. 4.

2. Simple Models Exhibiting Distortive Phase Transitions

Distortive phase transitions include structural, ferroelectric and antiferroelectric phase transitions in crystals. An essential property is the existence of an order parameter which represents a linear combination of particular displacements. Examples are the angular displacements of the oxygen octahedra in perovskite-type crystals and of hydrogen octahedra in ammonium halides. Such complicated systems have many other degrees of freedom. Consequently, a detailed theoretical description, taking all degrees of freedom into account, would lead to insurmountable difficulties. In this case, it is not surprising that the model systems considered so far only take into account the order parameter and those degrees of freedom strongly coupled to it [15].

Here we consider two 3-d models with a scalar ($n = 1$) order parameter. Model I corresponds to a discrete Φ^4 system having one particle per unit cell. Model II is an extension with two particles per unit cell which reduce in a certain limit to Model I.

Model I

The Hamiltonian of the ferrodistortive model is [10]

$$\mathcal{H} = \sum_{\ell} \frac{M \dot{X}_{\ell}^2}{2} + \frac{A-12C}{2} \sum_{\ell} X_{\ell}^2 + \frac{B}{4} \sum_{\ell} X_{\ell}^4 + \frac{C}{2} \sum_{<\ell,\ell'>} (X_{\ell} - X_{\ell'})^2. \tag{1}$$

ℓ labels the particle with mass M in the ℓ-th unit cell. $M\dot{X}_\ell$ and X_ℓ are momentum and displacement with respect to a rigid, cubic primitive reference lattice with lattice constant a. M, A, B and C are model parameters chosen as [10]

$$A = -1, \quad B = 1/3, \quad C = 1/6, \quad M = 1. \tag{2}$$

$<\ell,\ell'>$ denotes nearest-neighbor interactions. This choice of the model parameters guarantees that at $T = 0$ the order parameter given by

$$X^2 = \frac{12C-A}{B} \tag{3}$$

does not vanish. Consequently, the system will undergo a ferrodistortive phase transition at some $T = T_c > 0$, where $<X_c> = 0$.

The equation of motion reads

$$- M \ddot{X}_\ell = (A - 12C)X_\ell + B X_\ell^3 + 2C \sum_{\ell'} (X_\ell - X_{\ell'}) \ . \tag{4}$$

Introducing the scalar displacement field $f(\vec{R},t)$ this equation becomes, in the continuum limit, identical to the Φ^4 equation of motion

$$- M \ddot{f}(R,t) = (A - 12C)f + B f^3 - 2C a^2 \Delta f. \tag{5}$$

Linearization of (3) and (5) appears to be adequate at low temperatures where, at most, all particles perform small amplitude oscillations in one of the double wells defined by

$$- \frac{12C-A}{2} f^2 + \frac{B}{4} f^4; \tag{6}$$

which double well is preferred depends on the preparation of the system at T = 0. The frequencies of the small amplitude oscillations are given by

$$M \omega^2 = -(12C - A) + 3B <f^2> + 2 C a^2 q^2 , \tag{7}$$

where $\vec{q}$ is the wave vector. As pointed out in this context by KRUMHANSL and SCHRIEFFER [12], Eq.(5) also has other important particular solutions, namely the kink and anti-kink solutions

$$f(R_x,R_y = 0, R_z = 0,t) = \pm U_0 \tanh \left(\frac{R_x - U_x t}{\xi_x \sqrt{2}} \right) , \tag{8}$$

where

$$U_0 = \sqrt{\frac{12C-A}{B}} , \quad \xi_x^2 = \frac{2Ca^2 - MV_x^2}{12C-A} \ . \tag{9}$$

This kink takes f from U_0 to $-U_0$; and antikink takes $-f$ to f. Between $+f$ and $-f$ there is a wall which we call the cluster wall of approximate thickness $2\sqrt{2}\,\zeta_x$. The cluster wall moves with velocity V_x.

The partition function

$$Z = \int_{-\infty}^{+\infty} d\dot{X}_1 \cdots d\dot{X}_N \int_{-\infty}^{+\infty} dX_1 \cdots dX_N \exp - \frac{1}{k_B T} \mathscr{H} \tag{10}$$

of this system reduces in the limit

$$A \to -\infty, \quad B \to +\infty, \quad \frac{A}{B} = -1 \tag{11}$$

to that of the 3-d Ising model [13].

Approaching this limit, the phonon frequency (7) tends to infinity because $\langle f^2 \rangle \to 1$ [see (3)] while the cluster wall thickness tends to zero and U_0 to 1 [see (9)]. This behavior reveals that by approaching the Ising limit, the phonons cease to be of relevance. What remains are kink-like large amplitude motions giving rise to the formation of clusters separated by thin walls.

There is another interesting limiting case, i.e., the displacive limit [16] defined by $12C = A$ where the order parameter vanishes even at $T = 0$ [see (3)]. By approaching this limit the cluster wall thickness increases, the amplitude of the kinks decreases (9), and the phonon frequency (7) remains finite. Accordingly, near this limit both small and large amplitude motions seem to play an important role.

The model considered here and specified by the parameters in (2) is quite far from the two limits. In fact, according to (2),(3) and (7), the phonon frequency is at low temperature and small q, given by

$$\omega^2(q) = 6 + \frac{1}{3} a^2 q^2, \tag{12}$$

and

$$A/12C = -1/2 \tag{13}$$

being far from both the value at the Ising limit $(A/12C = -\infty)$ and the displacive limit $(A/12C = 1)$.

Model II

The Hamiltonian of this model is

$$\mathscr{H} = \sum_{\ell,\kappa} \frac{M_k \dot{X}_{\ell\kappa}^2}{2} + \frac{A}{2} \sum_{nn} (X_{\ell\kappa} - X_{\ell'\kappa'})^2 + \frac{B}{4} \sum_{nn} (X_{\ell\kappa} - X_{\ell'\kappa'})^4$$
$$+ \frac{C}{2} \sum_{nnn} (X_{\ell\kappa} - X_{\ell'\kappa})^2. \tag{14}$$

ℓ,κ label the κ-th particle with mass κ in the ℓ-th unit cell. $M_\kappa \dot{X}_{\ell\kappa}$ and $X_{\ell\kappa}$ are momentum and displacement with respect to a rigid cubic reference lattice. There are two particles with mass M_1 and M_2 per unit cell. nn denotes nearest-neighbor and nnn next-nearest-neighbor interactions. The arrangement of the particles is sketched in Fig. 1.

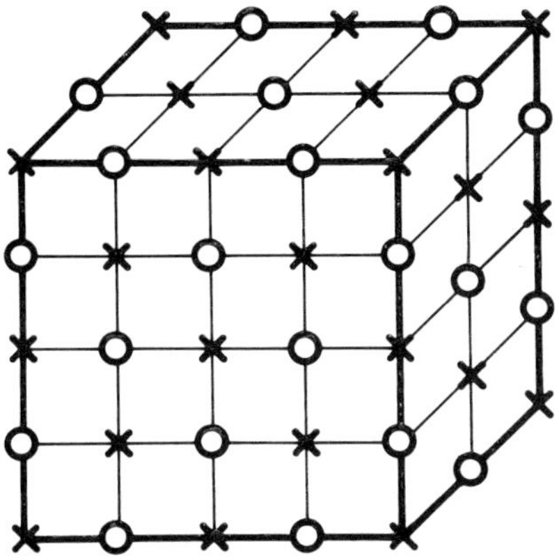

Fig.1 Schematic sketch of the high temperature phase. X denotes particles with mass M_1 and O particles with mass M_2

At zero temperature, the force acting on a given particle has to vanish. This condition determines the $T = 0$ order parameter given here by

$$(X_{\ell 1} - X_{\ell 2})^2 = -\frac{A}{B} . \tag{15}$$

The ordered phase will be antiferrodistortive because the mean displacement in sublattice 1 is positive and in sublattice 2 negative. Some additional insight might be obtained by treating the model in the Hartree approximation where the quartic term is linearized according to

$$\frac{B}{4} \sum_{nn} (X_{\ell\kappa} - K_{\ell'\kappa'})^4 \rightarrow \frac{3B}{2} \langle (X_{\ell 1} - X_{\ell 2})^2 \rangle \sum_{nn} (X_{\ell\kappa} - X_{\ell'\kappa'})^2 . \tag{16}$$

In this approximation, the Hamiltonian can be diagonalized in terms of two normal coordinates. One corresponds to the order parameter, describing the counter-motion within the cell and giving rise to an optic-phonon branch. The other normal coordinate is associated with an acoustic branch (Fig. 2). The expectation that large amplitude motions will also occur can be substantiated as follows: i) in the limit where the mass of the particles in one sublattice, say $M_{\kappa=1}$, tends to infinity, Model II obviously reduces to Model I; ii) moreover, it can be shown that the partition function of the potential energy reduces in the continuum limit to Model I. This is achieved by reducing the unit cell by a factor of two and relabeling the particles from 1 - 2N. Accordingly, the static critical properties will be identical to those in Model I. To describe the dynamic properties, however, the kinetic energy involving the two different masses must be taken into account. This is easiest seen in the Hartree approximation where the phonon frequencies are given by

$$\omega^2(q) = \frac{(M_1+M_2)\beta \pm \sqrt{(M_1-M_2)^2\beta^2 + 4M_1M_2\alpha^2\cos^2\frac{a}{2}q}}{2M_1M_2} , \tag{17}$$

where

$$\begin{aligned} \alpha &= 6\,A + 18\,B \;\langle(X_{11} - X_{12})^2\rangle \\ \beta &= \alpha \;\; + 6\,G\,(1 - \cos a\,q) \\ \vec{q} &= (1,\,1,\,1)\,q\;. \end{aligned} \tag{18}$$

The model parameters have been chosen as

$$A = -1, \quad B = 1, \quad C = 1, \qquad M_1 = 2M_2 = 1. \tag{19}$$

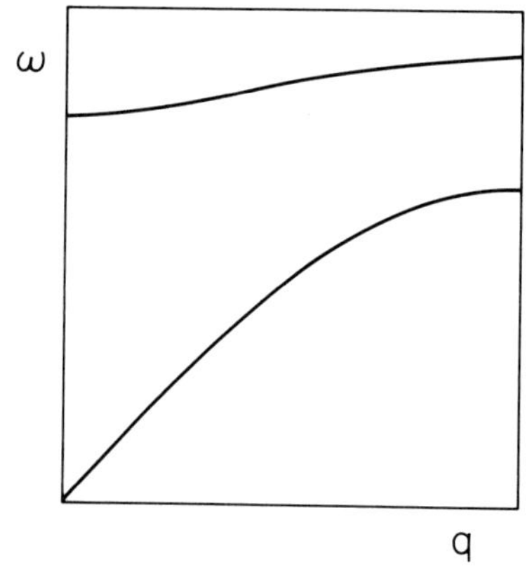

Fig.2 Sketch of the phonon-dispersion law in Model II in the Hartree approximation

Dynamic Variables and Spectral Densities

To investigate the excitation spectrum, we consider the spectral densities of the dynamic variables of interest

$$\hat{S}_{AA}(\vec{q},\omega) = \int_{-\infty}^{+\infty} dt\; e^{-i\omega t}\, \frac{\langle A(-\vec{q},t)\; A(\vec{q},0)\rangle}{\langle A(-\vec{q},0)\; A(\vec{q},0)\rangle} . \tag{20}$$

Variables A of particular interest are the displacement (order parameter) and energy fluctuations, defined by

Model I

$$X(\vec{q}) = \frac{1}{\sqrt{N}} \sum_{\ell} \delta X_{\ell}\; e^{i\vec{q}\cdot\vec{R}_{\ell}}, \qquad \delta X_{\ell} = X_{\ell} - \langle X_{\ell}\rangle \tag{21}$$

$$\mathscr{H}(\vec{q}) = \frac{1}{\sqrt{N}} \sum_{\ell} \delta \mathscr{H}_{\ell}\; e^{i\vec{q}\cdot\vec{R}_{\ell}}, \qquad \delta \mathscr{H}_{\ell} = \mathscr{H}_{\ell} - \langle \mathscr{H}_{\ell}\rangle \tag{22}$$

$$\mathscr{H}_\ell = \frac{M\dot{X}_\ell^2}{2} + \frac{A-12C}{2} X_\ell^{\,2} + \frac{B}{4} X_\ell^{\,4} + \frac{C}{2}\sum_{\ell'} (X_\ell - X_{\ell'})^2 . \tag{23}$$

$\vec{q}$ is the wave vector and the vectors $\vec{R}_\ell$ define the direct lattice.

Model II

$$X(\vec{q}) = \frac{1}{\sqrt{2N}} \sum_{\ell\kappa} \delta X_{\ell\kappa}\, e^{i\vec{q}\cdot\vec{R}_{\ell\kappa}} \; ; \; \delta X_{\ell\kappa} = X_{\ell\kappa} - \langle X_{\ell\kappa}\rangle \tag{24}$$

$$\mathscr{H}(\vec{q}) = \frac{1}{\sqrt{2N}} \sum_{\ell\kappa} \delta\mathscr{H}_{\ell\kappa}\, e^{i\vec{q}\cdot\vec{R}_{\ell\kappa}} \; ; \delta\mathscr{H}_{\ell\kappa} = \mathscr{H}_{\ell\kappa} - \langle \mathscr{H}_{\ell\kappa}\rangle \tag{25}$$

$$\mathscr{H}_{\ell\kappa} = \frac{M_\kappa \dot{X}_{\ell\kappa}^{\,2}}{2} + \frac{A}{2}\;{}_{\ell',\kappa'} (X_{\ell\kappa} - X_{\ell'\kappa'})^2 + \frac{B}{4}\;{}_{\ell',\kappa'} (X_{\ell\kappa} - X_{\ell'\kappa'})^4$$

$$+ \frac{C}{2}\sum_{\ell'} (X_{\ell\kappa} - X_{\ell'\kappa})^2 . \tag{26}$$

It is important to note in this context that the energy density $\mathscr{H}(\vec{q})$ is a conserved variable because

$$\frac{d}{dt}\mathscr{H}(\vec{q} = 0,t) = \dot{\mathscr{H}}(\vec{q} = 0,t) = 0, \tag{27}$$

as it should be for a Hamiltonian system.

3. The Molecular-Dynamics Technique

In the conventional molecular-dynamics technique [17,18] , one solves the set of coupled Newton's equations associated with a given Hamiltonian according to a set of difference equations with a time increment. This set of difference equations approximates Newton's equations. Starting from given initial conditions for the positions and velocities, the particles are then allowed to move, and the time evolution of their canonical variables $(X_\ell, \dot{X}_\ell)$ are calculated. Assuming the system is ergodic, estimates for microcanonical ensemble averages may be obtained in terms of time averages.

It would be preferable to have a molecular-dynamics technique, simulating a canonical ensemble, as most experiments are performed at constant global temperature. This was achieved by considering, in place of the Newton's equations, the coupled set of Langevin equations [10]

$$M\ddot{X}_\ell = -\frac{\partial\mathscr{H}}{\partial X_\ell} - \Gamma M \dot{X}_\ell + \eta_\ell(t), \tag{28}$$

where

$$\langle\eta_\ell(t)\,\eta_{\ell'}(t')\rangle = 2\,M\,\Gamma\,k_B T\,\delta(t-t')\,\delta_{\ell\ell'} . \tag{29}$$

Here, it is assumed that the particles suffer collisions with much lighter ones which represent the heat bath defining the temperature T. The collisions are described by the friction $\Gamma M \dot{X}_\ell$ and a random force $\eta_\ell(t)$. It may be shown that the stationary solution of the associated Fokker-Planck equation is the canonical distribution function

$$P_{eq}(\dot{X}_1, \ldots, \dot{X}_N, X_1, \ldots, X_N) = \exp - \frac{1}{k_B T} \mathcal{H}. \tag{30}$$

Starting from initial values for positions and velocities, the particles are then allowed to move under the influence of the computer-generated random force. The temporal evolution of the variables are then calculated with a set of difference equations approximating the Langevin equations (28). On this basis, one obtains

$$X_\ell(t), \dot{X}_\ell(t), \ddot{X}_\ell(t), \text{etc.} \tag{31}$$

For a detailed description of the algorithm and the random force generation, we refer to [10]. The system is then allowed to age, or, in other words, to reach equilibrium. After this interval, the subsequent 10^5 steps are used to perform time averages representing canonical ensemble averages.

From the Langevin equations (28), it is obvious that the dynamic properties will be modified, in particular, owing to the damping term. To reduce this modification, Γ must be chosen in such a way that

$$\frac{1}{\Gamma} \gg \tau_c, \tag{32}$$

where τ_c denotes the characteristic time of the dynamics. This guarantees that the excitations do not become overdamped owing to the friction term. Another important constraint on Γ evolves from the energy conservation of a Hamiltonian system. Since our system evolves according to the Langevin equation, it follows that

$$\frac{d\mathcal{H}}{dt} = \sum_\ell \frac{\partial \mathcal{H}}{\partial M \dot{X}} M \ddot{X}_\ell + \frac{\partial \mathcal{H}}{\partial X_\ell} \dot{X}_\ell$$

$$= - \sum_\ell [\Gamma M \dot{X}_\ell^2 - \dot{X}_\ell \eta_\ell(t)]. \tag{33}$$

Consequently, energy is not conserved because the Hamiltonian system is in contact with the heat bath. To avoid artificial features due to the random-noise pulses, the mean time between two pulses must be small compared to τ_c. In this case, we may average (33) over some pulses. This leads to

$$\frac{d\mathcal{H}}{dt} = - \Gamma[2 E_{kin}(t) - N k_B T] = - \delta E_k(t). \tag{34}$$

With the Ansatz

$$\delta E_k(t) = \alpha e^{-t/\tau} \tag{35}$$

and

$$\frac{d\mathscr{H}}{dt} = \frac{d\mathscr{H}}{d\,\delta E_k}\ \frac{d\,\delta E_k}{dt} = \frac{d\mathscr{H}}{dT}\ \frac{dT}{d\,\delta E_k}\ \frac{d\,\delta E_k}{dt} \approx \frac{2\,C_v}{k_B}\ \frac{d\,\delta E_k}{dt}\ , \tag{36}$$

we find

$$\tau = \frac{C_v}{k_B}\ \frac{1}{\Gamma}\ , \tag{37}$$

where C_v is the specific heat. Accordingly, energy is almost conserved within the characteristic time τ_c, provided that

$$\tau = \frac{C_v}{k_B}\ \frac{1}{\Gamma} >> \tau_c . \tag{38}$$

Moreover, owing to the fact that the system evolves according to the Langevin equation, the time interval τ_{ch} over which an evolution is followed must be larger than τ so that

$$\tau << \tau_{ch} . \tag{39}$$

Combining inequalities (38) and (39), we finally obtain

$$\tau_{ch} >> \frac{C_v}{k_B\ \Gamma} >> \tau_c . \tag{40}$$

From this relation, it becomes evident that energy can be almost conserved provided Γ and the chain length τ_{ch} are appropriately chosen. An exception is very close to T_c, where the characteristic time τ_c becomes very long.

In the calculations presented here, we have considered systems of 8000 particles defined by Hamiltonians (1) or (14) and model parameters specified in (2) and (19), respectively. The systems were subjected to periodic boundary conditions. In the time interval, where time averages were performed, Γ was chosen as

$$\Gamma = 0.005. \tag{41}$$

4. The Excitation Spectrum: Numerical Results

Before examining the explicit results, it is helpful to summarize those features of the excitation spectrum which will be expected on general grounds. At low temperatures $T << T_c$, the particles will oscillate about a mean position. The displacements are small and anharmonic perturbation theory should be adequate. We therefore expect an excitation spectrum exhibiting phonons and, by virtue of energy conservation, heat diffusion or second sound. The resulting Rayleigh or second-sound peaks may occur in both the energy and displacement spectral densities, due to the coupling between order parameter and energy fluctuation below T_c.

With increasing T, anharmonic effects become more important and a softening of the optic branch will occur for small $\vec{q}$, as expected from the

Hartree approximation. Moreover, large amplitude motions will become more important, because it is much easier for the particles to overcome the potential barrier. This will result in the formation and dynamics of clusters. A cluster represents particles connected by nearest-neighbor bonds, having displacements with a sign opposite to that expected from zero temperature. A cluster is surrounded by a cluster wall where the displacements change sign. This phenomenon will illustrate the importance of the particular kink and anti-kink solutions of the underlying equation of motion in the continuum limit. The formation and dynamics of these clusters will certainly affect the excitation spectrum producing, in particular, a central peak. This peak will be superimposed, at least below T_c and for $\vec{q} \neq \vec{0}$, on the heat-diffusion peak. Above T_c, energy and displacement fluctuations are no longer directly coupled. Nevertheless, central peak will be expected in the displacement spectral density due to the cluster dynamics and in addition, phonons, due to the small amplitude motions.

To summarize, the excitation spectrum is expected to be rather rich and complicated due to the presence of large and small amplitude oscillations and the implications of energy conservation. In fact, by approaching T_c from below we expect in $\hat{S}_{xx}(\vec{q},\omega)$ a softening of the optic-phonon resonance for $\vec{q} = \vec{0}$, a heat-diffusion central peak for small but $\vec{q} \neq \vec{0}$, and a central peak due to the cluster dynamics. Above T_c, the excitation spectrum of the displacement spectral density should be simpler since the heat-diffusion central peak cannot occur by means of a direct coupling between energy and displacement fluctuations. Accordingly, by approaching T_c from above, we expect in $\hat{S}_{xx}(\vec{q},\omega)$ a softening of the optic-phonon resonance for $\vec{q} = 0$ and a central peak due to the cluster dynamics.

Model I

For a detailed discussion of the numerical results we refer to [10]. Here we only present some of the crucial results. As mentioned above, we expect the formation of clusters separated by walls. Figure 3 shows snapshots of the particles in a plane perpendicular to (1,0,0). Only those particles are marked where $\operatorname{sgn} X_\ell = -\operatorname{sgn} \langle X \rangle_{T=0}$.

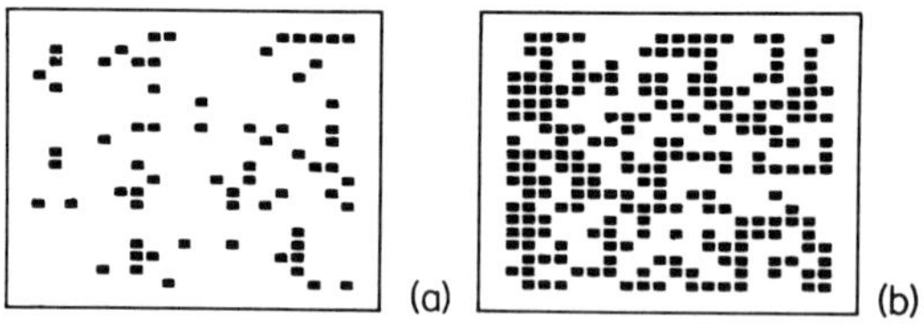

Fig.3 Snapshots of instantaneous cluster configurations (a) $k_BT = 6$; (b) $k_BT = 8$

Above $T_c \approx 7.2$ the clusters of positive and negative X_ℓ must, of course, be equal on the average so that the order parameter vanishes. The formation of these clusters and the associated cluster walls has important consequences. Their formation implies the presence of large amplitude motions which will certainly affect the excitation spectrum. Here it might be appropriate to start with an overview of the excitation spectrum.

In Fig. 4 we have sketched the essential features appearing in $\hat{S}_{xx}(\vec{q},\omega)$ and $\hat{S}_{\mathcal{H}\mathcal{H}}(\vec{q},\omega)$ for various temperatures at wave vectors $\vec{q} = (0,0,0)$ and $\vec{q} = (\pi/10a,0,0)$, respectively. At $k_B T = 15$ and 30, there is a broad phonon peak in S_{xx}, but no central peak (CP) occurs. $S_{\mathcal{H}\mathcal{H}}$ only exhibits a CP peak with half-width $\Delta\omega \sim q^2$, due to heat diffusion. This phenomenon

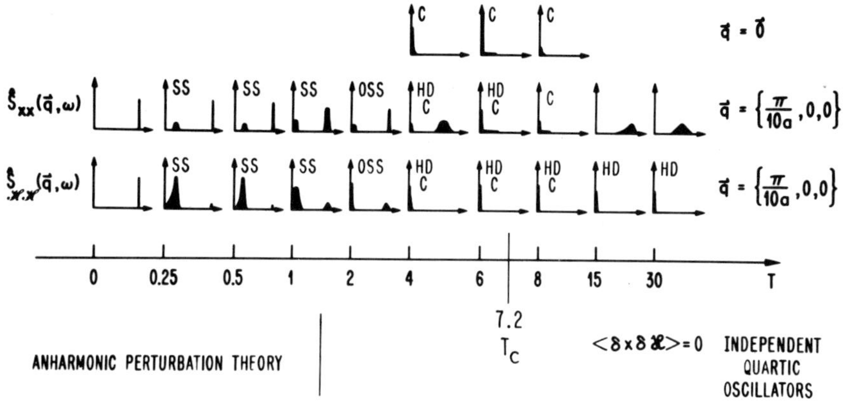

Fig.4 Sketch of the essential features of the excitation spectrum in $\hat{S}_{xx}(\vec{q},\omega)$ and $\hat{S}_{\mathcal{H}\mathcal{H}}(\vec{q},\omega)$ at various temperatures for wave vectors $\vec{q} = \vec{0}$ and $\vec{q} = (\pi/10a,0,0)$. HD denotes the central peak due to heat diffusion, and C the central peak arising from the cluster dynamics. ss denotes the second-sound peak, and oss overdamped second sound

does not occur in S_{xx}, as expected, because the displacement-energy coupling vanishes above T_c. Moreover, the broad phonon resonance shifts to lower frequencies with decreasing temperature. At $k_B T = 8$, which is close to $k_B T_c \simeq 7.1$, new features appear: a CP at $\vec{q} = (0,0,0)$ in S_{xx} and the soft mode becomes overdamped. At $\vec{q} = \vec{0}$, heat diffusion does not exist. Consequently, the $\vec{q} = \vec{0}$ CP must be attributed to the cluster dynamics. In $S_{\mathcal{H}\mathcal{H}}$, it is superimposed by the Rayleigh peak.

Below T_c, the coupling between displacement and energy fluctuations is nonzero. The Rayleigh peak will appear, therefore, in both S_{xx} and $S_{\mathcal{H}\mathcal{H}}$. Accordingly, we have at $k_B T = 6$ and $\vec{q} = 0$, the Rayleigh and cluster central peaks superimposed. The cluster CP also appears, as is seen from $\vec{q} = \vec{0}$. The soft mode is overdamped. At $k_B T = 4$, the soft mode becomes underdamped and the height of the cluster CP is reduced, as expected. In fact, for the formation of clusters, it is necessary that particles overcome the potential barrier. This process becomes less probable as the temperature is lowered. This fact is illustrated at $k_B T = 2$, where the cluster CP no longer appears $(\vec{q} = \vec{0})$. At $\vec{q} \neq \vec{0}$ there is a CP due to the energy fluctuations in both S_{xx} and $S_{\mathcal{H}\mathcal{H}}$, but it is weak in S_{xx}. The $\vec{q}$ dependence of the half width reveals, however, that this CP cannot be explained in terms of heat diffusion. Since this peak splits at lower temperatures ($k_B T = 1$), it must be attributed to overdamped second sound. The phonon resonance is now defined, indicating that we enter the regime where anharmonic perturbation theory should work. At $k_B T = 1$, the phonon is very well defined. It dominates $\hat{S}_{xx}$, but at small ω the weak second-sound resonance appears. This peak dominates $S_{\mathcal{H}\mathcal{H}}$, where the phonon peak is very weak. At the lower temperature $k_BT = 0.5$, the features are similar to those at $k_BT = 1$. The only difference is a zero-frequency tail of the second-sound resonance in $S_{\mathcal{H}\mathcal{H}}$. This tail must be attributed to two phonon processes. Finally, at $T = 0$, where the spectrum can be calculated exactly, only the phonon resonance at the frequency given by (12) survives.

An essential feature of the $\vec{q} = 0$ CP is, that by approaching T_c from above or below, it appears in addition to the optic soft-mode resonance. This feature, illustrated in Fig. 5, clearly reveals that the CP at $\vec{q} = \vec{0}$ does not evolve from the small amplitude oscillations but is associated with the large amplitude motions originating from the formation and the dynamics of clusters and cluster walls.

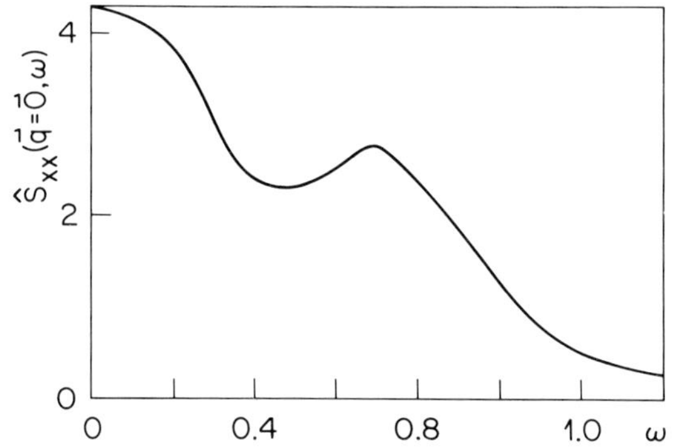

Fig.5 $\hat{S}_{xx}(\vec{q} = \vec{0},\omega)$ at $k_BT = 11$

These results reveal that the excitation spectrum appearing in $\hat{S}_{xx}(\vec{q} = \vec{0},\omega)$ is close to T_c dominated by a central peak originating from the large amplitude motions associated with the dynamics of clusters and cluster walls. Moreover, the strength of the soft-mode resonance evolving from small amplitude motions, decreases in this regime by approaching T_c. Accordingly, what remains are the motions reminiscent of the Ising limit.

Model II

As pointed out in Sec. 2, the partition function of this system can also be reduced to the Φ^4 model. Accordingly, here we also expect the formation of clusters and cluster walls. This expectation is confirmed in Fig. 6 showing snapshots of the particles in a plane perpendicular to (1,0,0). Only those particles are marked where $\mathrm{sgn}\, X_{\ell\kappa} \neq \mathrm{sgn} \langle X_{\ell\kappa} \rangle_{T=0}$.

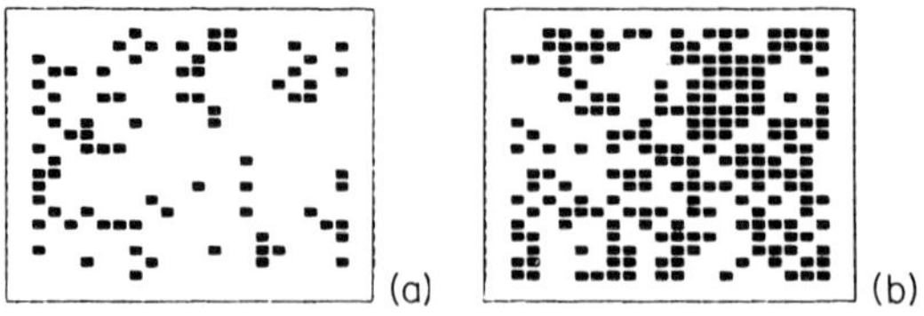

Fig.6 Snapshots of instantaneous cluster configurations at $k_BT = 1.75$ (a) and $k_BT = 2.1$, where $T_c \approx 2$

We expect that the large amplitude motions associated with the formation of the cluster walls will, in analogy to Model I, affect the excitation spectrum close to T_c.

To test this expectation, we next discuss the temperature dependence of the excitation spectrum above T_c. Figures 7,8 and 9 show the ω dependence of $S_{xx}(\vec{q} = 0,\omega)$ at $k_BT = 2.3$, 2.2 and 2.1, respectively. It should be emphasized that acoustic-branch resonances no longer occur at $\vec{q} = \vec{0}$. At $k_BT = 2.3$, the spectrum is dominated by the optic-phonon resonance at $\omega_{max} \approx 0.62$.

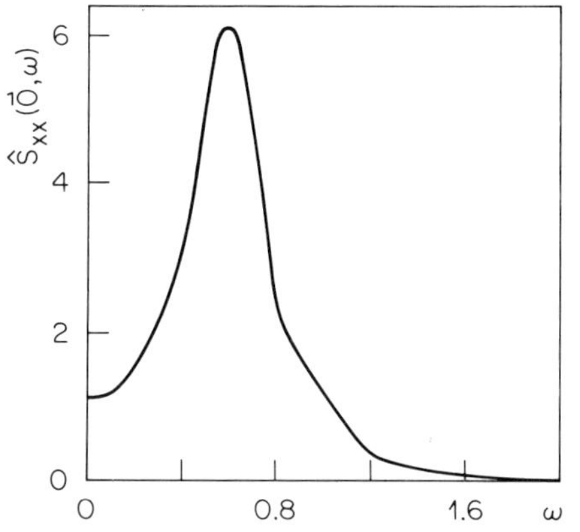

Fig.7 $\hat{S}_{xx}(\vec{q} = \vec{0},\omega)$ at $k_BT = 2.3$

By lowering the temperature to $k_BT = 2.2$ (Fig. 8), the peak position of the soft optic mode remains essentially the same, but the strength of the resonance is considerably lowered and a new feature appears in addition, namely a CP. By lowering the temperature further to $k_BT = 2.1$ (Fig. 9), the strength of the soft optic-mode resonance becomes so small that the spectrum is dominated by the CP.

These results again reveal that by approaching T_c the excitation spectrum of $\hat{S}_{xx}(\vec{q} = 0,\omega)$ is exhausted by a central peak originating from motions which are distinct from the small amplitude oscillations, giving rise to the soft optic-mode resonance. In fact, the strength of the optic-mode resonance is seen to decrease by approaching T_c and over a certain temperature range, both this resonance and the CP are present. In view of Fig. 6, we are then led to conclude that the large amplitude motions associated with the dynamics of clusters and cluster walls dominate the $\vec{q} = \vec{0}$ excitation spectrum close to T_c.

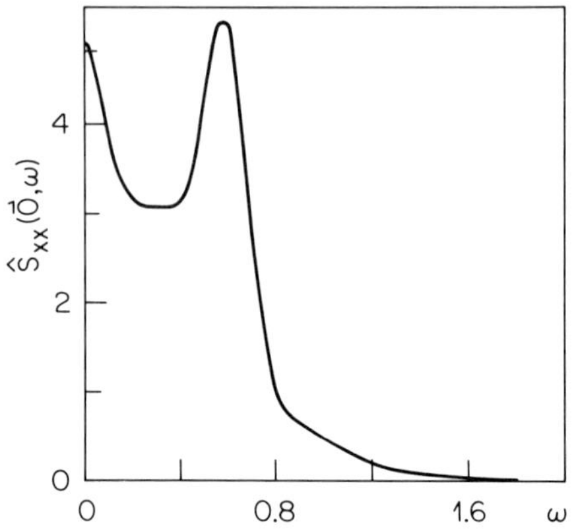

Fig.8 $\hat{S}_{xx}(\vec{q} = \vec{0},\omega)$ at $k_BT = 2.2$

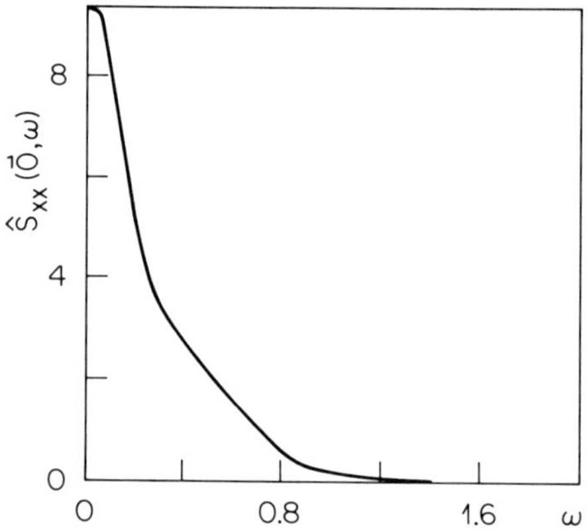

Fig.9 $\hat{S}_{xx}(\vec{q} = \vec{0},\omega)$ at $k_BT = 2.1$

5. Summary and Conclusions

The present discussion has revealed that in 3-d Φ^4 systems, nonlinearity, dispersion and energy conservation led to a rather rich excitation spectrum. Phonon resonances, second sound, heat diffusion and a CP due to large amplitude motions, associated with the formation and the dynamics of clusters do occur. At zero wave vector and close to T_c, however, the excitation spectrum of the displacement spectral density becomes simple. It is exhausted by the central peak arising from the cluster dynamics. The numerical results on 2-d Φ^4 systems indicate that this conclusion even holds close to the displacive limit. Accordingly, the cluster displacement patterns and the associated large amplitude motions seem to dominate the partition function and the zero wave-vector dynamics of the displacement spectral density close to T_c. If one accepts this conclusion, then it offers a pictorial explanation of universality according to which the Φ^4 system belongs to the Ising universality class [14]. Conversely, it does not suggest trivial critical behavior for $d \geq 4$, where the critical exponents are expected to be classical [14].

References

1. A.C. Scott, F.Y.F. Chu, D.W. McLaughlin: Proc. IEEE 61, 1443 (1973)
2. R.K. Bullough, R.K. Dodd: In "*Synergetics*", ed. by H. Haken (Springer, Heidelberg 1977)p.92
3. E. Fermi, J.R. Pasta, S.M. Ulam: LASL Report No.LA-1940, 1955; To be published
4. M. Toda: Progr. Theoret. Phys. (Kyoto) Suppl. 59, 1 (1976)
5. T. Schneider, E. Stoll: Phys. Rev. Lett. 31, 1257 (1973); Phys. Rev. B 13, 1216 (1976)
6. M.L.P. Bartolome, W.C. Kerr: Solid State Commun. 21, 253 (1977)
7. S. Aubry: J. Chem. Phys. 62, 3217 (1974); 64, 3392 (1976)
8. T. Schneider, E. Stoll: Phys. Rev. Lett. 35, 296 (1975)
9. T.R. Koehler, A.R. Bishop, J.A. Krumhansl, J.R. Schrieffer: Solid State Commun. 15, 1515 (1975)
10. T. Schneider, E. Stoll: Phys. Rev. B 17, 1302 (1978)
11. T. Schneider, E. Stoll: Phys. Rev. Lett. 36, 1501 (1976)
12. J.A. Krumhansl, J.R. Schrieffer: Phys. Rev. B 11, 3535 (1975)
13. St. Sarbach, T. Schneider: Phys. Rev. B 13, 464 (1976)
14. M.E. Fisher: Rev. Mod. Phys. 46, 597 (1974)
15. A.D. Bruce: Ferroelectrics 12, 21 (1976)
16. T. Schneider, E. Stoll: Phys. Rev. B 10, 2004 (1974)
17. A. Rahman: Phys. Rev. 136, A405 (1964)
18. A. Verlet: Phys. Rev. 159, 98 (1967)

Computer Simulation of Structural Phase Transitions

William C. Kerr

Centre Européen de Calcul Atomique et Moléculaire
Université de Paris XI, Orsay, France, and

Department of Physics, Wake Forest University
Winston-Salem, NC 27109, USA[1,2]

Phenomena associated with structural phase transitions (SPTs) have been intensively studied in recent years, ever since the discovery of the "central peak" in the neutron and light scattering spectra of the soft mode in many diverse materials [1]. These lattice dynamical systems lend themselves to study by molecular dynamics (MD) and several such investigations have been carried out. The results presented here are from a MD study of a two-dimensional 1600 particle system which exhibits a structural transition.

The high temperature phase is a square lattice with parameter a. At the transition the unit cell doubles, giving a square structure for the low temperature phase also. At the transition each atom shifts parallel to the [01] direction and adjacent atoms move in opposite directions, giving an antiferrodistortive transition with soft mode wavevector $\vec{q}_c$ at the corner of the Brillouin zone [2]. The mechanism producing the transition is that the mode with wavevector $\vec{q}_c$ and polarization parallel to [01] is unstable in the bare harmonic approximation. The lattice is then stabilized by the anharmonic potential energy, which has the form

$$\Phi^{(a)} = \tfrac{1}{4}(M\omega_o^2/a^2)\Sigma_{\vec{R}}\{u_o[u_x^2(\vec{R}) + u_y^2(\vec{R})]^2 + v_o[u_x^4(\vec{R}) + u_y^4(\vec{R})]\}, \qquad (1)$$

where $\vec{u}(\vec{R})$ is the atomic displacement at site $\vec{R}$, M is the particle mass, $M\omega_o^2$ is a nearest neighbor harmonic force constant, and u_o and v_o are constants. This form for $\Phi^{(a)}$ has been used for renormalization group analyses of the phase diagrams of perovskite crystals [3].

Some features of the atomic motions in this system can be understood on the basis of the continuum approximation to the equations of motion. To obtain slowly varying quantities for which continuum equations are appropriate, the staggered displacement field $\vec{S}(\vec{R},t)$, defined by reversing the displacements on alternate lattice sites, is introduced. The equation for S_y is

$$\begin{aligned}&(\omega_o a)^{-2}\partial^2 S_y/\partial t^2 - [-(1/2) - (3/8)r_2 + (1/8)\overline{r}_4]\nabla^2 S_y + (2 + r_2 - \overline{r}_4 - g)S_y \\ &- (1/8)(r_2 - \overline{r}_4)(\nabla_x^2 S_y - \nabla_y^2 S_y) - g\nabla_x^2 S_y - r_5\nabla_{xy}^2 S_x \qquad (2) \\ &+ [u_o(S_x^2 + S_y^2)S_y + v_o S_y^3] = 0.\end{aligned}$$

There is a similar equation for S_x. The constants $r_2, \overline{r}_4, r_5$ characterize the harmonic part of the potential energy; g is an anisotropy parameter arising from an external stress on the system (the same as used in [3]). The terms of the first line of (2) form the Klein-Gordan equation; the coefficient $(2+r_2+\overline{r}_4-g)$ is negative, which gives the instability of the harmonic approximation. The terms of the second line enter because the symmetry of the lattice is

[1] Permanent address

[2] Work at Wake Forest University supported by a Research Corporation grant

square rather than isotropic, there is an external stress on the system, and there is a coupling between the components of $\vec{S}$. The terms of the third line describe the non-linear forces. For the special conditions of propagation parallel to the y-axis with only a y-component to the motion, these equations reduce to equations of one-dimensional models studied by other authors [4,5], and thus this model possesses the propagating domain wall solutions found by those authors.

Several thermodynamic functions and correlation functions have been obtained from the MD calculations on this model. The most interesting results therefrom are presented here; a complete report is in preparation. The results for the specific heat, order parameter, and susceptibility show that the model has a SPT with the critical temperature at $T_c = (5.90 \pm 0.03) \times 10^{-3}$.

The instantaneous order parameter, its correlation function, and frequency spectrum are defined by

$$P(t) = N^{-1} \Sigma_{\vec{R}} S_y(\vec{R},t); \quad D(t) = \langle P(t)P(0) \rangle; \quad D(\omega) = \int dt\, e^{-i\omega t} D(t). \qquad (3)$$

Some of the results obtained for $D(\omega)$ are shown in Fig.1. At $T = 7.357 \times 10^{-3}$ the soft mode peak and a central peak of comparable intensity are evident. As T decreases, the central peak grows rapidly; the soft mode peak keeps about the same intensity while decreasing in frequency. Two curves are shown at $T = 6.022 \times 10^{-3}$ in order to bring out the soft mode peak at $\omega = 0.012$. At this temperature the central peak intensity is about 450 times that of the soft mode. Over the temperature range shown in the figure, the unnormalized value of $D(\omega=0)$ increases by a factor of 8000. As T goes below T_c, the sequence of changes of $D(\omega)$ reverses: the central peak intensity decreases and the soft mode moves out to higher frequency.

For T outside an interval between 5.4×10^{-3} and 6.2×10^{-3} the square of the soft mode frequency follows a linear dependence on temperature, $\omega^2(\vec{q}_c) \propto$

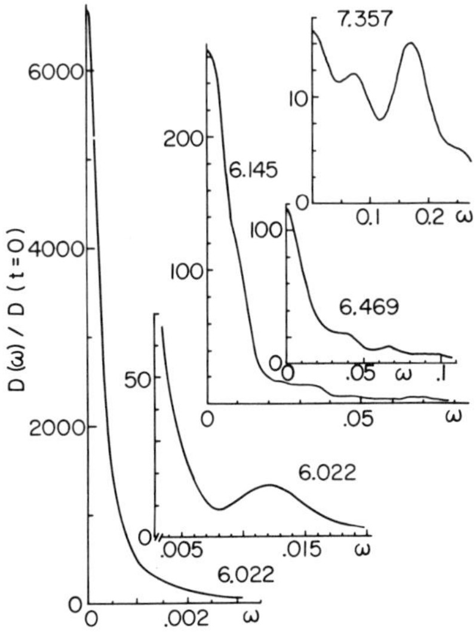

Fig.1 Spectral function of the order parameter correlation function. The numbers identifying each curve are 10^3T, where T is the temperature. Throughout the paper temperature is expressed in units of $M\omega_0^2 a^2/k_B$ and frequency is expressed in units of ω_0

$|T - T_o|$, with $T_o > T_c$. For temperatures within this interval around T_c, $\omega^2(\vec{q}_c)$ deviates from this linear behavior, and it appears to remain finite at T_c. This behavior is in agreement with the observed behavior of strontium titanate [6] and lead germanate [7].

The width of the central peak is a rapidly varying function of T. Excluding the point at $T = 6.022\times10^{-3}$, the values of the width can be fit by a $\sqrt{|T - T_c|}$ dependence, both above and below T_c. These are nearly the same temperatures where $\omega^2(\vec{q}_c)$ follows a linear dependence on T. However, the width at $T = 6.022\times10^{-3}$ is considerably narrower than is given by this square root dependence, so there may be a crossover between two different dependences here.

The correlation function for the spatial Fourier components of the staggered displacement field is defined by

$$D(\vec{q},t) = N^{-1}\langle S_y(\vec{q},t)S_y(-\vec{q},0)\rangle; \quad S_y(\vec{q},t) = \Sigma_{\vec{R}} e^{-i\vec{q}\cdot\vec{R}} S_y(\vec{R},t), \tag{4}$$

and its spectral function $D(\vec{q},\omega)$ is defined as in (3). For $\vec{q} = 0$, $D(\vec{q},t)$ reduces to the order parameter correlation function, so this function gives the wavevector dependence of the central peak characteristics. The results for $D(\vec{q},\omega)$ at the temperature at which $D(\omega)$ has the most intense central peak are shown in Fig.2. As $\vec{q}$ moves away from zero, the central peak splits so that the maximum is at a small finite frequency. A similar splitting has been found by SCHNEIDER and STOLL [8]. This splitting occurs in a smaller temperature interval around T_c than the interval for which there is a central peak in $D(\omega)$. As $\vec{q}$ increases further, the central peak intensity decreases, and $D(\vec{q},\omega)$ reverts to familiar anharmonic lineshapes.

The final results are for the energy density correlation function. The energy density $h(\vec{R},t)$ is the function whose sum over the lattice is the total energy. Its spatial Fourier transform $h(\vec{q},t)$, correlation function $E(\vec{q},t)$, and spectral function $E(\vec{q},\omega)$ are defined analogously to (4) and (3), respectively. Results for $\vec{q} = q_c$, and for $T > T_c$ are shown in Fig.3. The most

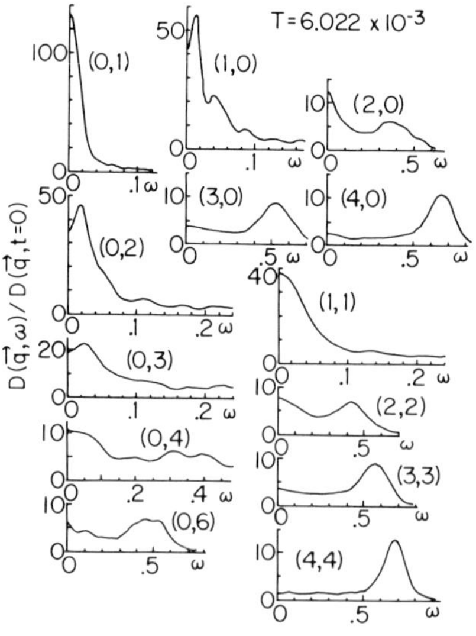

Fig.2 Spectral functions for the correlation functions of the Fourier components of the staggered displacement field. The integer labels with each curve identify the wavevector in the usual scheme for periodic boundary conditions

noticeable feature in $E(\vec{q}_c,\omega)$ is the growth of a well-defined high frequency peak near $\omega = 0.9$ as T approaches T_c from above. This feature is unlikely to be a second sound peak because it occurs in a high temperature regime and it is a short wavelength phenomenon. It is more likely to be a propagating two phonon resonance.

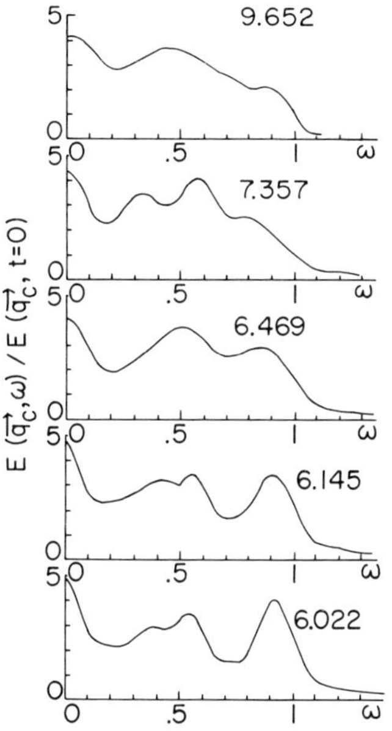

Fig.3 Spectral functions for the energy density correlation function for wavevector $\vec{q}_c$. The numbers with each curve identify the temperature in the same way as for Fig.1

In conclusion these MD calculations have shown that it is possible to obtain a central peak which is an intrinsic anharmonic effect and which has the same features as those found experimentally. In addition they have found a splitting of the central peak for non-zero wavevectors and a high frequency propagating mode in the energy density correlation function.

References

1. G. Shirane: Revs. Mod. Physics 46, 437 (1974)
2. M. L. P. Bartolome and W. C. Kerr: Solid State Commun. 21, 253 (1977)
3. A. D. Bruce and A. Aharony: Phys. Rev. B 11, 478 (1975)
4. S. Aubry: J. Chem. Phys. 62, 3217 (1975); ibid. 64, 3392 (1976)
5. J. A. Krumhansl and J. R. Schrieffer: Phys. Rev. B 11, 3535 (1975)
6. S. M. Shapiro, J. D. Axe, G. Shirane, T. Riste: Phys. Rev. B 6, 4332 (1972)
7. K. B. Lyons and P. A. Fleury: Phys. Rev. B 17, 2403 (1978)
8. T. Schneider and E. Stoll: Phys. Rev. B 13, 1216 (1976)

Soliton-Like Features in a Two-Dimensional XY Model with Quartic Anisotropy

E. Stoll and T. Schneider

IBM Zurich Research Laboratory, CH-8803 Rüschlikon, Switzerland

1. Introduction

Recently [1,2] we have shown that in a two-dimensional XY model with quartic anisotropy, soliton-like features do occur. Here we extend this study to include a variable anisotropy. Our results include: (i) the cluster wall thickness multiplied by T_c seems to remain constant, (ii) the rather sharp cluster walls prevent the formation of vortices [3], and (iii) the phase diagram agrees quite well with predictions as obtained by the real-space renormalization group [4].

2. The Model

The Hamiltonian of this model reads

$$\mathcal{H} = \frac{M}{2} \sum_{\ell} (\dot{X}_{\ell}^2 + \dot{Y}_{\ell}^2) + \frac{A}{2} \sum_{\ell} (X_{\ell}^2 + Y_{\ell}^2) + \frac{B}{8} \sum_{\ell} (X_{\ell}^2 + Y_{\ell}^2)^2$$

$$+ \frac{B_1}{4} \sum_{\ell} (X_{\ell}^4 + Y_{\ell}^4) - C \sum_{\ell,m} (X_{\ell} X_{\ell+m} + Y_{\ell} Y_{\ell+m}). \quad (1)$$

The momentum and displacement vector $\vec{X}_{\ell}$ has two components, $\dot{\vec{X}}_{\ell} = (\dot{X}_{\ell}, \dot{Y}_{\ell})$ and $\vec{X}_{\ell} = (X_{\ell}, Y_{\ell})$, respectively. We consider nearest-neighbor interactions only. For $B_1 = 0$, the model would be rotationally invariant. The rigid reference lattice is assumed to be a simple square lattice with lattice constant a. The equations of motion read for the X_{ℓ} component

$$-M \ddot{X}_{\ell} = (A - 8C) X_{\ell} + \frac{B}{2} (X_{\ell}^2 + Y_{\ell}^2) X_{\ell} + B_1 X_{\ell}^3 + 2C \sum_{m} (X_{\ell} - X_{\ell+m}). \quad (2)$$

Introducing polar coordinates, the equations of motion (2) become in the continuous limit for the angular field

$$\nabla^2 \varphi - \frac{M}{2C a^2} \ddot{\varphi} = - \frac{B_1 \rho^2}{8C a^2} \sin 4\varphi. \quad (3)$$

The model parameters have been chosen as follows

$$A = -1, \quad B = 2/3 - 2 B_1, \quad 1/240 \leq - B_1 \leq 2/3, \quad C = 1/4. \quad (4)$$

Of particular interest close to T_c is the kink solution [5] of the sine-Gordon equation (3)

$$\varphi(R_x, R_y = 0, t) = \pm \tan^{-1} \exp(\pm \alpha \xi), \tag{5}$$

which is a soliton (+) or antisoliton (-), respectively. Moreover,

$$\alpha^2 = \frac{-B_1 \rho^2}{2C a^2} \left(1 - \frac{V^2 M}{2C a^2}\right)^{-1}, \quad \xi = R_x - V_x t \; . \tag{6}$$

Of particular interest is the case where $|B_1|$ becomes small, because in the limit $|B_1| = 0$ the system reduces to the two-dimensional XY model. Although evidence from high-temperature series expansions [6] suggests that the susceptibility of the isotropic model diverges at some finite T_c, the lack of long-range order at any temperature has been proven rigorously [7]. KOSTERLIZ [3] suggested that below T_c bound vortex-antivortex pairs should occur, which dissociate above T_c. Recently, JOSE *et al.* [4] studied the effect of symmetry-breaking perturbation, such as quartic anisotropy using the Migdal recursion scheme. Their results suggest that quartic anisotropy is a strongly relevant variable, leading to a *conventional* continuous phase transition but with non-universal exponents.

3. Molecular-Dynamics Results

For quantitative discussions we solved (2) by means of molecular dynamics [1,8]. Systems of 1600 particles subjected to periodic boundary conditions were studied. We found that in a temperature window below and above T_c, clusters are formed, built up of particles connected by a nearest-neighbor bond with positions lying in the same quadrant. Figure 1 shows a snapshot of cluster configurations in the model with parameters given by (4) and $B_1 = -1/240$ and T close above T_c.

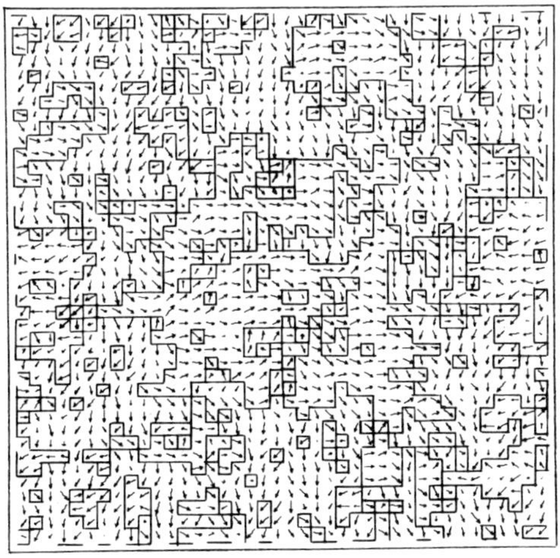

Fig.1 Snapshot of a cluster pattern for $T/T_c = 1.3$ and $B_1 = -1/240$. The arrows mark the positions of the particles, and the clusters are separated by lines

Following the arrow directions, it is seen that the clusters are separated by rather narrow $\pi/2$ walls. In fact, Fig. 2 reveals that for clusters consisting of more than five particles the cluster wall thickness $1/\alpha$ is less than two lattice constants, even for small $|B_1|$.

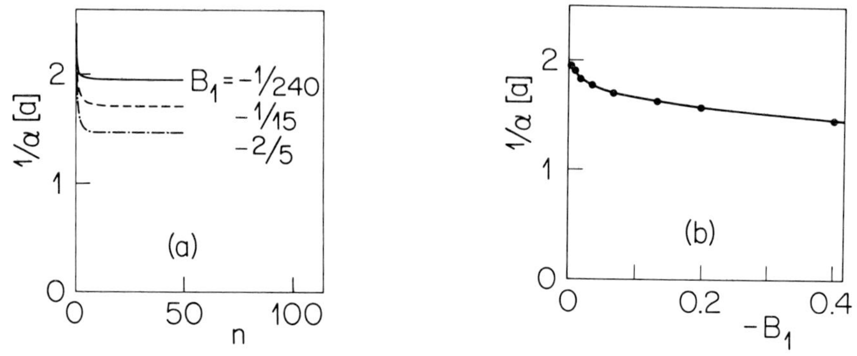

Fig.2 Cluster wall thickness $1/\alpha$ depending on (a) the cluster size n and (b) for a cluster consisting of more than five particles on the anisotropy $-B_1$, in units of lattice constants a

These results do not agree with the predictions of (6), where $\alpha \sim \sqrt{|B_1|}$.

We note, however, that according to Fig. 3 the product $T_c \times (1/\alpha)$ tends to a constant for small $|B_1|$.

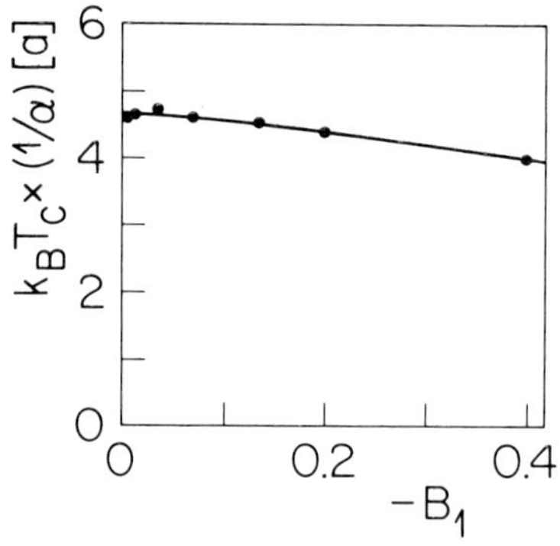

Fig.3 Product of cluster wall thickness $1/\alpha$ and T_c depending on the anisotropy $-B_1$

From these features and the absence of vortices (Fig. 1), we expect that for the anisotropies considered here (4) Ising-like critical behavior will dominate close to T_c. The non-universal behavior of the critical exponents as proposed by JOSE *et al.* [4] appears to be restricted to very small anisotropies. Nevertheless, their predictions of the anisotropy dependence of T_c

$$|B_1|\rho^4 \sim T_c \exp(-b\, T_c^{\,2} \exp g/T_c) \qquad (7)$$

agrees quite well with our numerical estimates (Fig. 4).

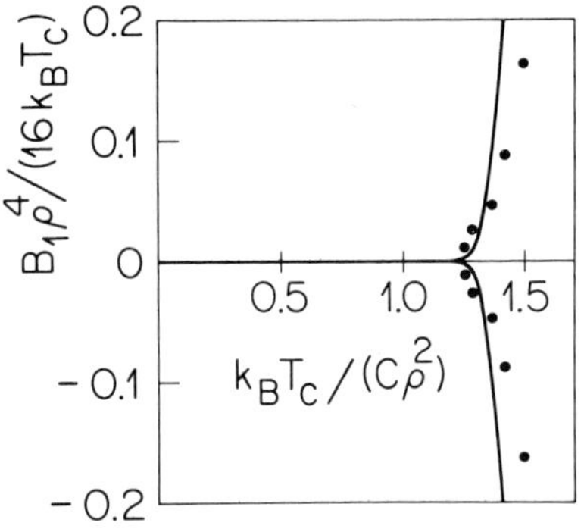

Fig.4 Dependence of T_c on B_1. ——— after (7) [4]. • molecular-dynamics results

4. Conclusions

To summarize, we have presented numerical evidence that close to T_c kink and anti-kink solitons are relevant even in a two-dimensional discrete lattice model. We find sharp cluster walls preventing the system from forming vortices. Finally, the anisotropy dependence of the critical temperature agrees quite well with the prediction of a real-space renormalization-group study.

References

1. T. Schneider, E. Stoll: Phys. Rev. Lett 36, 1501 (1976)
2. E. Stoll, T. Schneider: In *"Proceedings of the International Conference on Lattice Dynamics"*, ed. by M. Balkanski (Flammarion, Paris 1978) p.609
3. J.M. Kosterlitz, D.J. Thouless: J. Phys. C 6, 1181 (1973); J.M. Kosterlitz: J. Phys. C 7, 1046 (1974)
4. V. José, L.P. Kadanoff, S. Kirkpatrick, D.R. Nelson: Phys. Rev. B 16, 1217 (1977)
5. A.C. Scott, F.Y.F. Chu, D.W. McLaughlin: Proc. IEEE 61, 1443 (1973)
6. H.E. Stanley, T.A. Kaplan: Phys. Rev. Lett. 17, 913 (1966); H.E. Stanley: Phys. Rev. Lett. 20, 589 (1968); M.A. Moore: Phys. Rev. Lett. 23, 861 (1969)
7. N.D. Mermin, H. Wagner: Phys. Rev. Lett. 17, 1133 (1966); P.C. Hohenberg: Phys. Rev. 158, 383 (1967); D. Jasnow, M.E. Fisher: Phys. Rev. Lett. 23, 286 (1969)
8. T. Schneider, E. Stoll: Phys. Rev. B 17, 1302 (1978)

Behavior of a ϕ^4-Kink in the Presence of an Inhomogeneous Perturbation

N. Theodorakopoulos, S. Hanna, and R. Klein

Fachbereich Physik der Universität Konstanz
D-7750 Konstanz, Fed. Rep. of Germany

1. Introduction

In recent years, the ϕ^4 field theory became of interest as a one-dimensional model for structural phase transitions. In particular, it was shown that in a linear chain of harmonically coupled particles, which in addition sit in local double minimum potentials, ordered clusters exist. The walls separating ordered clusters are the kinks of ϕ^4-theory [1]. The possible relevance of an ensemble of independently moving kinks to the spectrum of scattered neutrons and light has been investigated [1,2].

In this contribution the response of the system to an external field of wave number K and frequency Ω in the presence ofconstant damping η is studied. We are mainly interested in the polarisation of a kink at rest (calculated in section 2) and in the absorption spectrum of an ideal gas of moving kinks, treated in section 3.

2. Polarisation of a Kink at Rest

In properly chosen units, the equation of motion for the displacement field ϕ is

$$\frac{\partial^2\phi}{\partial\tau^2} - \frac{1}{2}\frac{\partial^2\phi}{\partial\xi^2} - \phi + \phi^3 + \eta\frac{\partial\phi}{\partial\tau} = E\exp\left[i(K\xi - \Omega\tau)\right] . \tag{1}$$

The unperturbed ϕ^4 field is known to pocess excited states (j) which are solutions to (1) (with η=0=E) linearized in the neighborhood of a static kink solution $\phi_k(\xi) = \tanh\xi$. These excited states are characterized by $f_j(\xi)$ $\exp(i\omega_j\tau)$, where $f_j(\xi)$ form a complete orthonormal set of functions. These represent the translation mode (j=T), the localized mode (j=L) and a continuum (j=c) of scattering states {q}; the corresponding energies are $\omega^2_T=0$, $\omega^2_L = 3/2$ and $\omega^2_q = 2+q^2/2$ [3].

In the presence of damping and external field we follow a linearization procedure similar to that employed by FOGEL et al. [4] for a homogeneous time dependent field in the case of the Sine-Gordon (SG) equation. Accordingly,

$$\phi(\xi,\tau) = \phi_k(\xi) + \psi(\xi,\tau) \tag{2}$$

$$\psi(\xi,\tau) = \sum_j B_j(\tau)\, f_j(\xi) = \sum_j \psi_j(\xi,\tau) \tag{3}$$

The expansion coefficients $B_j(\tau)$ can be calculated in the framework of linear response theory and are given by

$$B_j(\tau) = \left[\omega^2{}_j - \Omega^2 - i\eta\Omega\right]^{-1} F_j(K)\ E\exp(-i\Omega\tau) \tag{4}$$

$$F_j(K) = \int d\xi\ \exp(iK\xi)\ f_j^*(\xi) \tag{5}$$

The $F_j(K)$'s are given in Table 1 for both the SG and ϕ^4 cases.

Table 1 The Fourier coefficients $F_j(K)$ of $f_j(\xi)$ for SG and ϕ^4 fields. For K=0 the SG results reduce to those derived in [4]. P stands for principal value

Mode j	SG	ϕ^4
T	$\frac{2}{\cosh\pi\ K/2}$	$\sqrt{3}\ \frac{\pi\ K/2}{\sinh\pi\ K/2}$
L	-	$\sqrt{\frac{3}{2}}\ \frac{i\ \pi\ K}{\cosh\pi\ K/2}$
{q}	$\frac{1}{\sqrt{2\pi}}\frac{1}{\omega_q}\left[2\pi q\ \delta(q-K) + P\frac{\pi}{\sinh\pi(K-q)/2}\right]$	$\frac{1}{\left[2\pi(q^4+5q^2+4)\right]^{1/2}}\left[2\pi(2-q^2)\delta(q-K) + 3\pi q\ P\frac{1}{\sinh\frac{\pi}{2}(q-K)} + \frac{3\pi(K-q)}{\sinh\pi\ (q-K)/2}\right]$

From (3) - (5) and Table 1 $\psi(\xi,\tau)$ can be calculated. In the SG case, as well as in the ϕ^4 case, the scattering states produce a contribution ψ_{vac} to $\psi(\xi,\tau)$, which is plane-wave-like:

$$\psi_{vac}(\xi,\tau) = \left[\omega^2{}_K - \Omega^2 - i\eta\Omega\right]^{-1} E\ \exp(iK\xi - i\Omega\tau) \tag{6}$$

The vacuum corresponds in this context to a crystal with optical phonons but no kink. This part is subtracted out in the following definition of a measure for the kink polarizability. The total kink polarization is taken as

$$P = \sum_j P_j \tag{7}$$

In the ϕ^4 case the various contributions can be shown to be

$$P_c = \int_{-\infty}^{\infty} d\xi(\psi_c - \psi_{vac}) = -\ 3\ \frac{\pi\ K/2}{\sinh\pi\ K/2}\left[\omega_o{}^2 - \Omega^2 - i\eta\Omega\right]^{-1} E\ \exp(-i\Omega\tau) \tag{8}$$

$$P_T = \int_{-\infty}^{\infty} d\xi\ \psi_T = 3\ \frac{\pi\ K/2}{\sinh\pi\ K/2}\left[-\Omega^2 - i\eta\Omega\right]^{-1} E\ \exp(-i\Omega\tau) \tag{9}$$

$$P_L = \int_{-\infty}^{\infty} d\xi\ \psi_L = 0 \tag{10}$$

The definition used in (8) ensures that only effects are considered, which are directly connected with the presence of the kink. It represents a generalization of the definition used in [4]. Note that the localized mode does not contribute.

3. Absorption Spectrum of Moving Kinks

We may obtain additional information about the dynamics of the nonlinear elastic continuum underlying (1) by considering the absorption spectrum of a moving kink. At a first stage we may then average over all possible kink positions, assumed uniformly distributed over all space; at a second stage we can average over kink velocities, introducing on appropriate distribution. The resulting spectrum contains, apart from a "vacuum" contribution due to absorption by optical phonon modes, new features due to the presence of a kink. To the extent that kinks may be considered as non-interacting entities such "corrections" to the spectrum will be additive and thus proportional to the total number of available kinks.

Let ξ_α and v_α be the (initial) position and velocity of the αth kink in the laboratory coordinate system. We may then introduce coordinates in the kink rest frame via a Lorentz transformation:

$$x_\alpha = \gamma_\alpha(\xi - \xi_\alpha - v_\alpha\tau) \quad ; \qquad t_\alpha = \gamma_\alpha[\tau - 2v_\alpha(\xi - \xi_\alpha)] \tag{11}$$

where $\gamma_\alpha = [1 - 2v_\alpha^2]^{-1/2}$ (Note that the maximum kink velocity in the units prescribed by (1) is $1/\sqrt{2}$). We shall now make the assumption (a) that the perturbing field is a scalar under Lorentz transformations and (b) that a dispersion relation of the type Ω=cK holds, where c is of the order of 10^5, corresponding to the magnitude of the light velocity in the dimensionless units chosen. We may now compute the quantity ψ in the kink's rest frame using (3)-(5) and considering that on account of the scalar property of the external field the kink now sees a field, in its rest frame, with wavevector $K_\alpha = \gamma_\alpha(K-2v_\alpha\Omega)$ and frequency $\Omega_\alpha = \gamma_\alpha(\Omega-v_\alpha K)$. Upon transforming back to laboratory coordinates and averaging over all positions $\{\xi_\alpha\}$ we obtain

$$<\psi^\alpha(\xi,\tau)> = \chi^\alpha(K,\Omega)\ E\ \exp(i(K\xi - \Omega\tau)) \tag{12}$$

$$\mathrm{Im}\chi^\alpha(\Omega) = \frac{1}{L_\alpha}\sum_j \frac{\eta\,\Omega_\alpha}{(\omega_j^2-\Omega_\alpha^2) + \eta^2\Omega_\alpha^2}\,|F_j(K_\alpha)|^2 \equiv \sum_j \mathrm{Im}\chi_j{}^\alpha(\Omega) \tag{13}$$

where $L_\alpha = \gamma_\alpha L$, L being the length of the chain. In the following we shall consider the limit $\eta\to 0$. The contributions of the various modes are:
(i) j=T becomes relevant only in the "pathological" case $\Omega=v_\alpha K$ which is excluded by the assumption (b) made above; the latter acts as a selection rule; note, however, that for a static kink the above condition corresponds to Ω=0, and this is precisely the case in which energy can be absorbed by the static kink - as predicted by (9) above.
(ii) j=L gives rise to absorption at $\Omega_\alpha=\omega_L$; note again that the oscillator strength $F_L(K)$, which would be zero for a static kink ($K^2\sim 10^{-10}$ for $\Omega=\omega_L$) is in fact non-negligible if we consider a distribution of velocities $P(v) = A\gamma^3\exp(-\gamma E_0/T)$, where E_0 is the rest energy of the kink, T the temperature and A a normalisation constant [5]. Upon performing the thermal average over the velocity distribution we obtain an absorption band for $\Omega<\omega_L$

$$\mathrm{Im}\chi_L^\alpha(\Omega) = (A\,\pi^3/L)\,\omega_L\,\frac{(\omega_L^2 - \Omega^2)^{3/2}}{\cosh^2\frac{\pi}{\sqrt{2}}(\omega_L^2-\Omega^2)^{1/2}}\exp(-\frac{E_0\omega_L}{2T\Omega}) \tag{14}$$

(iii) j=c. The continuum modes give rise, besides the vacuum absorption (obtainable from (6) as well), to two distinct kink-related contributions.

Firstly we obtain absorption at frequency ω_K and secondly a diffuse backround which vanishes if we go far from ω_K.

$$\mathrm{Im}\chi_c^\alpha(\Omega) = + \frac{1}{L_\alpha} B(K_\alpha)\pi\ \delta(\Omega^2 - \omega_K^2)$$

$$+ P \frac{1}{L_\alpha} \frac{9\pi^2 K_\alpha^2}{(K_\alpha^2+1)(K_\alpha^2+4)} \frac{\pi}{Q_\alpha} \left[\frac{1}{\sinh^2 \frac{\pi}{2}(Q_\alpha - K_\alpha)} + \frac{1}{\sinh^2 \frac{\pi}{2}(Q_\alpha + K_\alpha)}\right] \quad (15)$$

where $Q_\alpha = [2(\Omega_\alpha^2 - 2)]^{1/2}$ and $B(K_\alpha)$ turns out to be, for long-wavelength phonons, equal to -3.

The qualitative structure of the absorption spectrum is shown in Fig. 1.

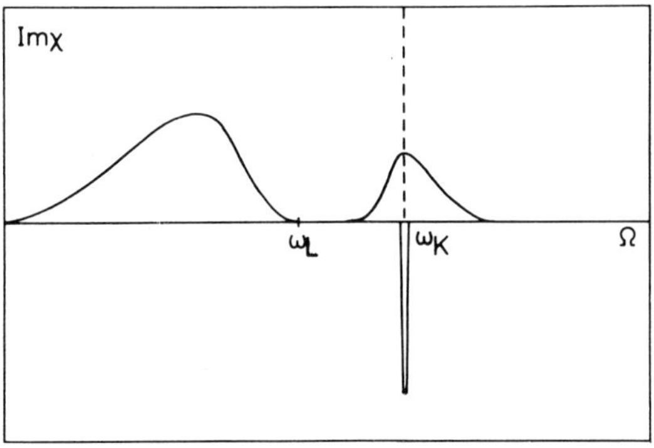

Fig.1 Schematic plot of the kink absorption spectrum. The dotted line represents the vacuum contribution; also shown are the band arising from the localised modes and the δ-peak as well as the diffuse backround due to the scattering modes

4. Conclusion

In section 2 we have considered the total polarization of a single static kink. The approach of section 3 should be regarded as complementary, since it averages over all kink positions and therefore restores translational invariance. It should be pointed out, as a further common characteristic of the two approaches, that they both explicitly exhibit the number of phonon modes removed by the kink. This can be seen by considering the $K\to 0$ limit of (8) and (15) where the strength of the phonon absorption peaks is in both cases equal to $\lim_{K\to 0} d\delta/dK = -3$, where $\delta(K)$ is the phase shift of a phonon with wavevector K [3].

References

1 J.A. Krumhansl and J.R. Schrieffer, Phys.Rev. B11, 3535 (1975)
S. Aubry, J.Chem.Phys. 62, 3217(1975); 64, 3392(1976)

2 W. Hasenfratz, R. Klein, N. Theodorakopoulos, Solid State Comm.18,893(1976)

3 See for instance R. Rajaraman, Physics Reports 21,229(1975) or
W. Hasenfratz and R. Klein, Physica 89A, 191(1977)

4 M.B. Fogel, S.E. Trullinger and A.R. Bishop, Phys.Lett.59A, 81(1976)

5 L.D. Landau and E.M. Lifshitz, Statistical Physics, (Pergamon Press 1969)

Solitary Wave Solutions in a Diatomic Lattice

H. Büttner

Physikalisches Institut, Universität Bayreuth
D-8580 Bayreuth, Fed. Rep. of Germany, and

H. Bilz

Max-Planck-Institut für Festkörperforschung
D-7000 Stuttgart, Fed. Rep. of Germany

Recently, it has been shown that the non-linear and anisotropic oxygen polarizability explains successfully the unusual dynamical properties of oxidic perovskites [1]. In particular the temperature dependence of the ferroelectric (transverse optical) soft mode and its coupling to the transverse acoustical modes in incipient ferroelectrics ($SrTiO_3$ and $KTaO_3$) has been described by a single quartic electron-phonon coupling parameter (localised at the oxygen lattice site) in the renormalised harmonic approximation (RHA [2]). This coupling leads, using the adiabatic condition, to a long-range anharmonic interionic potential along the (100)-direction where we have diatomic chains of alternating transition metal and oxygen ions. In view of the well-known limits of the RHA [3] we are interested in more general solutions of the problem. The structure of the Hamiltonian suggests a simplified version which allows analytical solutions while keeping the essential features of the underlying physics. The simplest model consists of a quasi-one dimensional diatomic lattice with harmonic and quartic nearest-neighbour interactions. It is shown that this lattice has at least two kinds of solitary wave solutions differing in their acoustical or optical-mode character. The low-energy acoustical excitation obeys a modified Korteweg-de Vries equation similar to the soliton in a monoatomic chain [4]. The optical mode is the solution of a ϕ^4-type wave equation and represents a 'kink-soliton'; it exhibits a close relation to the Krumhansl-Schrieffer model [5] for structural phase transitions. The group velocity of this solitary wave function is determined by the dispersion of the (harmonic) optical-phonon branch. This is clearly seen in a lattice with additional next nearest-neighbour interaction which makes closer contact to the experimental phonon dispersion curves.

The problem of nonlinear lattices has been studied in recent years quite extensively [6]. The lattices studied so far have, to our knowledge, only one atom per unit cell. In the following we study a diatomic chain with two different atoms of masses M_1 and M_2. The nearest-neighbour harmonic force constant is denoted by k_2 and the nonlinear one by k_4. The classical equations of motion for the displacements u_{1n} and u_{2n} of atoms M_1 and M_2 are:

$$M_1\ddot{u}_{1n}=k_2(u_{2n}-u_{1n})+k_2(u_{2n-1}-u_{1n})+k_4(u_{2n}-u_{1n})^3+k_4(u_{2n-1}-u_{1n})^3 \qquad (1)$$

$$M_2\ddot{u}_{2n}=-k_2(u_{2n}-u_{1n})-k_2(u_{2n}-u_{1n+1})-k_4(u_{2n}-u_{1n})^3-k_4(u_{2n}-u_{1n+1})^3 \qquad (2)$$

In the following we focus on long-wavelength solutions and therefore we use the continuum-limit approximation in both sublattices (i=1,2):

$$u_{in\pm 1} = u_{in} \pm 2d\, u'_{in} + (4d^2/2)u''_{in} \pm \ldots \qquad (3)$$

A general solution of the resulting partial differential equations would lead to a definite relation between u_{1n} and u_{2n}. The main point in our solution procedure is to try a general ansätz for u_{2n} as a function of u_{1n} which, for $k_4 \to 0$, reduces to the harmonic limit and, for equal masses, describes the known solitary wave solution in a monatomic chain. This ansatz reads:

$$u_{2n,\nu} = a_\nu \{ u_{1n,\nu} + \beta_{1,\nu} d u'_{1n,\nu} + \beta_{2,\nu}(d^2/2)u''_{1n,\nu} + \ldots \} \quad ,\nu = 1,2 \qquad (4)$$

with $a_1 = 1$ (acoustical mode) and $a_2 = -M_1/M_2$ (optical mode). The coefficients $\beta_{i,\nu}$ are determined by the condition that both equations, (1) and (2), should result in the same differential equation for $u_{1n,\nu}$.

For the acoustical mode, we obtain a modified Korteweg-de Vries equation which is identical to the one found in [4] for the monoatomic chain except for the different meaning of the parameters:

$$2\lambda M_1 \dot{v}_1 + k_2(\beta_{4,1}/12)d^4 v_1''' + 3k_4\, \beta_{2,1}\, d^4 v_1^2 v_1' = 0 \qquad (5)$$

with $v_1(x,t) \equiv \delta u_{1n,1}(x,t)/\delta x$ and the parameters: $\beta_{2,1} = 2M_1/(M_1+M_2) = 2\mu/M_2$; $\beta_{4,1} = 4\beta_{2,1}(1 - 3\mu/(M_1+M_2))$; $\lambda^2 = 2k_2 d^2/(M_1+M_2)$. A propagating wave solution of (5) is given by

$$v_1(x-ct) = \pm\, (2\lambda^{\pm} c/b)^{1/2} \operatorname{sech}\left((2\lambda^{\pm} c)^{1/2}(x-ct) + \Theta\right) \qquad (6)$$

where $\lambda^{\pm} = 12\lambda/(k_4 \beta_{2,1} d^4)$; $b = 3k_4\beta_{2,1}/2k_2\beta_{4,1}$, and c,Θ determined by boundary conditions.

For the optical mode, $u_{1n,2}$, we find (up to second order in d) the following equation:

$$\ddot{u}_1 + (2k_2/\mu)u_1 + \lambda^2 u_1'' + 2(k_4/\mu)(M_1/\mu)^2 u_1^3 =$$
$$= -6(k_4/\mu)(M_1/M_2)^2\, d^2 \left[u_1^2 u_1'' + (M_1/M_2)(u_1')^2 u_1\right] \qquad (7)$$

The l.h.s. is identical with the continuum approximation of the Krumhansl-Schrieffer-mode [5] for displacive phase transitions. The relation to their parameters is given by (see (6) of [5]):

$$A = 2k_2 M_1/\mu\, ; \quad B = (k_4/4k_2^3)A^3, \quad c_o^2 = \lambda^2. \qquad (8)$$

Note that k_2 is a negative constant in our model. The condition for a displacive phase transition given in (5) of [5] reads, with our parameters,

$$M_1 + M_2 < 16\ \mu\ . \qquad (9)$$

This means that, in this simple model, the phase transition is determined by the mass ratio, only. In our calculation we find additional terms which, for not too large amplitudes u_1, define renormalized harmonic force constants and sound velocities:

$$\tilde{k}_2 = k_2 + 3k_4 \, (M_1/M_2)^2 d^2 \, (u_1')^2 \, , \tag{10}$$

$$\tilde{\lambda}^2 = \lambda^2 + 6 \, (k_4/\mu) \, (M_1/M_2) d^2 u_1^2 . \tag{11}$$

These equations may be regarded as the first step in a self-consistent procedure reminiscent of the RHA used in (1) . In the 'paraelectric' regime above the phase transition, the values of the second terms, in the r.h.s. of (10) and (1) or their self-consistent analogues, overcompensate the negative values of the first terms thus leading to positive effective force constants and sound velocities. In a more general solution these renormalising terms correspond to a residual coupling between the solitary optical and acoustical solutions. The result of this coupling near the phase transition deserves particular interest.

We note that the solitary solutions which are usually visualized as originating from double-well potentials at the lattice sites of a monoatomic lattice (5) may be derived from the translationally-invariant quartic interaction potential of a diatomic chain.

Furthermore, we have investigated the effect of additional next nearest-neighbour harmonic force constants, α_1 and α_2. With these force constants we simulate the compensation between the strong attractive Coulomb forces which, in the bare harmonic approximation, would lead to an instable ferroelectric transverse mode, (k_2<o), and the effective repulsive forces. We can then readjust the effective phonon frequency, ω_{TO}, of this mode so that it is in agreement with realistic values, say, for $SrTiO_3$ or $KTaO_3$ in the paraelectric phase at a given temperature. Repeating the above calculations for this new case leads to the following equation for the acoustic mode:

$$2\bar{\lambda} \, M_1 v_1 + k_2 (\bar{\beta}_{4,1}/12) d^4 v_1''' + 3k_4 \, {}_{2,1} d^4 v_1^2 v_1' = 0 \tag{12}$$

with the new parameters

$$\bar{\lambda}^2 = \lambda^2 \left(1 + 2(\alpha_1 + \alpha_2)/k_2\right) \tag{13}$$

$$\bar{\beta}_{4,1} = \beta_{4,1} \left(1 + (2M_2/k_2)(\alpha_2/M_2 - \alpha_1/M_1) \right) \tag{14}$$

In contrast to the ansatz in (4), here we have used a non-linear relation between $u_{2n,1}$ and $u_{1n,1}$:

$$u_{2n,1} = \sum_i \bar{\beta}_{i,1} u_{1n,1}^{(i)} + \delta \, (d^4/24)(u_1'^2) u_1'' \, , \tag{15}$$

where

$$\delta = - \, (6k_4/k_2^2) \, \mu \, (\alpha_2/M_2 - \alpha_1/M_1) . \tag{16}$$

This extension of (4) re-establishes the form of the original modified Korteweg-de Vries equation except for a renormalization of the acoustic sound velocity and of the parameter $\beta_{4,1}$.

For the <u>optical</u> mode we obtain instead of (7), with (10) and (11),

$$\ddot{u}_1 + (2k_2/\mu) \, u_1 + \bar{\lambda}^2_{opt} \, u_1'' + 2(k_4 M_1^2/\mu^3) \, u_1^3 = 0, \tag{17}$$

where $$\bar{\lambda}^2_{opt}=\lambda^2 \left[1- (2/k_2)((M_2/M_1)\alpha_1+(M_1/M_2)\alpha_2)\right] \qquad (18)$$

Here, we have not considered renormalising terms analogous to those in (10) and (11). (18) is different from (13) for the sound velocity λ. Thus, it becomes obvious that, for the optical soliton, λ has no physical relation to the sound velocity of the system but merely describes the dispersion of the optical-soliton branch at very long wavelengths for small ratios k_4/k_2:

$$\omega^2_{opt}(q) = 2k_2/\mu - \bar{\lambda}^2_{opt}\, q^2+... \qquad (19)$$

With an arbitrary choice of the force constants $k_2(<o)$, $\alpha_1(>o)$ and $\alpha_2(>o)$ we may, for example, describe a rather strong dispersion of $\omega_{opt}(q)$ $(\bar{\lambda}^2_{opt}>\lambda)$ while the sound velocity could be pretty flat $(\bar{\lambda} > \lambda)$.

We note that the condition (9) for a displasice phase transition is now modified:

$$M_1+M_2 < 16\mu\, (\bar{\lambda}_{opt}/\bar{\lambda})^2. \qquad (20)$$

This condition indicates that strong mass differences (with $M_1+M_2>>\mu$) or flat optic dispersion $(\bar{\lambda}_{opt}<<\lambda;\alpha_1\alpha_2>o)$. favors order-disorder transitions <u>independent</u> of the strength of the non-linear coupling parameter k_4.

As our discussion has shown, one might expect that in the acoustical and optical branches of diatomic crystals with short-range forth-order anharmonicity solitary wave solutions may appear simultaneously, one acoustical of the Korteweg-de Vries type (pulse-soliton) and a second 'optical' one of the Krumhansl-Schrieffer type ('kink'-soliton). As the calculation has exhibited, the two solutions are not completely decoupled but, in particular for the optical soliton, terms remain which may be treated as renormalisation effects only for small values of k_4, as has been done in (10) and (11).

In summary, it has been shown that a simple one-dimensional diatomic lattice with quartic nearest-neighbour anharmonicity leads to the simultaneous appearance of acoustical (pulse) and optical-mode (kink) solitary waves. Since the model keeps several important features of the dynamical behaviour of displasive ferroelectrics, such as $SrTiO_3$, it is argued that the results obtained in this paper may be of importance for the understanding of phase transitions in those crystals.

The authors are grateful for discussions and critical comments to A. Bishop, W. Dieterich. R. Klein, W. Kleppmann and R. Zeyher.

<u>References</u>

1 R. Migoni, H. Bilz and D. Bäuerle, Phys. Rev. Lett. 37, 1155 (1976),
2 R. Migoni, H. Bilz and D. Bäuerle, Proc. Int. Conf. Lattice Dynamics, Paris 1977, Ed. M. Balkanski, Flammarion, p. 650
3 N.S. Gillis, in: Dynamical Properties of Solids, Vol.2., Ed. G.K.Horton and A.A. Maradudin,(North-Holland, Amsterdam, 1975) p. 105
4 M. Wadati, J. Phys. Soc. Jap. 38, 673 (1975) and 38, 681 (1975)
5 J.A. Krumhansl and J.R. Schrieffer, Phys. Rev. B11, 3535 (1975)
6 For recent reviews see: M. Toda, Phys Reports C18, 1 (1975), A. R. Bishop in: Springer Tracts Mod. Phys., Vol. on Solitons,1978, Heidelberg

Lattice Models of High Velocity Dislocation Motion*

N. Flytzanis

University of Virginia, Charlottesville, Virginia, USA

Abstract

The high velocity motion of a screw dislocation is critically affected by the discreteness of the medium. The motion is accompanied by the emission of sound waves following the defect as a wake. The dynamic Peierls stress s, required to maintain the dislocation in uniform motion is a structured function of velocity for $v \leq 0.5c$ but increases monotonically from $v \simeq 0.5c$ up to a critical $v \simeq 0.9c$. The external stress needed depends very strongly on the interatomic force law and can correspond to a strain as low as 10^{-5} for a straight screw dislocation with a piecewise linear force law. The interaction of the dislocation with phonons is strongly non-linear for phonons with phase velocity equal to the dislocation velocity. They can form a phonon-dislocation complex moving with no external stress.

1. Introduction

Plastic flow takes place through the propagation of dislocations which can be ingrown in the crystal or can be nucleated under the influence of the applied stress [1,2]. While many flow phenomena can be understood by phenomenological models using continuum elasticity, plastic deformation is controlled by processes happening on the atomic level, in particular near the core of the defects where continuum elasticity theory is inadequate. A realistic description, therefore, of a dislocation in a crystal must take into account the discrete nature of the medium.

It is clear in principle how to construct a realistic model of a dislocation in a pure crystal. We can separate the crystal in two regions:

*Supported by the National Science Foundation Grant DMR-76-01059

(i) the core of the defect that consists of a finite number of atoms interacting with non-linear forces and (ii) the rest of the crystal. Outside the core the atomic motions are described using lattice dynamics [3]. All the necessary information is in the frequency spectrum $\omega(\vec{k},\alpha)$ and the polarization vectors $\vec{e}_\alpha$. The frequency spectrum can be evaluated directly from the elastic constants and neutron scattering data. In the core where the atomic displacements are outside the region of linearity we must solve the equations of motion with full interatomic forces or use some simple phenomenological model. For static dislocations many realistic calculations have been made using various methods [4,5,6]. From these one can determine the core energy of the dislocation, the static Peierls barrier that must be overcome to move a straight dislocation segment from valley to valley in the periodic structure of the crystal, and the atomic configuration of the defect. These parameters depend very strongly on the type of interatomic forces used.

In most experimental measurements for the dislocation velocity as a function of the applied stress one can distinguish three regions [2,7,8]: i) the low velocity region $v<10^{-4}c$ (c is the appropriate speed of sound) where the motion occurs by kink nucleation and diffusion, ii) an intermediate region where the velocity increases rapidly with applied stress, and iii) the high velocity region $v>0.1c$. It is in this region that steady state motion is possible and we will concentrate on it in the rest of the paper. The appropriate experiments to study high velocity motion are those using the tensile test, where dynamic effects due to long range dislocation propagation are observed in the form of bands of deformation [9]. The most direct method to study single dislocation motion is the use of the etch pit technique [2,10] on a pulse-stressed crystal, but this measures an average velocity over the duration of the pulse. On the other hand ultrasonic attenuation experiments [11,12] sample small displacements of the dislocation lines from equilibrium and are more relevant to study damping mechanisms at low dislocation velocities.

There are many mechanisms that can damp the propagation of dislocations. In fact almost 95 percent of the plastic work done on aluminum crystals for a large range of strains at room temperature is converted into heat [13]. These processes include [14]: damping due to the emission and scattering of lattice waves, electronic damping, anelastic relaxation as well as interaction with impurities and other defects. In a pure crystal at low temperature the dominant mechanism is expected to be the creation of phonons because the lattice presents in general a high barrier that the dislocation must overcome. This viscous force depends very non-linearly on the dislocation velocity.

The theoretical study of moving dislocations is complicated by the fact that the core of the dislocation is being displaced. There are techniques to deal with this. We assume an arbitrary motion for the non-linear part for a given interaction force law. We find the linear motion of the rest of the crystal, and then we determine the non-linear motions in the core self consistently.

Here we shall limit ourselves to some simple interatomic force laws in order to study the effect of the discrete nature of the crystal. In section 2 we will introduce the snapping bond model that has been used in one dimensional [15] and simple three dimensional dislocation models [16,17]. In the next section this will be extended to include any force law consisting of linear segments. The influence of phonons will be examined and the last section will include a summary of the results as well as a discussion for future work.

2. Snapping Bond Model (SB)

We generalized the model used for a static screw dislocation in a discrete lattice [18] to the case of a moving dislocation with the appropriate corrections for the nonlinearity of the forces in the core region. The crystal is simple cubic (not a necessary restriction) with nearest neighbor interactions. The only component of the atomic displacements is parallel to the dislocation axis and will be denoted by $W_{m,n}$ where m and n are half integers, labeling the positions of the atoms in the lattice. The force law between nearest neighbor atoms (actually rows of atoms) is assumed to be a piecewise linear function of the relative displacement D i.e.

$$F(D) = -A(D-\nu b) \tag{2.1}$$

where b is the lattice spacing in the direction parallel to the dislocation (also the magnitude of the Burgers vector) and ν an integer such that

$|D-\nu b|\leq\frac{b}{2}$. As the dislocation moves in the crystal from one valley A to the next A', it breaks all the bonds connecting the two rows between A and A' across the slip plane and the atoms of these rows are connected to new equilibrium positions displaced by a lattice spacing. In the process the dislocation creates a disturbance which asymptotically consists of two main contributions [19] (i) a contribution from the $k\simeq 0$ waves that is equal to that of continuum elasticity and has a $1/r$ dependence for the strain, (ii) a contribution due to the emission of phonons (from the core) that have phase velocity matching that of the dislocation. This radiation field falls off like $1/r^{\frac{1}{2}}$ and creates oscillations within a Cerenkov angle behind the dislocation, while along the caustics the field falls off slowly like $1/r^{1/3}$. A plot of the atomic displacement at different planes as a function of μ-vt is shown in Fig. 1.

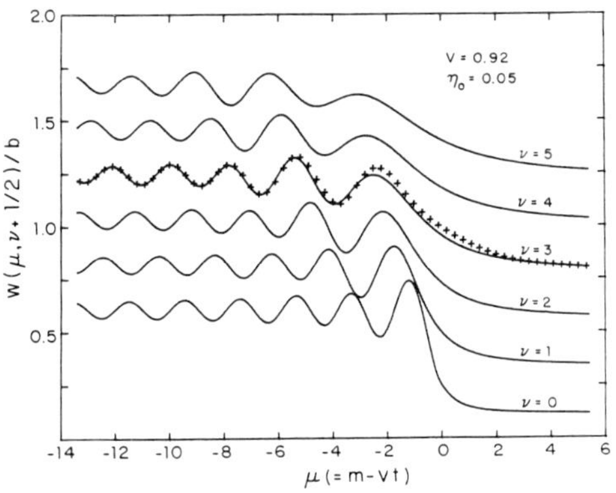

Fig. 1 Atomic displacements $W_{m,n}$ for different planes parallel to the slip plane. The crosses (+++) represent the asymptotic approximation

The solution can be written in terms of the Green's function for phonons with momentum $\vec{k}$ perpendicular to the dislocation axis, whose Fourier transform is:

$$G(\vec{k},\omega) = \left[\omega^2 - f^2(\vec{k})\right]^{-1} \tag{2.2}$$

where $f(\vec{k})$ is the dispersion relation for two dimensional transverse waves. Due to the coherence of the motion only those phonons will be most effective that obey the relation

$$f(\vec{k}) = \vec{v}.\vec{k} \tag{2.3}$$

The Fourier transform of the displacement across the slip plane as a function of $\mu=m-vt$ is

$$D^\circ(k_x) = \frac{1}{k_x-i\varepsilon} G(k_x) = \frac{1}{k_x-i\varepsilon}\int dk_y\, G(k_x,k_y) \qquad (2.4)$$

with

$$G(k_x,k_y) = G(\vec{k},\omega)|_{\omega = \vec{v}.\vec{k}}. \qquad (2.5)$$

and the factor $(k_x-i\varepsilon)^{-1}$ is the Fourier transform of the step function $\theta(m-vt)$ (since the sources of the radiation (the snapping bonds) are in front of the dislocation.) The solution for the displacement can be easily written down

$$W_{m,n} = sbn + \frac{b}{2\pi^2}\int_{-\infty}^{\infty} \frac{dk_x}{k_x-i\varepsilon}\int_{-\pi}^{\pi} dk_y\, e^{i\mu k_x} \frac{\sin(k_y/2)}{f^2(\vec{k})-v^2k_x^2} \qquad (2.6)$$

where s is the strain due to the externally applied stress and can be obtained from the energy balance condition, i.e. the work done by the external stress is equal to the energy emitted in the lattice waves. It is given by

$$s = \frac{1}{\pi^2}\int \frac{dk_x}{k_x}\, \mathrm{Im}(iG(k_x)) \qquad (2.7)$$

where $\mathrm{Im}(iG(x))$ is the density of states of possible emitted phonons which satisfy eq. (2.3) and comes from the vanishing of the denominator in (2.6). For details of this derivation the reader is referred to reference [16].

For the simple SB model the strain s can be calculated from the phonon spectrum of the perfect crystal. A plot of the strain as a function of the dislocation velocity for a cubic crystal is shown in Fig. 2. The main feature is that the stress is a nonlinear function of the velocity. There is a broad minimum near $v\simeq.5c$. For low velocities there are many unstable regions, and because of the high strains required it is more likely that the motion is of the thermally activated type or with large oscillations in the average velocity. For $v\geq c$ the strains are finite but enough to cause a breakdown of the uniform motion. In this phenomenon there are new dislocations being created in the wake as was first evidenced in computer simulations of the one-dimensional modified Frenkel-Kontorova model [20] (MFKM).

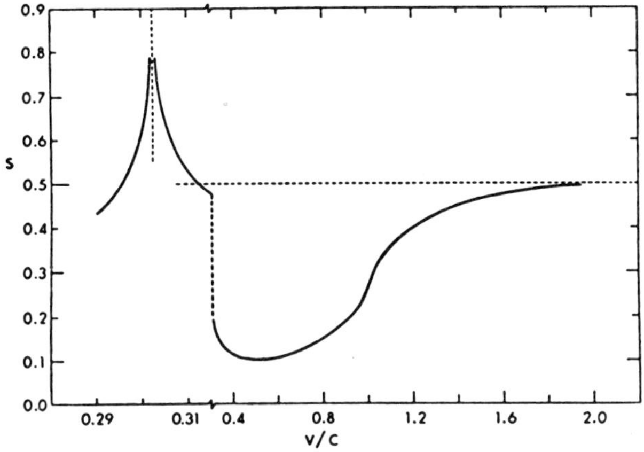

Fig. 2 Plot of the strain s (in the SB model) as a function of the dislocation velocity V (in units of the speed of sound c)

This is shown in Fig. 3 where the bonds behind the dislocation snap again for the velocity v=1.2c (i.e. $|D-1|>\frac{1}{2}$). While the strains of Fig. 1 are much higher than possible applied strains (the static Peierls stress is also high), many of the features of the plot will survive in the extension to a more general force law to be discussed in the next section. The advantage from this calculation is that we have the Green's function for the moving defect, from which in principle we can solve analytically the problem for any piecewise linear force law.

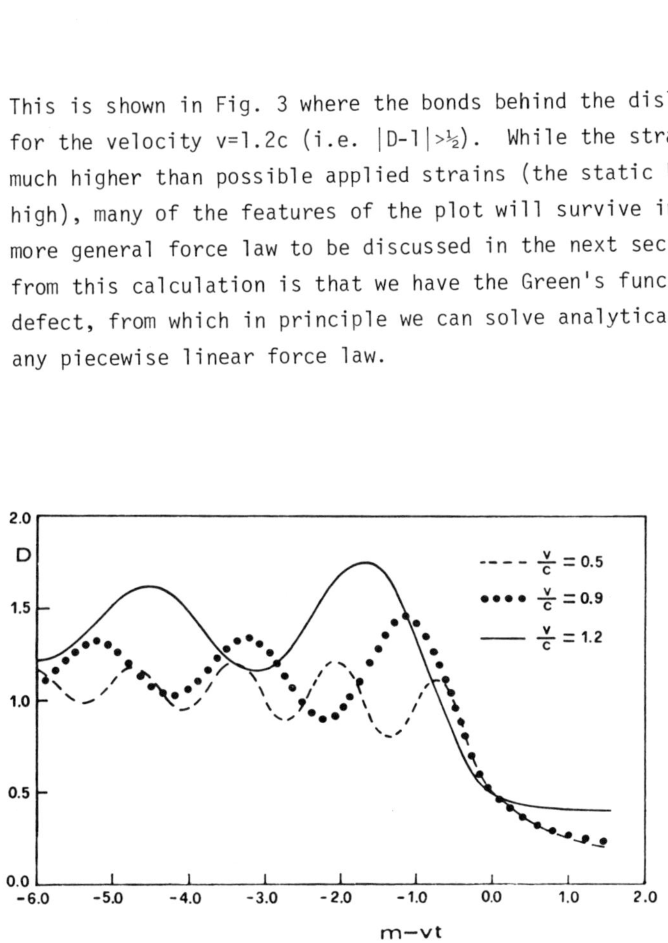

Fig. 3 Plot of D vs μ(=m-vt) for three dislocation velocities

3. Piecewise-Linear (PWL) Force Law

The extension of the lattice dynamical theory of dislocation motion to allow for arbitrary interatomic force laws in the core region, is a non-trivial problem if one wants to treat self-consistently the coupling between the highly non-linear motions in the core of the dislocation and the motions of the rest of the crystal, which are adequately described by discrete linear elasticity theory. The problem has been treated numerically for the one dimensional Frenkel-Kontorova model by ISHIOKA [21] for PWL and sinusoidal substrate forces. More extensive computer simulation studies were performed by EARMME and WEINER [22] for the same model with emphasis on breakdown phenomena, and by WEINER and PEAR [23] for a simplified two dimensional model of an edge dislocation. Using a method similar to that applied by WEERTMAN [24] to a smeared out dislocation in a continuum, we can treat the problem analytically in terms of a shape function h(u) [25,26](analogous to Weertman's distribution function) that will be defined below.

For simplicity we limit the discussion to the PWL force law, shown in Fig. 4a, with only 3 linear segments, where βb is the maximum displacement for which Hooke's law is valid. The problem is solved once we can determine the stress-strain relationship across the slip plane as a function of time. To accomplish this we write the force for the bonds that are stretched beyond βb in terms of a dislocation shape function h(u) (different for each non-equivalent soft bond):

$$F\left[D(\mu)\right] = AbD(\mu) + Ab\int_{-z}^{z} du\,h(u)\,\theta(u-\mu) \qquad (3.1)$$

where μ=m-vt is the distance of the dislocation at time t from the m-th atom on the slip plane. The force (3.1) is identical to the one in Fig. 4a, once h(u) is normalized to unity over the range (-z,z) (2z is the dislocation width) and for $|\mu|<z$ satisfies the self consistency condition

$$Eh(u)=\int_{-z}^{z} du'g(u-u')h(u') \quad ,|\mu|<z \qquad (3.2)$$

where $E=-(1-2\beta)(\pi/2)$ and g(u-u') can be obtained from the Green's function, and in fact equals $(\pi/2)$ times the derivative of the SB displacement across the slip plane. Eq. (3.2) is solved numerically by expanding h(u) in polynomials [25]. Once h(u) is known we can obtain the atomic displacements from the convolution of the source function h(u) with the Green's function for the defect, $D_0(u-u')$

$$D(u)=(s-s_0)bn + \int_{-z}^{z} h(u')\ D_0(u-u')\ du' \tag{3.3}$$

where $D_0(u-u')$ and s_0 are obtained from the solution of the SB model. From energy balance we can evaluate the external strain s required to maintain uniform motion,

$$s = \frac{1}{\pi^2}\int \frac{dk_x}{k_x}\ |H(k_x)|^2\ Im(iG(k_x)) \tag{3.4}$$

where $H(k_x)$ is the Fourier transform of h(u). The dependence of the strain s on β comes through $|H(k_x)|^2$. In particular, s will be small if $|H(k_x)|^2$ is small for a wave vector k_x which corresponds to a high density of states for emitted phonons, i.e. to a large value of $Im(iG(k_x))$.

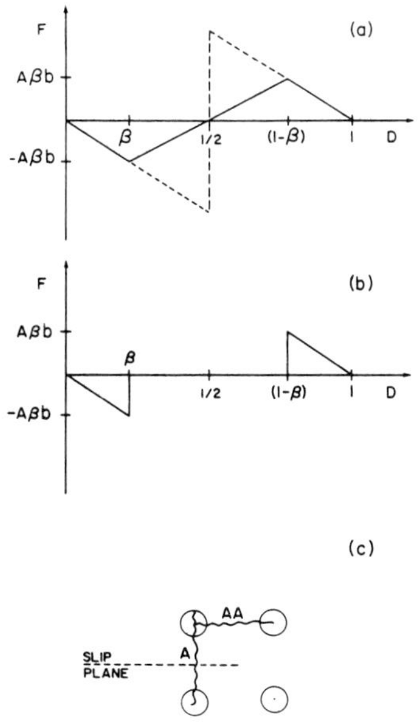

Fig. 4 Interatomic force laws and bond configurations. (A) The solid line is the PWL (piecewise linear) force law; the dashed line is the SB (snapping bond) force law. (B) The DB (dangling bond) force law. (C) Schematic diagram of the bonds near the slip plane

Calculations reported previously [25] show that the stress field of a moving dislocation for a PWL interatomic force law in the velocity range $0.7c<v<0.92c$ is not markedly different from the stress field for the snapping-bond model. Thus the analysis of high-speed dislocation phenomena in this range can be based on the much simpler SB model with the expectation that the results will be at least qualitatively correct. It is important that the calculations be carried out self-consistently within the model chosen; it is not permissible to choose a velocity - independent dislocation shape function $h(u)$ from continuum elasticity as done by ISHIOKA [28] and implicitly by BOFFI et al. [29]. However, this simplifying assumption may be justified at low velocities. The major effect of the softening of the bond law is a phase shift of the oscillatory displacement field in the wake of the moving dislocation; however the height of the oscillations and the magnitude of the external strain required to maintain the dislocation in uniform motion are not much changed from the predictions of the snapping bond model.

For $0.92c<v<0.99c$ it is difficult to achieve complete self-consistency by the methods used in this paper, since the bonds across the slip plane are repeatedly stretched beyond the linear limit in the wake of the dislocation. The most interesting questions in this region concern the critical velocity at which breakdown will occur [20] and which bond will be responsible for the breakdown [28]. The results indicate that breakdown occurs first for bond A (see Fig. 4c) and that the critical velocity is in the range of velocities indicated, depending nonmonotonically on the value of β, because of the phase shift of the oscillatory wake field.

For $v<0.7$ there are noticeable deviations from the SB model. In Fig. 5 we show a graph of s for a velocity $v = 0.5$; the points computed for several values of $1/\beta$ are connected by straight line segments. In curve A1 only the bonds across the slip plane are treated self consistently. We see that the strain decreases by orders of magnitude from the SB value ($\beta = 0.5$) and displays sharp minima, which correspond to almost symmetric core configurations for $h(u)$. The structure, however, disappears when the bond parallel to the slip plane are treated self-consistently (curve A2 in Fig. 5). We have also considered a crystal with preferred slip anisotropy, with the bonds perpendicular to the slip plane weaker by a factor K than the bonds parallel to the slip plane [27]. For $K = 1/3$ the minima in s are even sharper (curve B) and the motion near the minima is self-consistent. In Fig. 6 we show a plot of

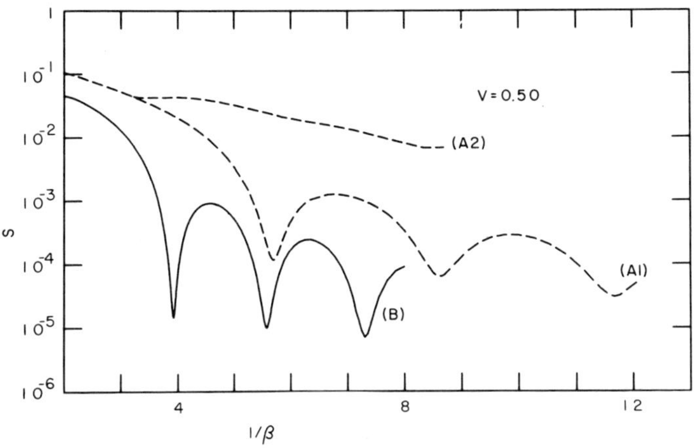

Fig. 5 The external strain s vs. the parameter $1/\beta$ for a dislocation velocity v = 0.5 and two values for the anisotropy parameter K: (A1) K = 1, only bonds across the slip plane treated consistently; (A2) K = 1, all bonds treated consistently; (B) K = 1/3, only bonds across the slip plane need to be treated consistently

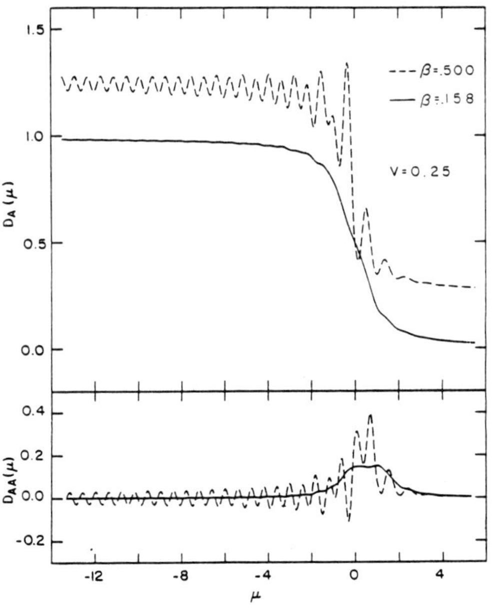

Fig. 6 Plot of the strain $D_A(D_{AA})$ of bonds A(AA) (Fig. 4C) vs μ for v = 0.25 with β = .158 (solid line) and β = .500 (dashed line)

the displacement of the bonds A and AA (see Fig. 4c) for the SB model and for the PWL force law with β corresponding to a minimum in the s vs $1/\beta$ plot, and we see that there are no residual oscillations behind the defect.

The same analysis was also made for a "dangling bond" force law (DB) [27] (see Fig. 4b) as an approximation to interatomic forces that are negligibly small outside a small region from the atomic core. The results are similar to those in the PWL force law. In this way we can bracket a reasonable range of interatomic force laws with the DB and PWL force laws. The PWL method can also be extended to fit any desired potential with a linearized force law containing an arbitrary number of segments, since the computational effort involved increases slowly with the number of segments used.

We also examined the width of the dislocation core as the force law shape was modified. The core width (2z) increases almost linearly with β^{-1}. The slope depends on the velocity and the anisotropy constant K (for $v=0.5$: $K=1$, $2z \approx 1/(2\beta)$; $K=1/3$, $2z=1/(3\beta)$). Thus the s vs $1/\beta$ plot (Fig. 5) can also be interpreted as an s vs z plot. The minimum shear stress required to advance a straight dislocation in a crystal has been examined by several authors [30,34] with respect to the associated width w of the dislocation. In general one sees an exponential decrease for the strain-width relationship $s \alpha \exp(-\lambda w)$ although λ depends on the particular model. Discrete models also show a superimposed non-monotonic decrease in the strain-width relationship for both short range [35] and long range [36] interatomic forces.

Most of the discussion above was at a fixed velocity. But as seen in Fig. 7 for a fixed interatomic force law there are preferential velocities for dislocation propagation. Motion at velocities slightly higher than the one corresponding to the minimum requires a significantly higher stress.

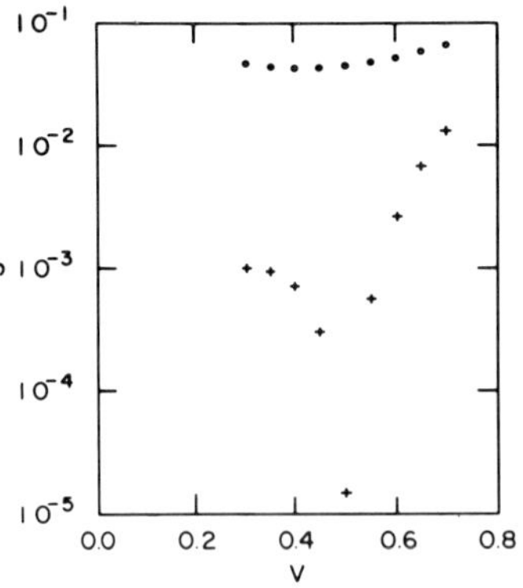

Fig. 7 The external strain s in an anisotropic crystal with $K = 1/3$ as a function of the dislocation velocity v. The circles (0 0 0 0) are for $\beta = .500$ ($1/\beta = 2.00$ and the crosses (+ + + +) for $\beta = .255$ ($1/\beta = 3.92$)

4. Dislocation-Phonon Interaction

Various phonon mechanisms of dislocation damping have been studied: thermoelastic losses [37,38], phonon viscosity [39], phonon wind [40,41] and flutter effect [42]. A review on this subject gives a critical discussion of calculations based on these different mechanisms [43]. Calculations of phonon scattering from the dislocation in a continuum approach taking into account the phonon relaxation which includes as special cases the effects of the phonon wind, phonon viscosity and thermoelastic losses have been performed [44,45] and it was concluded that at low temperatures relaxation processes can be neglected as compared to scattering processes. At high velocities, however, the scattering mechanism must be included along with the emission of phonons (since this is the dominant) necessitating thus the study of a discrete model.

In the framework of perturbation theory FLYTZANIS and CELLI [46] studied the propagation of a dislocation in a crystal at low temperature under the action of a constant applied stress to balance the frictional stress due to the emission and scattering of phonons, which causes a nonuniformity in the motion. It was found that the response of the dislocation to the scattered waves depends on the difference of $|v_p - v|$ between the phase velocity v_p of the waves and the velocity of the dislocation. If $|v_p - v|/v \leq 0.1$ the response of the dislocation can be treated as a perturbation and the frictional stress consists of two terms: the emission term modified by a Debye-Waller factor and a term proportional to the temperature, due to the scattering of the phonons which act as a damping mechanism. For phonons that are phase matched i.e. with $v_p = v$, however, the response is divergent in the perturbation approach and must be considered in a fully nonlinear treatment [47]. The set of phase-matched phonons is identical with the set that is radiated by the dislocation, carrying away energy in the wake behind the dislocation. An analysis of the contributions to this radiative field is given in [19]. The predominant phonon field inside the wake for v = 0.50 and K=1/3 has wave number $(k_x, k_y) = (4.175, \pi)$, which will be used for the remainder of this discussion. These phonons can inhibit or help the dislocation propagation depending on their phase. This could be particularly important near the minima found in the previous section for the applied strain ($s \sim 10^{-5}$) which could be supplied by lattice waves having an amplitude $\sim$ s, obtaining thus a dislocation phonon complex. The problem is treated along the lines of section 3, except that now the displacement across the slip plane D also includes that due to the lattice waves (ΔP),

$$\Delta P=-4|A_k| \sin(\mu k_x+\phi_k) \sin(k_y/2) \tag{4.1}$$

where $|A_k|$ and ϕ_k are the amplitude and phase of the lattice wave (k_x,k_y). Then we must solve the consistency conditions for h(u), which includes implicity, but in a correct way, the nonlinear interaction between the wave and the dislocation. Some sample calculations are presented for $1/\beta=3.33$($s=0.0052$ at $|A_k|=0$). Fig. 8 shows that for fixed phase, the external strain s depends

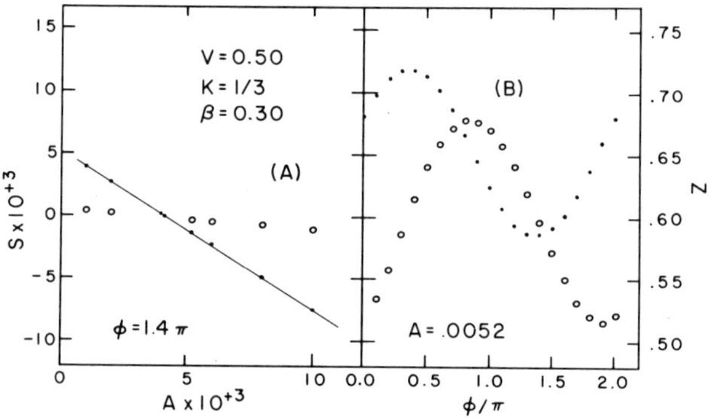

Fig. 8 Plot of the strain s and the half-width z vs phonon amplitude (with fixed phase) and vs phase (for fixed amplitude)

linearly on the phonon amplitude and for fixed amplitude the strain has a roughly sinusoidal dependence on ϕ and can be driven to zero if $|A_K|$ is large enough. A similar dependence is seen for the core half-width z. Table 1 shows that the influence of the phonon as measured by $\Delta s/2$ and $\Delta z/2$ is strongly affected by the parameter β [$\Delta s(\Delta z)$ is the maximum difference in strain (half-width) values generated by varying ϕ from 0 to 2π]. A phonon of fixed amplitude $|A_K|$ = 0.001 can change the strain by $\Delta s/2=127.47 \times 10^{-5}$ for β =.300, but by only $\Delta s/2=8.18 \times 10^{-5}$ for β = .255, corresponding to the first minimum in the solid curve (K = 1/3) of Fig. 5. In this preliminary work, considerations are limited to the single dominant phonon, but the method is easily extended to any superposition of phonons of various amplitudes, phases and wave vectors (as long as $\omega = vk_x$). The combination of perturbative and non linear methods described leave only the treatment of near resonance phonons to provide a complete picture of the dislocation-phonon interaction.

Table 1 Effect of phonon on external strain

$1/\beta$	$S\times10^5$ (no phonon)	z (no phonon)	$\frac{\Delta s}{2} \times 10^5$	$\frac{\Delta z}{2}$
2.00	4494.7	.0000	400.00	.0000
3.33	520.11	.60425	127.47	.0159
3.77	27.55	.83540	22.59	.0069
3.85	7.23	.8814	7.53	.0026
3.88	3.19	.9007	3.07	.0011
3.92	1.47	.9299	8.18	.0030
4.00	7.78	.9793	21.00	.0082
4.55	91.32	1.2517	54.61	.0245
5.00	51.99	1.4430	36.57	.0237
5.88	11.48	1.8954	20.93	.0270
6.35	24.63	2.1008	23.07	.0363

5. Discussion

Linear continuum elasticity predicts no radiation loss for a velocity $v<1$ (in units of the speed of sound), but this is simply because all effects due to the discreteness of the medium (including the Peierls stress) are absent in this theory. It is not surprising therefore, that attempts by ESHELBY [48] using the Peierls-Nabarro model with the inclusion of dispersion predicted the effect is small. The Peierls-Nabarro model allows for the periodic nature of the force across the slip plane that joins two elastic half-spaces held together by non-Hookean forces. When the dislocation width is small the defect behaves like a particle whose potential energy varies periodically. Calculations to evaluate the damping due to the non-uniformity of the motion [49,50] are also based on continuum elasticity. The relevant conclusion is that for velocities $v>0.1$ this radiation is small.

The radiation losses of a moving dislocation in a perfect crystal at zero temperature are due to the emission of phonons resulting from the rearrangement of atoms in the core of the dislocation. We find that the radiation losses depend very strongly on the velocity (for $v < 0.5$) and on the interatomic force law. We can conclude that for a wide range of interatomic forces there are velocities where nearly stress-free modes of motion can occur which are stable and will be preferentially selected by the dislocation, if high enough velocities can be attained. The one-dimensional Frenkel-Kontorova model gives rise to the Sine-Gordon equation, if displacement differences are replaced by derivatives; this equation has "soliton" solutions [51,52] that represent a dislocation in uniform, loss-free motion at zero stress for any $v < 1$. Keeping finite differences, EARMME and WEINER [53] discovered that loss-free motion can still occur in a modified Frenkel-Kontorova model. Wake-free motion is also observed in the computer simulations of HOOVER et al. [54] for the motion of a dislocation in a 2-dimensional triangular lattice.

Even the single phonon considered has a pronounced non-linear effect on the resulting motion. The results presented indicate that a phonon can have a "lubricating" effect, carrying along the dislocation in the absence of an applied stress or even against the direction of motion favored by such a stress. (In such a case, the moving DPC absorbs energy from the phonon field, leaving behind an "antiwake".) Such results might be expected on simple intuitive grounds, but have never been shown to follow in a self-consistent way from a lattice model. One can describe the DPC as a type of elementary excitation of the non-linear, discrete medium, analogous to soliton solutions in non-linear continuous media. Recent computer simulations show that temperature can have an enhancing effect on the mobility of a dislocation segment, perhaps due to the phenomenon of thermal energy trapping by the moving dislocation. [55]

It should be pointed out that there is a characteristic difference of the damping mechanisms in the low velocity and high velocity regions of dislocation motion. In general it is not easy to demonstrate the relative importance of a given mechanism and in most cases more than one mechanism might be contributing to the resistance to uniform motion. For a real crystal one must use the macroscopic description of kinetic rate theory with phenomenological parameters obtained from microscopic model calculations. Whether the high velocity motion studied here occurs will ultimately depend on the structure of a real crystal that can favor activation of highly correlated slip processes or complex dislocation interactions due to cross slip. The materials to look for high velocity motion are the high yield alloys [9] . To understand these experiments, however, we must study the interaction of dislocations in moving arrays and the influence on their motion of substitutional impurities. The former has been studied extensively in continuum models but some discrete models have also been considered [56]. In [56] it was found that it is much easier to move two dislocations because one can ride on the wake of the leading dislocation. Several studies of the impurity damping have been based on the string model, which considers the dislocation line as a flexible string being pinned at the impurity sites. This model has met both successes [57] and failures [58] in the low velocity region. If the rate of unpinning, however, is high one must consider the dynamic interaction of the defect with the impurity [59,60].

References

1. W.T. Read Jr., Dislocations in Crystals (McGraw-Hill, New York, 1953).
2. W.G. Johnston and J.J. Gilman, J. Appl. Phys. 30, 129 (1959).
3. A.A. Maradudin, E.W. Montroll and G.H. Weiss, Solid State Physics Supplement 3, Theory of Lattice Dynamics in the Harmonic Approximation (Academic Press, New York 1963).
4. Z.S. Basinski, M.S. Duesbery and R. Taylor, Phil. Mag. 21, 1201 (1970).
5. V. Vitek, R.C. Perrin and D.K. Bowen, Phil. Mag. 21, 1049 (1970).
6. J.W. Christian and V. Vitek, Rep. Prog. Phys. 33, 307 (1970).
7. E. Yu. Gutmanans, E.M. Nadgornyi and A.V. Stepanov, Soviet Phys. Solid State 11, 3081 (1970).
8. H. Araki and T. Ninomiya, J. Phys. Soc. Japan 41, 1684 (1976).
9. R.B. Schwartz and J.W. Michell, Phys. Rev. B9, 3292 (1974).
10. K.M. Jassby and T. Vreeland, Jr., Acta Met. 20, 611 (1972).
11. A.V. Granato and K. Lücke, J. Appl. Phys. 27, 583 (1956).
12. A. Hikata, R.A. Johnson and C. Elbaum Phys. Rev. Letters 24, 215 (1970).
13. W.S. Farren and G.I. Taylor, Proc. Roy. Soc. 107, 422 (1925).
14. F.R.N. Nabarro, Theory of Crystal Dislocations (Clarendon Press, Oxford, 1964).
15. W. Atkinson and N. Cabrera, Phys. Rev. 138A, 763 (1965).
16. V. Celli, and N. Flytzanis, J. Appl. Phys. 41, 443 (1970).
17. S. Ishioka, J. Phys. Soc. Japan, 30, 232 (1971).
18. A.A. Maradudin, J. Phys. Chem. Solids 9, 1 (1958).
19. V. Celli, N. Flytzanis and S. Crowley, J. Phys. Chem. Solids 37, 1125 (1976).
20. Y.Y. Earmme and J.H. Weiner, Phys. Rev. Lett. 31, 1055 (1973).
21. S. Ishioka J. Phys. Soc. Japan, 34 462 (1973).
22. Y.Y. Earmme and J.H. Weiner, J. Appl. Phys. 45, 603 (1974).
23. J.H. Weiner and M. Pear, Phil. Mag. 31, 1055 (1975).
24. S. Weertman, Symp. Math. Theory Dislocation. American Society of Mechanical Engineers. Northwestern University (1969).
25. N. Flytzanis, S. Crowley, and V. Celli, J. Phys. Chem. Solids, 38, 539 (1976).
26. N. Flytzanis, S. Crowley and V. Celli, Phys. Rev. Lett. 39, 891 (1977).
27. S. Crowley, N. Flytzanis and V. Celli, J. Phys. Chem. Solids, to be published.
28. S. Ishioka, J. Phys. Chem. Solids 36, 427 (1935).

29. S. Boffi, G. Caglioti, G. Rizzi, and F. Rossitto, J. Appl. Phys., 45 3220 (1974).
30. A.J. Foreman, M.A. Jaswon, and J.K. Wood, Proc. Phys. Soc. (London) A64, 156 (1951).
31. A.H. Cottrell, Dislocations and Plastic Flow in Crystals (Oxford University Press, New York 1953) p. 64.
32. R. Hobart, Jour. Appl. Phys., 36, 1944 (1965).
33. V.L. Indenbom, Soviet Phys. - Cryst. 3, 193 (1958).
34. W.T.O. Sanders, Phys. Rev., 128, 1540 (1962).
35. S. Ishioka, J. Phys. Soc. Japan, 36, 187 (1974).
36. T. Kurosawa, J. Phys. Soc. Japan 13 153 (1958).
37. J.D. Eshelby, Proc. Roy. Soc., A197, 369 (1949).
38. J.H. Weiner, J. Appl. Phys. 29, 1305 (1958).
39. W.P. Mason, J. Appl. Phys. 35, 2779 (1964).
40. G. Leibfried, Zs. Phys. 127, 344 (1950).
41. J. Lothe, J. Appl. Phys. 33, 2116 (1962).
42. T. Ninomiya, J. Phys. Soc. Japan, 25,830 (1968).
43. V.I. Al'shitz and V.L. Indenbom Sov. Phys. Usp. 18, 1 (1975).
44. A.D. Brailsford, J. Appl. Phys. 43, 1380 (1972).
45. V.I. Al'shitz, A.G. Malshukov, ZETP 63, 1849 (1972).
46. N. Flytzanis and V. Celli, J. Appl. Phys. 43 3301 (1972).
47. S. Crowley, N. Flytzanis and V. Celli, submitted to J. Phys. Chem. Solids.
48. J.D. Eshelby, Proc. Phys.Soc. London A62, 307 (1949).
49. E.W. Hart, Phys. Rev. 98 1775 (1955).
50. V.I. Al'shitz, V.L. Indenbom and A.A. Shtol'berg JETP, 33, 1240 (1971).
51. A. Kochendorfer, A. Seeger and H. Donth Zeits. Phys., 127, 533 (1950); 130, 321 (1951); 134, 173 (1953).
52. A.C. Scott, Am. Journ. Phys. 37, 52 (1969).
53. Y.Y. Earmme and J.H. Weiner, Phys. Rev. Lett. 33, 1550 (1974).
54. W.G. Hoover, N.E. Hoover and W.C. Moss, Phys. Lett. 63A, 324 (1977).
55. J.H. Weiner, A. Hikata and C. Elbaum, Phys. Rev. 13B, 531 (1976).
56. N. Flytzanis and V. Celli, J. Appl. Phys. 45, 5176 (1974).
57. A.V. Granato, Phys. Rev. Lett. 27, 660 (1971).
58. A. Hikata, and C. Elbaum, Phys. Rev. B9 4529 (1974).
59. J. Takamura and T. Morimoto, J. Phys. Soc. Japan 18, Suppl. 1, 28 (1963).
60. A. Ookawa and K. Yazu, J. Phys. Soc. Japan 18, Suppl. 1, 36 (1963).

Grain Boundaries as Solitary Waves

Ralph J. Harrison and George H. Bishop, Jr.

Army Materials and Mechanics Research Center, Watertown, MA 02172, USA

and

Sidney Yip and Thomas Kwok

Nuclear Engineering Department, Massachusetts Institute of Technology
Cambridge, MA 02139, USA

In computer molecular dynamics "experiments" on high angle symmetric tilt grain boundaries we have discovered that these boundaries exhibit many features attributed to solitons. For example, the grain boundaries have a geometrically stable motion which is nonlinear in that it requires a finite temperature to activate and it persists over times long compared to other relaxation times. The analogy with dislocation motion naturally comes to mind since low angle grain boundaries may be regarded as rows of dislocations and there is extensive literature [1] on the soliton aspects of dislocations. If we had started out to look for soliton motion we might well have started out with low angle grain boundaries; however, we were originally interested in using dynamic simulation methods to obtain the thermodynamic properties of the metallurgically more important high angle grain boundaries [2]. We attempted to measure the excess boundary entropy by observing the change relative to the perfect crystal in the spectrum of the velocity autocorrelation function for atoms near the boundary. However, we found changes far from the boundary as well, and in tracking this down we discovered that the two crystals were undergoing a relative sliding motion which was coupled with migration of the boundary [3]. This is the motion we call soliton-like and we shall describe this motion in some detail in the following.

Figure 1 gives the arrangement of the two crystallographically distinct layers of atoms in our model bicrystal simulating a coincidence boundary in an fcc crystal. Coincidence boundaries are formed at those special angles of rotation such that a super lattice is formed by crystal lattice points common to a crystal lattice and the rotated lattice. The corresponding

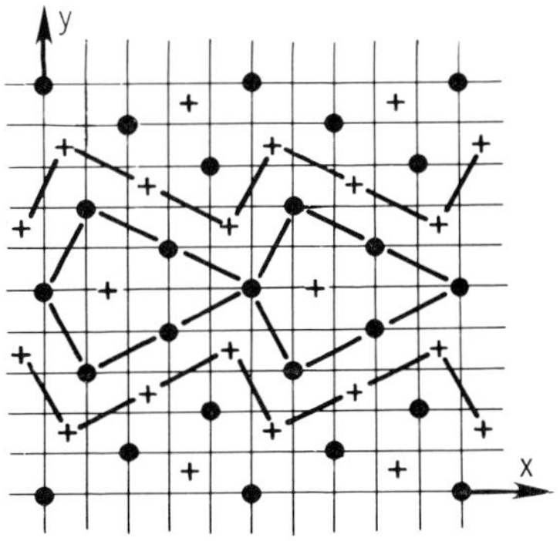

Fig. 1 Projection on a {100} plane of the {310} fcc twin grain boundary atomic sites in successive {100} layers are shown by dots and crosses. The line patterns drawn, the "kite" and zig-zag are mnemonics useful for recognition of the boundary structure

superlattice is the *coincidence site lattice* (CSL) [4]. The boundary is characterized by periodicity in the transverse x direction. Periodicity in the z direction is associated with the ABABA.. [001] stacking sequence of the fcc lattice. The grain boundary pattern of Fig. 1 is one which results from intuitive geometrical concepts supported by static energy minimization procedures [5]; it also resembles "bubble raft" models of boundaries. Fig. 2a represents essentially the same pattern except that circles are drawn to represent the atoms and the centers are at the actual positions obtained by the static relaxation method. In Fig. 2 one has an additional "pattern recognition" aid, namely the shapes of the interstices. Fig. 2b-d represent the time development of the motion after random velocities were imparted to each atom in Fig. 2a resulting in a simulated temperature of 0.2 energy units defined by the depth of the Lennard-Jones potential. Short time averaging has been done in order to eliminate some of the high frequency motions. One notes that individual atoms in the top crystal move to the right, while those in the bottom move to the left. In the course of this motion the boundary has moved downward, although individual atoms have not. That is, the two crystals have undergone *sliding* relative to each other, while the crystal has *migrated* downward. In this particular case the *coupled sliding and migration* has produced a migratory motion equal in magnitude to the sliding motion.

The fine grid drawn in Fig. 1 is the *DSC lattice* introduced by Bollman [6]. It is the coarsest lattice which contains both the crystal lattice and the rotated lattice as sublattices and it is relevant to the description of the periodic character of the observed grain boundary motions. When the two bicrystals have slid relative to one another by [100] DSC, the boundary has also migrated exactly [010] DSC. Fig. 2b shows the motion after about 1/2 [100] DSC as one may judge from the interstices near the boundary; if one would pinpoint the boundary it seems to be about half way between the two distorted kites. We have computed the static Peierls barrier to the boundary motion for Lennard-Jones potentials. The barrier exhibits the periodicity of the DSC lattice. It has a shallow minimum at the symmetric position, a deeper minimum at a displacement of 1/8 [100] DSC and a maximum at 1/2 [100] DSC. Apart from the exact details, the noteworthy features of the barrier are that the heights of the peaks and depths of the valleys are much less than for the barrier to shear in a perfect crystal because of the fact that the DSC lattice has a much finer grid than does the crystal lattice, so that the boundary atoms never have to move very far from their minimum energy positions. Another consideration is that there is some "free volume" at the boundary. The fact of thermal activation of boundary motion is consistent with the computed static Peierls barrier.

Figure 3 shows the time dependence in another computer run of the relative coordinates of the centers of mass of each crystal. The large sliding motion is in the x coordinate, whereas the y and z coordinates simply show oscillations associated with phonon-like modes of the computational cell. Other similar runs show: an incubation period before the boundary starts to move; the coupled sliding and migration mode with motion in a direction statistically chosen; the motion stops for a while; then starts again in the same or reverse direction. The velocity when moving is a few percent of the velocity of sound, with the temperature dependence not well established yet, although at very low temperatures there is no sliding or migration at all. The computational time steps shown in the figure are 0.015 using the dimensionless unit of time which is approximately the longitudinal speed of sound divided by the lattice parameter. Characteristic times for reestablishing "equilibrium" in computer molecular dynamics runs on single crystals after a parameter such

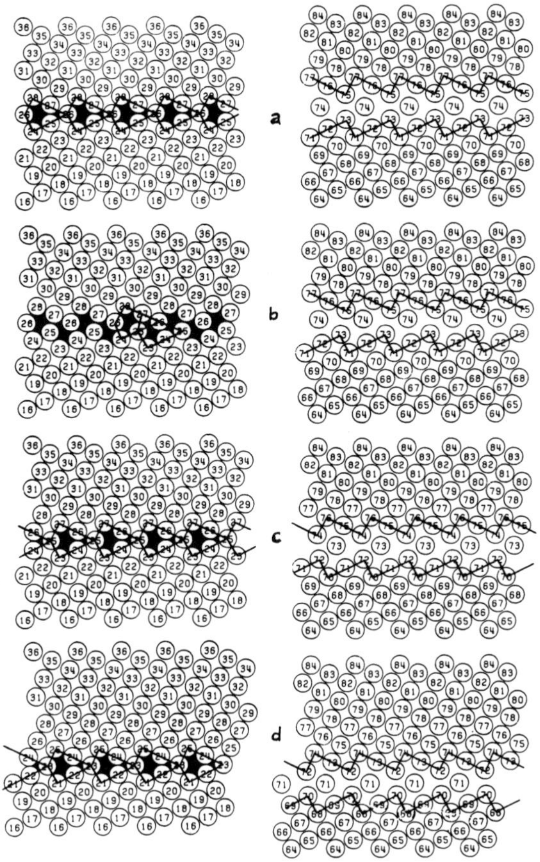

Fig. 2 Atom positions in the vicinity of the grain boundary. The two layers in <100> fcc stacking sequence are shown separately (left and right). (a) Starting structure. (b)-(d) Positions at selected later times

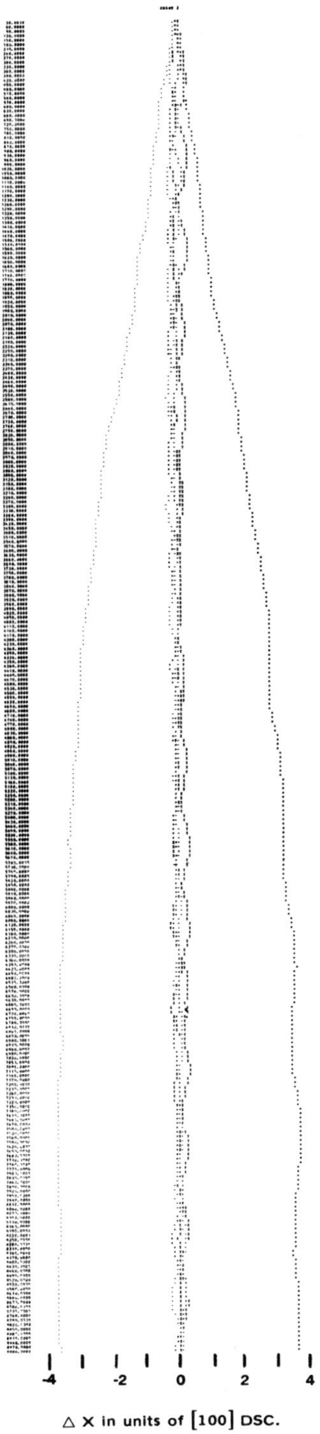

Fig. 3 The time variation in the x, y, and z coordinates of the center of mass of groups of 30 atoms in the center of each bicrystal. The sliding is revealed by the drift in the x coordinates. The y and z coordinates show ordinary thermal fluctuations

as volume or temperature is changed are typically several hundred computational time steps [7], i.e., much shorter than the typical time for persistence of the grain boundary motion. The boundary motions may be compared to Brownian motion [11], resembling also that of domain walls [12].

In addition to the boundary described above we have studied two-dimensional grain boundaries [8] which also exhibit thermally activated motions. We have also carried out some exploratory studies where boundary motion was initiated without random thermal motions simply by starting the crystals with relative sliding motions of various magnitudes. We may also mention that the motion we observed of the grain boundary between a pair of three-dimensional fcc crystals is crystallographically the same as in the formation of a deformation twin and similar to that in the martensitic transformation [9]. Special interfacial dislocations (e.g., the 1/6 [112] twinning dislocation) have been thought to be involved in effecting the associated lattice rotations locally. One would expect a martensitic interphase soliton to differ geometrically from the interphase soliton proposed by HOROVITZ et al. [10] for the bcc to ω phase transformation in Zr-Nb alloys. There are many new and fascinating phenomena to be explored by computer molecular dynamics techniques having relevance to soliton theory as well as to metallurgy.

Acknowledgment S.Y. and T.K. acknowledge support for part of their work by the United States Army Research Office under Grant No. DAA29-78-C-0006.

References

1. Some basic and some more recent papers in this area are: A. Kochendörfer and A. Seeger, Z. Phys., 127, 533 (1950); A. Seeger and A. Kochendörfer, ibid. *130*, 321 (1951); A. Seeger, H. Donth, and A. Kochendörfer, ibid. *134*, 173 (1953); N. Flytzanis, J. Crowley, and V. Celli, Phys. Rev. Levtt., *39*, 891 (1977); Y. Y. Earmme and J. H. Weiner, J. Appl. Phys., *48*, 3317 (1977).
2. R. J. Harrison, J. A. Cox, G. H. Bishop, Jr., and S. Yip, Nucl. Metal., *20*, 604 (1976); G. H. Bishop, Jr., G. A. Bruggeman, R. J. Harrison, J. A. Cox, and S. Yip, Nucl. Metal., *20*, 522 (1976).
3. G. H. Bishop, Jr., R. J. Harrison, T. Kwok, and S. Yip, Trans. Amer. Nucl. Soc., *27*, 323 (1977).
4. G. Friedel, *Lecons de Crystallographic*, Gauthier-Villars, Paris (1926); M. L. Kronberg and F. H. Wilson, Trans. AIME, *185*, 501 (1949).
5. R. J. Harrison, G. A. Bruggeman, and G. H. Bishop, Jr., *Grain Boundary Structure and Properties*, ed., G. A. Chadwick and D. A. Smith, p. 45, Academic Press, N. Y. and London (1976); G. A. Bruggeman, G. H. Bishop, Jr., J. A. Cox, and R. J. Harrison, Nucl. Metal., *20*, 450 (1976).
6. W. Bollman, *Crystal Defects and Crystal Interfaces*, Springer Verlag, New York, Heidelberg, Berlin (1970).
7. O. Deutsch, Ph.D. Thesis, Dept. of Nuclear Engineering, M.I.T. (1975).
8. T. Kwok, M.S. Thesis, Dept. of Nuclear Engineering, M.I.T. (1978).
9. S. Mahajan and D. F. Williams, Int. Metall. Rev., *18*, 43 (1973); J. W. Cahn, Act. Met., *25*, 721 and 1021 (1977).
10. B. Horovitz, J. L. Murray, and J. A. Krumhansl, Bull. Amer. Phys. Soc., *23*, 274 (1978).
11. R. J. Harrison, G. H. Bishop, Jr., S. Yip, and T. Kwok, Bull. Amer. Phys. Soc., *23*, 253 (1978).
12. T. R. Koehler, A. R. Bishop, J. A. Krumhansl, and J. R. Schrieffer, Solid State Comm., *17*, 1515 (1975).

The Relation of Solitons to Polaritons in Coupled Systems

D.F. Nelson

Bell Laboratories, Murray Hill, NJ 07974, USA

Soliton solutions in acoustic modes [1] and in optic modes [1] of solids have been studied in the past. Envelope soliton solutions of coupled exciton-electromagnetic mode systems [2] and coupled two-level atom-electromagnetic mode systems [1] have also been studied. In this paper we wish to present soliton solutions for coupled optic mode-electromagnetic mode systems (for which linear sinusoidal excitations are called *polaritons*) and for coupled optic mode-acoustic mode systems. Since the mathematical derivations are similar in the two cases, we will present only the former one in detail.

The physical origin of the soliton solution in either system can be described as follows. Consider an optic mode whose potential energy consists of a term quadratic in the optic mode amplitude and a term quartic in the amplitude. We assume each term produces a restoring force though the restoring force from the quadratic term is regarded as weak. In other words, we are considering a soft mode above the phase transition temperature. If the transition is a ferroelectric-paraelectric phase transition, we will consider an electromagnetic wave coupled to the optic mode; if the transition is a ferroelastic-paraelastic phase transition, we will consider an acoustic wave coupled to the optic mode. In each case the coupling term to the traveling wave can cause the quadratic term to reverse its sign and so become non-restoring. Thus during the interaction the effective potential energy of the optic mode has two minima. As is well known, such a double-well potential energy allows soliton solutions.

Optic Mode-Electromagnetic Mode Solitons

We model the interaction of an optic mode with an electromagnetic wave by the electric field wave equation for the scalar amplitude E of an electric field eigenmode,

$$\frac{\partial^2 E}{\partial z^2} - \frac{\kappa_h}{c^2}\frac{\partial^2 E}{\partial t^2} = \frac{q}{\epsilon_o c^2}\frac{\partial^2 y}{\partial t^2} \quad , \tag{1}$$

and by the force equation for the scalar normal coordinate y of an optic mode,

$$m\partial^2 y/\partial t^2 = -A_{20}y - A_{40}y^3 + qE \quad . \tag{2}$$

Here κ_h is the dielectric constant at frequencies high compared to the resonant frequency of the optic mode, q and m are the charge and mass densities associated with the optic mode, and A_{20} and A_{40} are constants characterizing the linear and nonlinear restoring forces of the optic mode. We assume A_{20} and A_{40} are both positive; hence the potential energy of the optic mode has a single minimum at $y = 0$ and the unperturbed state of the crystal has $y = 0$ and $E = 0$.

We now search for a stationary pulse solution by transforming to a traveling coordinate system characterized by $\xi = z - vt$. The wave equation can then be integrated twice to yield

$$E = v^2 qy/\epsilon_o(c^2 - \kappa_h v^2) \tag{3}$$

since the integration constants must be taken as zero to satisfy the boundary conditions,

$$E = y = \partial E/\partial\xi = \partial y/\partial\xi = 0 \quad (\xi = \pm\infty) \quad . \tag{4}$$

The optic mode equation then becomes

$$mv^2\partial^2y/\partial\xi^2 = -A'_{20}y - A_{40}y^3 \tag{5}$$

where

$$A'_{20} \equiv A_{20}v_h^2(v_l^2-v^2)/v_l^2(v_h^2-v^2) \tag{6}$$

$$v_h^2 \equiv c^2/\kappa_h\ ,\quad v_l^2 \equiv c^2/\kappa_l\ , \tag{7}$$

$$\kappa_l \equiv \kappa_h + q^2/\epsilon_o A_{20}\ . \tag{8}$$

Here κ_l is the dielectric constant at frequencies low compared to the optic mode resonance.

For soliton solutions of (5) to exist it is necessary for

$$A'_{20} < 0 \tag{9}$$

which requires

$$v_l < v < v_h\ . \tag{10}$$

This allowed range of soliton velocities coincides with the forbidden range of polariton phase velocities.

Eq. (5) can now be integrated twice with the aid of (4) and (9) to obtain the soliton solution

$$y = y_o \operatorname{sech} k(\xi-\xi_o)\ , \tag{11}$$

$$E = \frac{qy_o}{\epsilon_o(v^{-2}-v_h^{-2})} \operatorname{sech} k(\xi-\xi_o) \tag{12}$$

with the optic mode amplitude y_o given by

$$y_o \equiv Akv/\pi\ , \tag{13}$$

the *pseudo-wavenumber* k given by

$$k = (\omega_L/v)(v^2-v_l^2)^{1/2}/(v_h^2-v^2)^{1/2}\ , \tag{14}$$

and ξ_o being an integration constant. Here A is an area constant,

$$A \equiv \int_{-\infty}^{+\infty} y\,dt = \pi(2m/A_{40})^{1/2}\ , \tag{15}$$

depending on the nonlinearity parameter A_{40} and ω_L and ω_T are the longitudinal and transverse optic frequencies

$$\left\{\omega_L/\omega_T\right\}^2 = (v_h/v_l)^2 = \kappa_l/\kappa_h\ , \tag{16}$$

$$\omega_T \equiv (A_{20}/m)^{1/2}\ . \tag{17}$$

Eq. (16) is the Lyddane-Sachs-Teller relation. Ordinarily the velocities v_h and v_l are regarded as limiting velocities measured on the high and low frequency sides of the optic mode resonances which produces a transmission stop-band for frequencies ω such that $\omega_T < \omega < \omega_L$. Examination of the soliton solution, (11)-(14), reveals an alternate interpretation of the LST relation, namely, that v_h is the velocity of the largest amplitude, narrowest width soliton and that v_l is the velocity of the lowest amplitude, broadest width soliton.

Eq. (14) is the dispersion relation of these solitons; it relates the pseudo-wavenumber k and the velocity v. Surprisingly we find that this relation depends only on linear properties of the medium. *Thus the dispersion relation of an inherently nonlinear wave contains no nonlinear properties.*

The dispersion relation can be put into a more suggestive form if we define a *pseudo-frequency* by $\omega \equiv kv$. Eq. (14) can then be rearranged into

$$k^2 = (\omega/v_h)^2(\omega_L^2+\omega^2)/(\omega_T^2+\omega^2)\ . \tag{18}$$

This should be compared to the dispersion relation for single frequency plane wave solutions of (1) and (2) (*polaritons*) in the absence of nonlinearity ($A_{40} = 0$) which is

$$K_p^2 = (\dot{\Omega}_p / v_h)^2 (\omega_L^2 - \Omega_p^2)/(\omega_T^2 - \Omega_p^2) \tag{19}$$

where the wavenumber K_p and frequency Ω_p of the polariton have their usual meaning [$E \sim expi(K_p z - \Omega_p t)$]. It can be seen that the replacement $K_p \rightarrow ik$, $\Omega_p \rightarrow i\omega$ in the polariton dispersion relation yields the soliton dispersion relation. Thus, if K_p and Ω_p in (19) are regarded as complex, *the soliton dispersion relation corresponds to the imaginary wavenumber, imaginary frequency solutions of the polariton dispersion relation.* The figure illustrates the soliton and polariton dispersion relations.

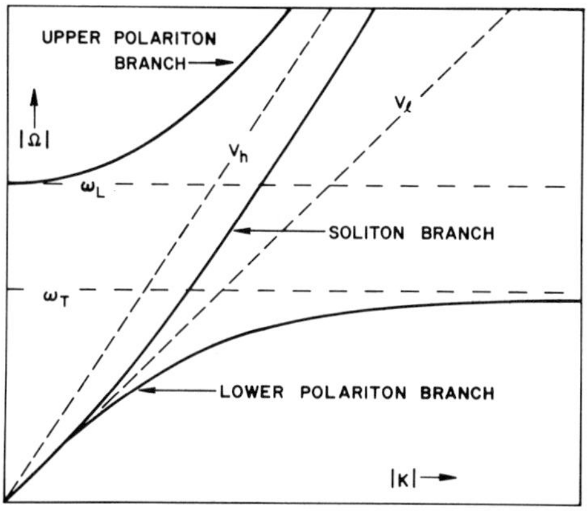

Plot of absolute values of complex wavenumber and complex frequency. For the polariton branches $|K| = K_p$, $|\Omega| = \Omega_p$ with K_p, Ω_p governed by (19); for the soliton branches $|K| = k$, $|\Omega| = \omega$ with k, ω governed by (18)

Optic Mode-Acoustic Mode Solitons

The interaction of an optic mode and an acoustic mode can be modeled by the acoustic wave equation,

$$\rho \partial^2 u/\partial t^2 = A_{02} \partial^2 u/\partial X^2 + A_{11} \partial y/\partial X \ , \tag{20}$$

and the optic mode force equation,

$$m \partial^2 y/\partial t^2 \equiv -A_{20} y - A_{40} y^3 - A_{11} \partial u/\partial X \ . \tag{21}$$

In these equations u is the scalar amplitude of an acoustic eigenmode, X is the material coordinate in the direction of propagation, ρ is the mass density of the medium, A_{02} is the elastic stiffness from all sources except the optic mode normal coordinate y, A_{11} is the coupling constant between y and the displacement gradient $\partial u/\partial X$, and all other quantities have the same meanings as in the last section. Piezoelectric coupling to an electric field E could be added to each of these equations along with the addition of the divergence of electric displacement equation. Since these additions make no qualitatively change in the conclusions, we omit them for simplicity.

By manipulations analogous to those in the last section the soliton solution of these equations is found to be

$$y = y_o \operatorname{sech} k (\xi - \xi_o) \tag{22}$$

$$\frac{\partial u}{\partial X} = \frac{A_{11} y_o}{\rho (v^2 - v_h^2)} \operatorname{sech} k (\xi - \xi_o) \tag{23}$$

with $\xi \equiv X - vt$ and (10), (13)-(16) (with the second equality in (16) dropped) holding once again. The definitions in (7) must be replaced by

$$v_h^2 \equiv A_{02}/\rho \ , \quad v_l^2 = (A_{02} - A_{11}^2/A_{20})/\rho \tag{24}$$

and the definition in (17) must be replaced by

$$\omega_L \equiv (A_{20}/m)^{1/2} \ . \tag{25}$$

The replacement of (17) by (25) is initially very puzzling because (a) $(A_{20}/m)^{1/2}$ was called the transverse optic frequency in the last section since it gave a pole in the dielectric constant and (b) here $(A_{20}/m)^{1/2}$ is called the longitudinal optic frequency even though it gives a pole in the elastic stiffness. The resolution of this follows from the fact that the elastic compliance (which is the inverse of the elastic stiffness) is analogous to the dielectric constant since the linear wave velocity is proportional to the inverse square root of each of these quantities in the respective cases. Hence (25) is the frequency at which the compliance is zero and so is rightfully identified as a longitudinal optic frequency.

All remarks concerning the soliton dispersion relation in the last section apply also to the coupled optic mode-acoustic mode system here.

References

1. See references contained in A. C. Scott, F. Y. F. Chu and D. W. McLaughlin: Proc. IEEE *61*, 1443-1483 (1973)
2. S. A. Moskalenko, V. A. Sinyak and P. I. Khadzhi: Sov. J. Quant. Electron. *6*, 464-465 (1976)

Solitons in $CsNiF_3$: Their Experimental Evidence and Their Thermodynamics

M. Steiner

Hahn-Meitner-Institut, Glienicker Str. 100
D-1000 Berlin 39, Fed. Rep. of Germany, and

J.K. Kjems

Research Establishment Risø, DK-4000 Roskilde, Denmark

Abstract

An inelastic neutron scattering study of the 1-D-ferromagnet $CsNiF_3$ in a magnetic field yields strong evidence for the existence of thermally activated solitons for 14 K $\geq T \geq$ 5 K and 2 kG $\leq H \leq$ 36 kG. Comparing the experimental data with a classical continuum-soliton theory developed by Mikeska for this particular system very good agreement was found, e.g. the activation energy was found experimentally as 28 K and theoretically as 34 K.

1. Introduction

In the last few years it has been shown, that soliton or kink solutions of the equations of motions of the constituents of a system undergoing a phase transition might be of importance [1,2,3,4,5]. Most of these theoretical treatments have been done for onedimensional systems since exact solutions exist in this case for the nonlinear equation of motions, the Sine-Gordon equation for example. It is possible to show that the dynamics of a 1-D-ferromagnet with planar anisotropy can be described by the Sine-Gordon equation in the classical continuum limit if a magnetic field is applied perpendicular to the chain direction (Mikeska 1978 [1]).

In a series of papers it has been shown that $CsNiF_3$ is an ideal model system to study the predicted solitons [6]. Its magnetic properties are determined by magnetic chains of Ni^{2+} running along the hexagonal c-axis, the intrachain interaction being a hundred times the interchain interaction. The magnetic moments are kept in the a-b-plane by a strong single site anisotropy. There is no anisotropy measurable in the a-b-plane. The following Hamiltonian is appropriate for $CsNiF_3$ in the temperature range $T \gtrsim 3$ K, T_N being 2.7 K:

$$\mathcal{H} = -2J \sum_i \vec{S}_i \cdot \vec{S}_{i+1} + A \sum_i (S_i^z)^2 - g\mu_B H^x \sum_i S_i^x \tag{1}$$

with the following constants as determined by inelastic neutron scattering: $J/k = 11.8$ K; $A/k = 9$ K and $S = 1$ [7].

2. Experiments

We studied a single crystal of about 1 cm^3 on a cold source - triple axis spectrometer at the DR3-reactor in Risø for 3.1 K $\leq$ T $\leq$ 14 K and 0 $\leq$ H $\leq$ 36 kG. In order to reduce incoherent background the crystal contained only the ^{58}Ni-isotrope. Measurements were performed in constant Q-scans around the (0,0,2) reciprocal lattice point especially at (0,0,1.9) or at 0.1 r.l.u. In order to extract the magnetic contribution one has to subtract the still existent incoherent background. The determination of this background is crucial. We used as background a measurement at 3.1 K and 36 kG at (0,0,1.9), under which conditions no solitons should exist. This background measurement was then subtracted from the different measurements at constant H = 5 kG for 3.1 K $\leq$ T $\leq$ 14 K and at constant T = 9.3 K and 0 $\leq$ H $\leq$ 20 kG. The remaining spectrum consists of three peaks: two spinwave peaks (creation and anihilation of spin waves) at $\pm$ 0.55 meV and a peak centered around zero energy. In order to extract the reliable parameters we tried to describe the experimental spectra by a lineshape of the following form: Two spinwave peaks as a convolution of the resolution and a Lorentzian line centered around the spinwave energy and in addition a Gaussian line centered around zero energy. The parameters like widths, position and intensity of the peaks were determined by a least squares fitting. Some of the results for H = 5 kG and T = 6.3 K and 10.1 K are shown in Fig.1 demonstrating the good agreement between the theoretical lineshape and the experimental results. It can be seen as well how strongly the peak around zero energy increases with increasing temperature.

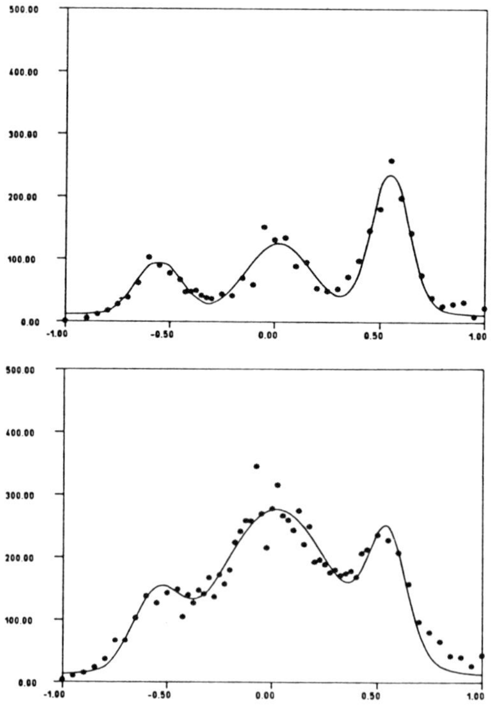

Fig.1 Experimental spectra after background subtraction for H = 5 kG, q = 0.1 r.l.u., T = 6.3 K (upper half) and T = 10.1 K (lower half). The abszissa and ordinates are given in units of meV and counts per 25 min respectively

3. Discussion

Since we want to check the soliton theory of Mikeska, we concentrate on the temperature and field dependence of the peak around zero energy. The neutron cross-section for scattering from solitons is given by Mikeska as

$$S(Q,\omega) \sim \frac{\beta e^{-8\beta m}}{2cq}\left(\frac{16}{\pi} e^{-\frac{2\beta m\omega^2}{c^2 q^2}} \frac{\pi q/2m}{\sinh \pi q/2m}\right)^2 \qquad (2)$$

with $m = (g\mu_B H/2J)^{1/2}$, $c = S\,(4AJ)^{1/2}$ and $g = 2.4$, $S = 1$, $J/k = 11.8$ K, $A/k = 4.5$ K (since the theory is a classical one).

This formula is analogous to scattering for neutrons from a gas of real particles with a thermal velocity distribution and does not describe the excitation of a single soliton. Thus the solitons show up in a neutron scattering experiment as a gas of noninteracting quasiparticles moving along the ferromagnetic chain with a velocity $v \lesseqgtr c$, (c being the slope of the spinwave dispersion). We see from formula (2) that the intensity is governed by the activation energy 8 m and the width for the peak can be calculated without adjustable parameter. The last term reflects the spatial extension of the solitons. Generally, all our results agree at least qualitatively with Mikeska's result: intensity J versus temperature at fixed field (H = 5 kG), J versus field at field temperature (9.3 K), halfwidth versus temperature and field. We want to present here the J(T)-results only , the other results will be published in another paper shortly. The intensity is simply given by $\sqrt{\beta}\ \exp^{-8m\beta}$ for H = 5 kG, q = 0.1 r.l.u.. Thus by plotting ln $(J\sqrt{T})$ versus 1/T one should find a straight line with slope 8 m. In Fig.2 the experimental results are shown together with a line representing the theoretical slope (full line). The best line through the experimental points is given by the broken line. It is obvious that the experiments show the expected exponential temperature dependence for $T \gtrsim 5$ K but the activation enery is slightly less then the theoretical value as calculated from the known parameters:

$$8\,m_{exp} = 27\ \text{K};\ 8\,m_{theor} = 34\ \text{K}$$

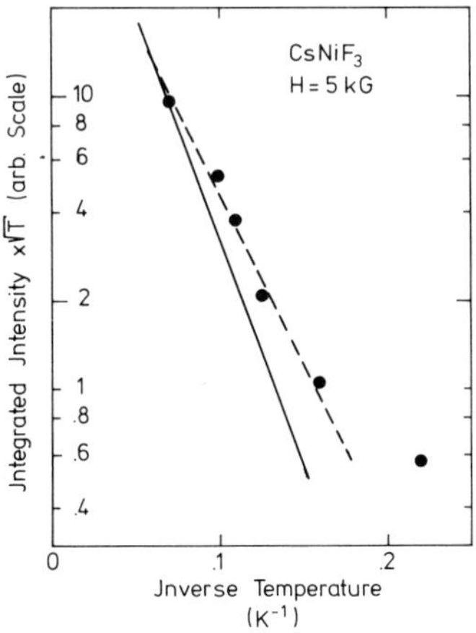

Fig.2 Semilog-plot of the integrated intensity times $\sqrt{T}$ versus 1/T. Full line: theory (scaled at 14 K); broken line: best straight line through experimental points

We thus conclude, that solitons are an important part of the dynamics in $CsNiF_3$ and can be regarded as noninteracting quasiparticles which are thermally activated. We were able to show that the thermodynamics of the solitons are as predicted. These first direct experimental studies of solitons hopefully stimulate theoretical treatments which look for the discrepancies we found between the theory and the experimental results.

References

1. H.J. Mikeska: J. Phys. C 11, L 29 (1978)
2. J.A. Krummhansl, J.R. Schrieffer: Phys. Rev. B 11, 3535 (1975)
3. N. Gupta, B. Sutherland: Phys. Rev. A 14, 1790 (1976)
4. R.A. Guyer, M.D. Miller: Phys. Rev. A 17, 1205 (1978)
5. J.F. Currie, M.B. Fogl, F.L. Palmer: Phys. Rev. A 16, 796 (1977)
6. M. Steiner, J. Villain, C. Windsor: Adv. Phys. 25, 87 (1976) and references therein
7. M. Steiner, J.K. Kjems: J. Phys. C 10, 2665 (1977)

Structure and Stability of Domain Walls – Phase Transition

J. Lajzerowicz and J.J. Niez*

Laboratoire de Spectrométrie Physique, B.P. 53
F-38041 Grenoble Cédex, France

*also with

Département de Récherche Fondamentale,
Centre d'Etudes Nucléaires de Grenoble, B.P. 85 X
F-38041 Grenoble Cédex, France

1. Introduction

We have shown [1] that a ferromagnetic domain wall can have two configurations, one corresponding to an Ising domain wall (for large uniaxial anisotropy) and the other corresponding to a Bloch type domain wall. This change of configuration is a two dimensional phase transition, the order parameter being the chirality. In this paper, we want to study the structure and stability of a ferroelectric domain wall when two types of order are competitive, for example Ferroelectric and Antiferroelectric phase transitions.

2. The Model

One uses a Landau Free Energy [2,3] to describe the occurence of Ferroelectricity and Antiferroelectricity. Let us call P_1 and P_2 the polarizations of the two sublattices and take an energy of the form :

$$\mathcal{F} = \frac{c}{2}\left[(\nabla P_1)^2 + (\nabla P_2)^2\right] + \frac{a}{2}(P_1^2 + P_2^2) + b P_1 P_2 + \frac{d}{4}(P_1^2 + P_2^2)^2 \quad , \quad d > 0 \qquad (1)$$

Let us introduce two new variables, the Ferroelectric and Antiferroelectric ones :

$$F = \frac{1}{\sqrt{2}}(P_1 + P_2) \quad ; \quad A = \frac{1}{\sqrt{2}}(P_1 - P_2) \qquad (2)$$

The new form of the Free Energy is :

$$\mathcal{F} = \frac{c}{2}\left[(\nabla F)^2 + (\nabla A)^2\right] + \frac{a+b}{2} F^2 + \frac{a-b}{2} A^2 + \frac{d}{4}(F^2 + A^2)^2 \qquad (3)$$

The phase diagram is given in fig. 1 where in the ferroelectric phase $A = 0$; $F = \pm\sqrt{-(a+b)/d}$ and where in the antiferroelectric one $F = 0$; $A = \pm\sqrt{b-a/d}$.

So if b varies we have a first order phase transition from a ferroelectric to an antiferroelectric at b = 0. Such a situation can be found in a crystal like $Pb(Zr_{1-x}Ti_x)O_3$ [2].

3. Structure of a Domain Wall

We can study it in the ferro or in the antiferroelectric phase but the situations are symmetric so we will study the ferroelectric domain wall. The EULER-LAGRANGE equations are :

$$\begin{aligned} c\,\nabla^2 F &- (a+b)F - d\,(F^2+A^2)\,F = 0 \\ c\,\nabla^2 A &+ (b-a)A - d\,(F^2+A^2)\,A = 0 \end{aligned} \qquad (4)$$

If A = 0, we have a ferroelectric domain wall

$$F_0(x) = \sqrt{-\frac{a+b}{d}}\ \tanh x\sqrt{-\frac{a+b}{2c}} \qquad (5)$$

4. Stability of the Domain Wall

We study small excitations around the static solution (5). If we associate an effective density for the ferroelectric and the antiferroelectric components, ρ_f and ρ_a (for simplicity we take $\rho_f = \rho_a = \rho$) and if we write

$$\begin{aligned} F &= F_0(x) + F_1(x,t) \\ A &= A_1(x,t) \end{aligned}$$

We have in first order

$$\begin{aligned} \rho\omega^2 F_1 &= -c\,\nabla^2 F_1 + (a+b)F_1 - 3(a+b)\,F_1 \tanh^2 x\sqrt{-\frac{a+b}{2c}} \\ \rho\omega^2 A_1 &= -c\,\nabla^2 A_1 + (a-b)A_1 - (a+b)\,A_1 \tanh^2 x\sqrt{-\frac{a+b}{2c}} \end{aligned} \qquad (6)$$

It is easy to find the spectra of these Shrodinger like equations. The system will be stable if the eigenvalues of these equations are positive. For the first one, they are all positive, for the second one we can easily write the lowest eigenvalue :

$$\rho\omega_0^2 = \frac{a-3b}{2} \qquad (7)$$

So the domain wall is stable against antiferroelectric fluctuations if a > 3b and is unstable against such fluctuations if 3 b > a. For ω_0, we have a soft mode behaviour near the threshold 3 b = a, therefore the domain wall changes its configuration in a cooperative way.

5. Structure of the Domain Wall in the New Configuration

We know an exact solution of the system (4) including non linear terms [4]. We have two cases

– α) $3b - a < 0$

$$F = \sqrt{-\frac{a+b}{d}} \tanh x \sqrt{-\frac{a+b}{2c}}$$

$$A = 0$$

– β) $3b - a > 0$

$$F = \sqrt{-\frac{a+b}{d}} \tanh x \sqrt{-\frac{2b}{c}}$$

$$A = \sqrt{\frac{3b-a}{d}} \; 1/ch \, x \sqrt{-\frac{2b}{c}}$$

We see that the amplitude of the antiferroelectric component is the order parameter in our problem, so the domain wall exhibits an antiferroelectric component.

It is interesting to note that when b goes to zero, near the ferroelectric-antiferroelectric transition the width of the antiferroelectric component diverges showing that, at this first order phase transition, the nucleation of the new phase occures in the domain wall.

Remark I

The situation is equivalent in the antiferroelectric phase where we have a ferroelectric component in the antiferroelectric domain wall. Such a structure must be sensitive when subjected to a field gradient. If we want to compare it with what happens in the ferroelectric it gives an effective field

$$\tilde{E} = (\text{grad } E) \times \delta$$

when δ is the width of the domain wall $\sqrt{c/-2b}$. This gives very low values for the effective field.

Remark II

Such a situation can, of course, be found for ferromagnetic-antiferromagnetic transition.

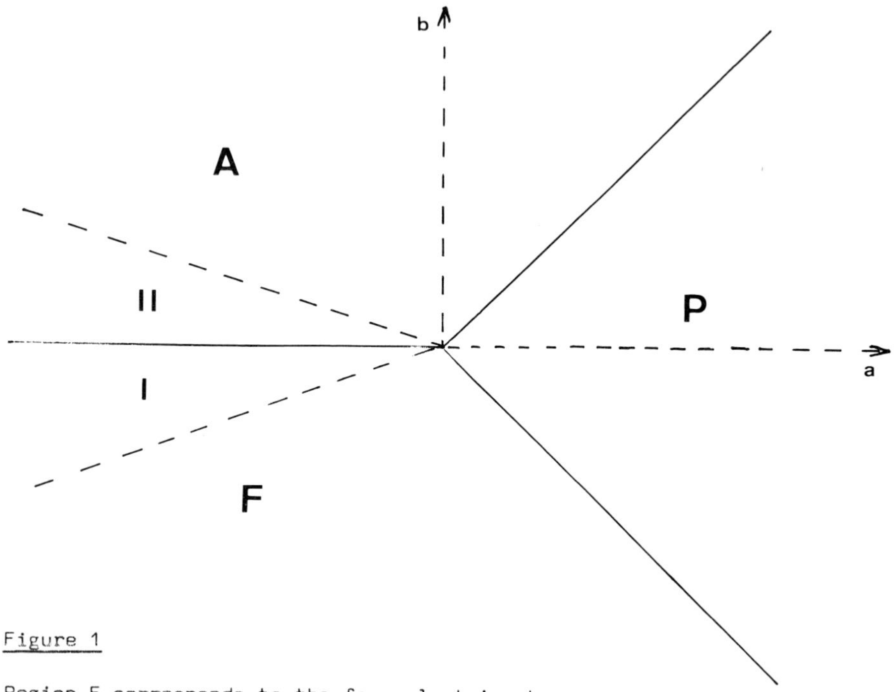

Figure 1

Region F corresponds to the ferroelectric phase
Region A corresponds to the antiferroelectric phase
Region P corresponds to the paraelectric phase.

In the region I the domain wall of the ferroelectric phase exhibits an antiferroelectric character.

In the region II, the domain wall of the antiferroelectric phase exhibits a ferroelectric character

References

[1] J. Lajzerowicz and J.J. Niez
Phase transition in a domain wall (to be published)

[2] M.E. Lines and A.M. Glass
Principles and Applications of Ferroelectrics and Related Materials
Clarendon Press, Oxford (1977)

[3] C. Kittel
Phys. Rev., 82, 729 (1951)

[4] S. Sarker, S.E. Trullinger, A.R. Bishop
Phys. Lett. A, 59, 255 (1976).

Periodic Lattice Distortions and Charge Density Waves in One- and Two-Dimensional Systems

R.H. Friend

Cavendish Laboratory, Cambridge, Great Britain

At low temperatures, low dimensional metals are commonly found to exhibit superlattices incommensurate with the primitive lattice (periodic lattice distortion coupled to charge density waves, PLD - CDW). Transitions to such states are driven by strong electron phonon coupling and favoured by the particular forms of the Fermi surfaces of these metals. Experimental investigations of the structural distortions, and of the changes in electronic structure around the Fermi level produced by the CDW - PLD state, will be discussed for the layered dichalcogenides, the linear chain platinum complex KCP, and the organic charge transfer salts.

The basic instability of a one-dimensional metal to a distortion with a wavevector that opens an energy gap at the Fermi level, the Peierls distortion, has been known for some time (1,2). However, it is only in the past few years that physical systems that appear to meet the conditions for such a distortion have been discovered. The study of these systems has revealed a fascinating wealth of phenomena, including the importance of one-dimensional fluctuations, the role of three-dimensional coupling, and the interaction between the lattice periodicity and that of the Peierls supperlattice.

In this review paper I shall first review the basic theoretical concepts and then discuss the properties of the three groups of materials that have been most intensively investigated, the linear chain platinum complexes, (KCP), the organic charge transfer salts (TTF-TCNQ) and the layered transition metal dichalcogenides. These compounds have been extensively reviewed (3-8).

In the case of a one-dimensional metal, for which the Fermi surface consists of two planes at wavevectors $-k_F$ and $+k_F$, an applied potential, v_q with wavevector $2k_F$, will introduce an energy gap in the electronic spectrum at the Fermi level. Since occupied states will have their energies lowered by this potential, and only unoccupied states are raised in energy, the overall electronic kinetic energy is lowered. When the periodic potential, v_q is derived from a periodic distortion of the lattice, at T=0 the lowering of the electronic kinetic energy will dominate the lattice distortion energy to give a distorted semiconducting state with a superlattice of wavevector $2k_F$.

Following the mean field treatment of RICE and STRASSLER (9) if the electron-phonon coupling is of the form

$$H_{e-p} = \frac{1}{\sqrt{N}} \sum_p \sum_q g(q)\, c^+_{p+q}\, c_p (b_q - b^+_{-q}) \tag{1}$$

in which the c's and b's denote annihilation and creation operators for the electronic Bloch states and phonon states, and g(q) denotes the electron-phonon coupling constant, the effect of the conduction electrons on the bare normal mode frequency $\omega_o(q)$ is to change it to

$$\omega^2(q) = \omega_o^2(q)\ (1 - 2\ (g^2(q)/\hbar\omega_o(q)\chi(q,T)) \tag{2}$$

where $\chi(q,T)$ is the Lindhard function,

$$\chi(q,T) = \sum_k \frac{f_k - f_{k-q}}{\varepsilon_{k-q} - \varepsilon_k} \tag{3}$$

(f_k is the Fermi function for Bloch state k and ε_k is the energy of state k).

For the case of a one-dimensional conductor, $\chi(2k_F,T)$ diverges as $N/E_F)\ \ln(E_F/kT)$ at low temperatures, where $N(E_F)$ is the density of states at the Fermi level. At a temperature T_p, the phonon frequency is renormalized to zero, and for $T<T_p$, a static distortion of the lattice will be present.

$$kT_p \sim E_F\ \exp\ (-1/\lambda) \tag{4}$$

where λ, the dimensionless electron-phonon coupling constant is given by

$$\lambda = \frac{g^2(q)\ N(E_F)}{\hbar\ \omega_o(q)} \tag{5}$$

Below T_p, the energy gap in the electronic spectrum builds up, following a BCS type temperature dependence, and at T=0,

$$2\ \Delta = 3.5kT_p \tag{6}$$

In this model the energy gap in the electronic spectrum is produced by a periodic lattice distortion (PLD). The conduction electrons will attempt to screen the periodic potential set up by the atomic displacements, creating a charge density wave (CDW) in the conduction electron density. As emphasised by FRIEDEL (10), the charge density wave stiffens the lattice and makes its instability less easy to produce. Thus if $v_q(\text{ext.})$ is the potential due to a phonon, the total potential acting on the electrons is

$$v_q = v_q\ (\text{ext.}) + v_q\ (\text{int}) \tag{7}$$

defining the dielectric constant, εq such that $v_q\varepsilon_q = v_q$ (ext.)

$$\varepsilon_q = 1 - \frac{4\ e^2}{q^2}\ \chi\ (q,T) \tag{8}$$

The effect of the CDW on the formation of the PLD-CDW state is to modify the stability criterion ($\omega(q) = 0$ in (2)) to

$$\frac{2\ g^2(q)}{\hbar\ \omega_o(q)}\ \frac{1}{\varepsilon_q} > \frac{1}{\chi(q,T)} \tag{9}$$

Including the effect of electron-electron repulsion for the conduction electrons further stiffens the phonon. If U is a short range Coulomb repulsion, equation (9) is modified to

$$\frac{2g^2(q)}{\hbar\omega_o(q)}\ \frac{1}{\varepsilon_q} - U > \frac{1}{\chi(q,T)} \tag{10}$$

A similar criterion, including also an exchange interaction between conduction electrons, has been obtained by CHAN and HEINE (11) using a variational technique.

Any real physical compound must of course be three dimensional, and even though extremely anisotropic electronic properties are possible in chain-like structures, as is the case for KCP and TTF-TCNQ, interchain coupling will distort the Fermi surface planes of a one-dimensional metal. The warping of the planar Fermi surface will tend to prevent any single superlattice wavevector introducing a gap over the whole Fermi surface, and will eventually suppress the Peierls transition (12). Characterising the interchain coupling by a transfer integral $t_{\perp}$, it is clear that if $kT_P > t_{\perp}$, the formation of the PLD-CDW state is unaffected by $t_{\perp}$. This criterion is satisfied for KCP and TTF-TCNQ, although other related charge transfer salts show distortions which are depressed by $t_{\perp}$, and are further weakened by pressure induced increases in $t_{\perp}$, (13).

The other class of compounds showing PLD-CDW formation that I shall discuss here is the layered structure transition metal dichalcogenides. These materials have effectively two-dimensional Fermi surfaces, with little band dispersion out of the plane of the layer. Large contributions to $\chi(q,T)$ in (3) can arise from areas of Fermi surface which are separated from each other by a single wavevector. This condition is termed Fermi surface "nesting". Calculations of $\chi(q,T)$ (15,16) based on band structure calculations for the layer compounds (16,17) indicate only weak nesting, and as stressed by DORAN et al (18), it is the strong electron phonon coupling in these compounds that is important in driving the distortion. It is no coincidence that many of these compounds which distort but remain metallic below the distortion are superconductors (19).

The mean field treatment of the Peierls distortion outlined above produced a phase transition at finite temperature in spite of the theorem that one-dimensional systems without long range forces cannot undergo phase transitions at finite temperatures. This is because, by considering one phonon mode only, the short range electron-phonon interaction was effectively turned into a long range sinusoidal force of wavevector $2k_F$. If all other Fourier components are included, the short range character of the force is restored and the phase transition is suppressed. This has been treated by LEE et al (20) using a Ginzburg-Landau expansion of the free energy. One-dimensional fluctuations are very sensitive to interchain, three dimensional coupling, and the effectiveness of interchain tunnelling coupling in suppressing fluctuations has been considered by RICE and STRASSLER (21) and HOROVITZ et al (12). Coulomb coupling between CDW's on adjacent chains has also been proposed as a means of establishing three-dimensional long range order (22). Of the three series of compounds discussed later, only in KCP is behaviour strongly influenced by fluctuations, although TTF-TCNQ remains controversial (HEEGER in (4) presents the case for their importance).

In this model of the Peierls distortion, the zero-temperature Peierls gap is destroyed at high temperatures by thermal excitation of electrons across the gap, giving the relationship between Δ and T_P in eqn. (6). For the layer compounds, where this relation is not obeyed (the factor of 3.5 in (6) is found to be much larger), McMILLAN (23) has developed a short coherence length model ($\xi \sim$ superlattice period), consistent with the poorly defined peaks calculated in $\chi(q)$ and the very broad Kohn anomaly in the phonon spectrum, extending over $\sim 1/3$ Brillouin zone (24). Since many modes are involved, the phonon entropy dominates the electron entropy at finite temperature and determines the value of T_p.

One of the interesting aspects of the Peierls distortion is that the superlattice wavevector is determined by the details of the Fermi surface, and may bear no simple relation to the lattice periodicity. If the superlattice wavelength Λ = na where a is the lattice dimension and n integer, the PLD-CDW is said to be a commensurate superlattice, and will have preferred positions with respect to the lattice and will be unable to move without surmounting an energy barrier. If however $\Lambda \neq$ na, the superlattice is termed incommensurate, and if the PLD-CDW is a simple sinusoidal distortion then it has no preferred position in the lattice, and if not impeded by too many impurities, it will be free to move. Such movement, which would carry current, is the collective mode envisaged by FROHLICH as a mechanism for superconductivity (2). It appears that any contribution to the conductivity from the Fröhlich collective mode is unimportant in the systems I shall discuss in this paper, however measurements of non-linear conductivity in the quasi one-dimensional system $NbSe_3$ (25), in which an incommensurate superlattice has recently been discovered (26) have rekindled interest in this conduction mechanism.

When Λ is close, but not equal to na, there will be competition between the electronic bandstructure energy which favours the best nesting wavevector, and the commensurability energy gained when the PLD-CDW sits in preferred positions in the lattice. The commensurate superlattice is favoured for large values of the distortion amplitude, and a sequence of transitions, first from the undistorted stated to an incommensurate superlattice, then at a lower temperature to a commensurate superlattice is commonly observed for the layered compounds, when the incommensurate superlattice is close to commensurability, and the weak Fermi surface nesting reduces the band structure energy penalty in moving away from the best nesting wavevector. For the quasi one-dimensional systems where nesting is better, and the distortion amplitudes smaller, a lock-in to the commmensurate superlattice is less likely, although it has recently been observed in TTF-TCNQ under pressure (27). This sequence of phase transitions has been modelled using a Landau theory by McMILLAN (28).

As an alternative to the incommensurate phase characterised by a simple sinusoidal distortion, McMILLAN (29) has proposed that when Λ is close to na the PLD-CDW may dissociate into regions of locally commensurate superlattice and regions where the phase changes rapidly with respect to the lattice, introducing harmonics of the superlattice period. The regions of rapidly changing phase, termed discommensurations by McMILLAN can be regarded as solitons. Evidence for their existence in the layer compounds will be discussed later.

$K_2(Pt(CN)_4)\ Br_{0.30} \cdot 3H_2O$

$K_2(Pt(CN)_4)\ Br_{0.30} \cdot 3H_2O$, usually abbreviated to KCP, is one of a family of non integral oxidation state platinum complexes of cyanide or oxalate investigated by Krogmann and co-workers. Their properties are extensively reviewed in (3).

A much simplified sketch of the backbone of the structure is shown in Fig.1.

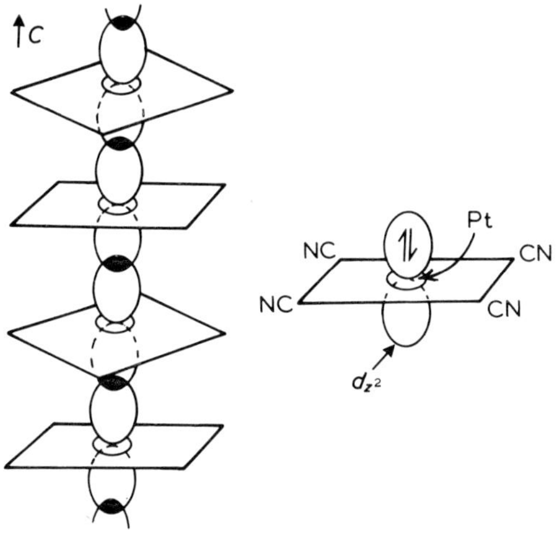

Fig. 1 Idealised sketches of a $Pt(CN)_4^{2-}$ group, and of the stacking of $Pt(CN)_4^{2-}$ groups in a KCP crystal, showing the overlap of the d_{z^2} atomic orbitals based on the Pt atoms

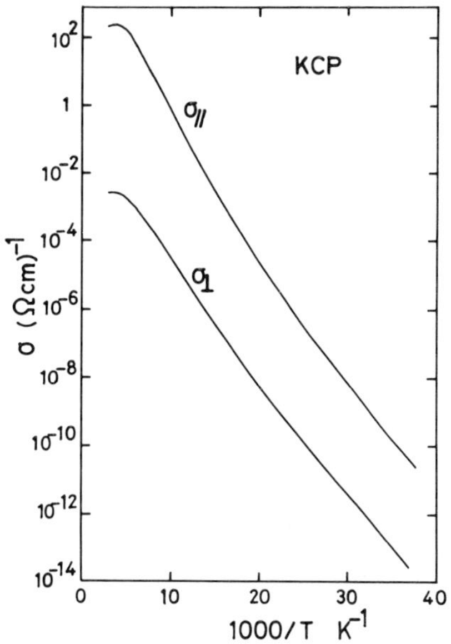

Fig. 2 Conductivity parallel and perpendicular to the chains for KCP, after (30)

The planar arrangement of cyanide anions around the platinum cations allows stacking along the c axis and overlap between the atomic $d_z{}^2$ platinum orbitals on adjacent ions. Not shown are the positions of the K^+, Br^- and H_2O; however the presence of 0.30 bromide ions per formula unit is crucial for the conduction properties, as they take electrons out of what would have been the filled platinum d_{z^2} band, and leave it with only 1.7 electrons per platinum ion. Thus in the absence of any distortion the material should be metallic along the chain axis. The conductivity measured parallel to the chains is several hundred $(\Omega\,cm)^{-1}$, and the anisotropy ratio $\sigma_{\parallel}/\sigma_{\perp}$ is about 5×10^4 at room temperature (30). Below room temperature the conductivity falls rapidly, as is shown in Fig. 2, and X-ray and neutron diffraction studies indicate the appearance of an incommensurate superlattice, with a chain axis repeat of 6.7 lattice units, which is the value of $2k_F$ expected for the 1.7 electron filled band (31). A large Kohn anomaly can be seen in the longitudinal acoustic phonon branch at room temperature, shown in Fig. 3.

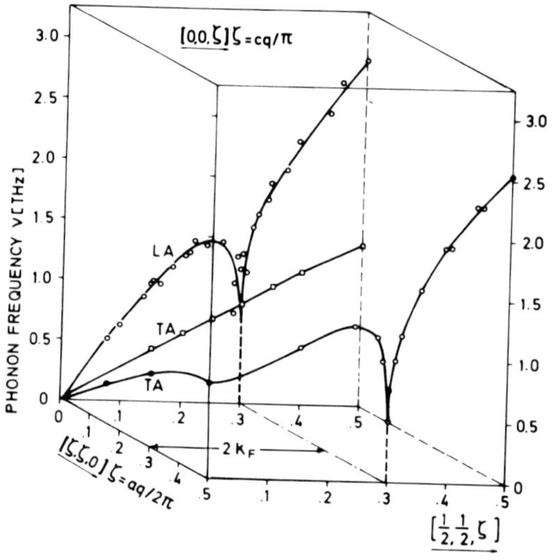

Fig. 3 Acoustic phonon branches in KCP at room temperature (31). An enhanced Kohn anomaly appears along the line $q=2k_F$=const

Elastic neutron scattering is also observed at all off symmetry directions studied which have a $2k_F$ component of the wavevector parallel to the chains, demonstrating the presence of a PLD together with the Kohn anomaly. Below 140K, the intensity of the superlattice point $(\pi/d_{\perp}, \pi/d_{\perp}, 2k_F)$ increases rapidly and then saturates at around 80K. The intensities of other $2k_F$ points, such as $(0,0,2k_F)$ fall in this temperature region, but do not reach zero. This indicates that whilst the coherence of the distortion with transverse period $2d_{\perp}$ increase rapidly below 140K, the system never attains long range order. X-ray experiments measure the low temperature coherence length to be some 60 Pt-Pt spacings. The choice of $(\pi/d_{\perp}, \pi/d_{\perp}, 2k_F)$ suits both Coulomb (22) and tunnelling (12) interchain energy considerations.

Measurements of the infra-red reflectivity show what is interpreted as the Peierls gap, at an energy of $\sim$ 0.2 eV (30). While this gap is most apparent at low temperatures, it is still present at room temperature. The value of 0.2eV implies a mean field transition of some 800K, from equation (6) and it is evident that KCP is a system where the transition is suppressed by one-dimensional fluctuations. The random potential produced by the partial occupation of the Br^- and H_2O lattice sites may be responsible for the absence of true long range order at low temperatures, but it is interesting to note that under pressure, when presumably interchain coupling is strengthened, a sharp anomaly in the resistivity as a function of temperature is observed (32). The anomaly is similar to that seen at the three-dimensional ordering transitions in the organic charge transfer salts (see Fig. 5), and indicates that there is a transition to a three-dimensional ordered state at low temperatures at pressures in excess of 20kbar.

ORGANIC CHARGE TRANSFER SALTS

The organic charge transfer salts, of which tetrathiofulvalenium tetracyanoquinodimethanide (TTF-TCNQ) is the most studied, are highly anisotropic, or quasi one-dimensional metals in which overlap between π orbitals of planar molecules stacked one on top of another results in band formation, and delocalisation of electrons along these stacks. TTF and TCNQ molecules are shown in Fig. 4a, and the three-dimensional structure of TTF-TCNQ is shown schematically in Fig. 4b. Segregrated stacks of TTF and TCNQ are formed along the b axis, and as determined from the value of Peierls distortion wavevector, there is charge transfer of .59 electrons per molecule from the TTF to the TCNQ stack.

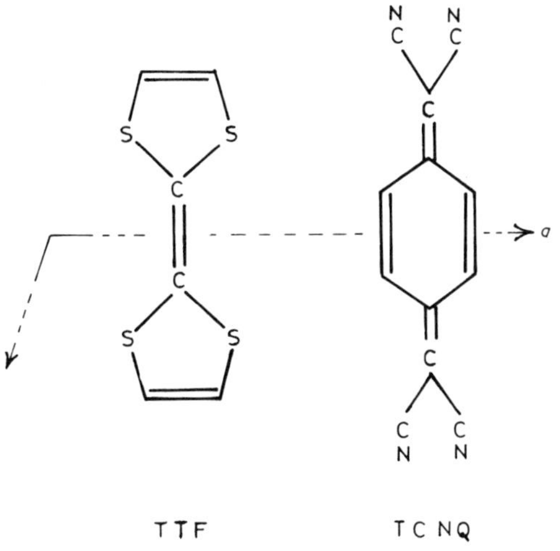

Fig 4a Structure of TTF & TCNQ molecules

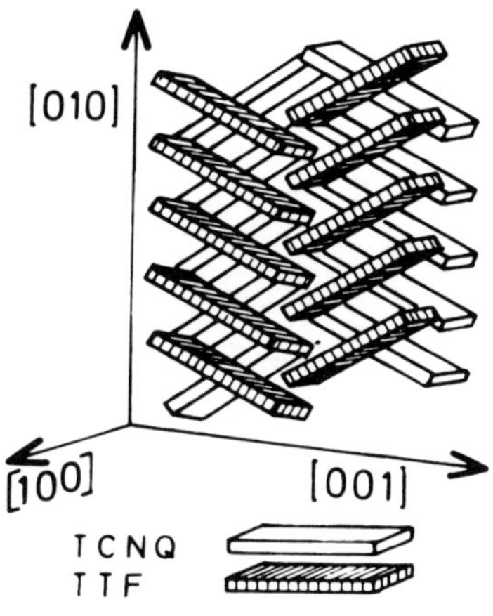

Fig 4b Schematic representation of the TTF-TCNQ structure, showing the tilted stacks of TTF & TCNQ molecules along the b axis

In a one electron one-dimensional band model, the Fermi wavevector, and therefore the charge transfer, is determined by the value of k at which the bonding TCNQ band crosses the antibonding TTF band. The transfer of charge puts .59 electrons into the TCNQ band, leaving the TTF band with

1.41 electrons, and with both bands partially filled, both stacks contribute to the metallic conductivity, at room temperature some 500 $(\Omega\ cm)^{-1}$ along the b axis (33). Coupling between stacks is weak, the conductivity anisotropy is of order 500 at room temperature (33), and the tunnelling integral between stacks is estimated from the NMR interstack relaxation time to be some 5meV (34). The conductivity increases on cooling, reaching a maximum for most crystals of between 10 and 25 times the room temperature value at around 56K (35). Larger measured increases have been interpreted in terms of a Fröhlich mode contribution to the conductivity (36), however these measurements have been criticised by Schafer et al (37), and Thomas et al (35) consider that the majority of conductivity measurements are consistent with single particle conduction mechanisms.

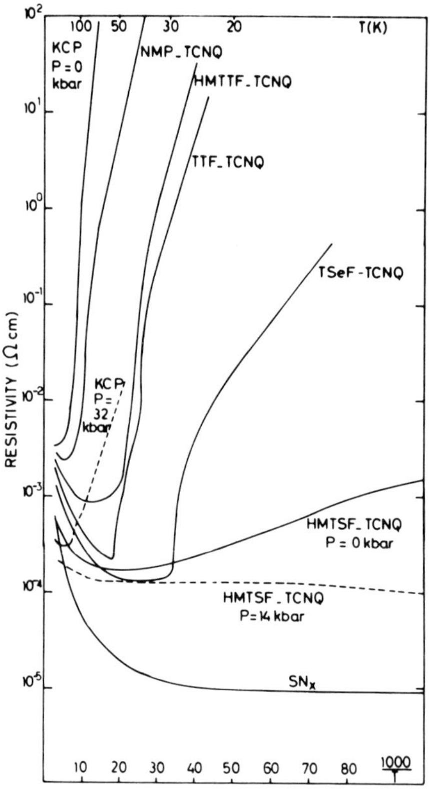

Fig. 5 Temperature dependence of the resistivity along the high conductivity axis for several 1d conductors (34)

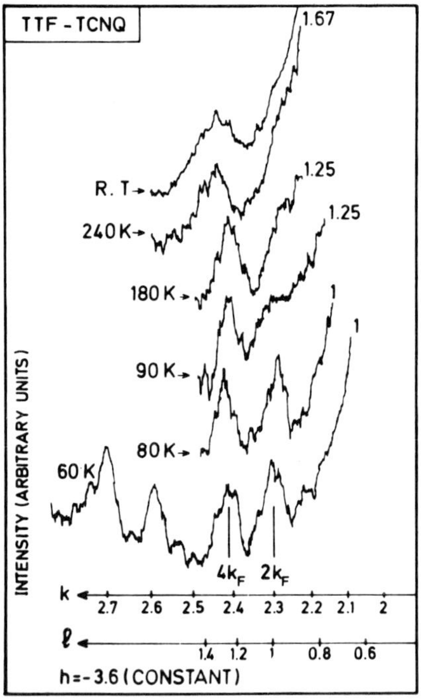

Fig. 6 Microdensitometer readings of a series of X-ray patterns between 60K and room temperature for TTF-TCNQ. Note the different temperature dependence of the $2k_F$ and $4k_F$ scattering (38)

Below the conductivity maximum, TTF-TCNQ becomes semiconducting, showing a series of phase transitions, two of which, at 53K and 38K are visible as peaks in the slope of the resistivity versus temperature curve shown in Fig. 5. X-ray and neutron scattering measurements have been hampered by the small size of crystals available, the elegant measurements of the structural distortions taking place are reviewed by Comès (38). Diffuse X-ray scattering at temperatures above the distortions show appreciable phonon softening at .295b*, which is interpreted as the $2k_F$ phonon. Scattering at twice this wavevector is also seen in the X-ray pattern shown in Fig. 6 (.59b*, or 41b* in the reduced zone). This $4k_F$ scattering has a different temperature dependence from the $2k_F$, being relatively stronger at room temperature, and it cannot be considered as a harmonic of the $2k_F$ distortion. The sequence of phase transitions seen at low temperatures is associated with varying stages of the three-dimensional ordering of the Peierls distortions on the two stacks. At 53K, the $2k_F$ scattering condenses to give a superlattice of (2a,3.4b,c). Many measurements including ESR (39) and C^{13}NMR (40) indicate that this corresponds to the formation of a Peierls distortion on the TCNQ stacks. The transverse period of 2a can leave the TTF stacks undistorted (41) and it is only below 49K that the distortion builds up on them, and the transverse period increases from 2a. Finally at 38K there is a first order transition where the transverse period locks in to a value of 4a.

A controversial issue for TTF-TCNQ is the validity of the free electron model. The maximum calculated values of the bandwidths of either stack is only $\frac{1}{2}$eV (42), whereas the value of the on site Coulomb repulsion, U is 10 times larger (43). Explanations for the $4k_F$ scattering are usually based on a large U model for one or both stacks (38,44,45). It should be noted that $4k_F$ scattering is not observed in most other TCNQ salts, such as the selenium analogue TSF-TCNQ (46), although it has been recently reported to be present in NMP-TCNQ (47) which is also considered to be a system where correlation effects are important.

At atmospheric pressure none of the charge transfer salts have a value of $2k_F$ close enough to a commensurate superstructure for there to be a lock-in transition. However, recent high pressure measurements (27) indicate that at 20kbar, charge transfer in TTF-TCNQ has increased to 2/3, giving a superlattice with period 3b. The range over which the PLD-CDW is commensurate is limited to $\Delta k/k < 1\%$, which is consistent with the almost perfect one-dimensional Fermi surface $(kT_p \sim t_\perp)$, and contrasts with the larger commensurability ranges seen in the layer compounds.

LAYERED TRANSITION METAL DICHALCOGENIDES

The group V transition metal dichalcogenides, MX_2 are layered structure materials, composed of close packed sheets of metal atoms sandwiched by layers of chalcogen atoms. Bonding between chalcogen layers in neighbouring sandwiches is weak, and these compounds, which are d band metals, show two dimensional metallic properties, although the anisotropy between the in-plane and out-of-plane conductivities is typically only 30 (48).

Coordination of the metal by the chalcogens can be either octahedral or trigonal prismatic; many ways of stacking the sandwiches, of either or both coordination types, are possible, and there are many polytypes found. Fig. 7 shows some of the more common of them, the simplest octahedral co-ordination polytype, with only one layer per unit cell, 1T, the simplest trigonal prismatic polytype with two layers per unit cell, 2H, and the mixed coordination polytype, $4H_b$.

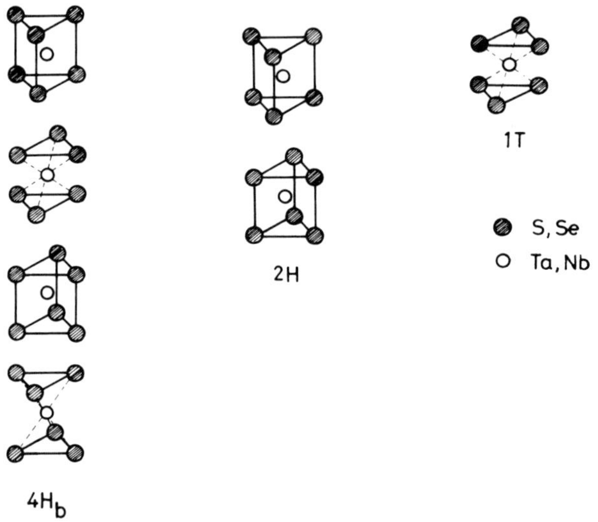

Fig. 7 Structures of some of the Group V dichalocogenide polytypes

Physical properties are mainly determined by the coordination type, in fact the properties of 1T and 2H materials are sufficiently different for it to be simplest to treat them separately. The Fermi level in 2H materials lies in the middle of a d subband of d_{z^2}, d_{xy} and $d_{x^2-y^2}$ character with a hybridisation gap between it and the rest of the d band. This results in a high $N(E_F)$ and a complicated Fermi surface, whereas the d band for the 1T polytypes is not split, $N(E_F)$ is lower and the Fermi surface simpler (16,17).

Structural distortions, not necessarily related to Fermi surface nesting, are a general feature of this class of materials; all the Te compounds are distorted from the hexagonal layer structure (49) and even the Group IV material $TiSe_2$ which has only a low carrier density from a small p-d band overlap exhibits a low temperature superstructure in which the in-plane and out-of-plane cell constants are doubled (50). The common feature for these systems is strong electron-phonon coupling, seen for instance in the high infra-red effective charge in $TiSe_2$ (51). Nevertheless the S and Se Group V compounds show a series of incommensurate and commensurate distortions for which it does seem possible to relate the superlattice period to Fermi surface nesting in the calculated band structures (14, 15).

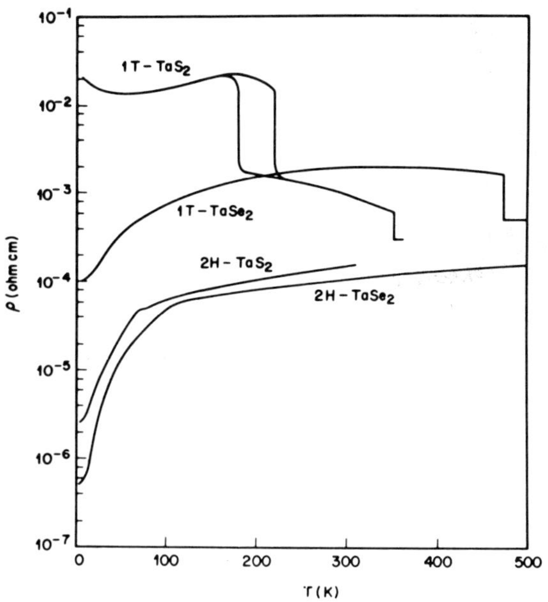

Fig 8. The electrical resistivity parallel to the layers of several layered conpounds (F.J. Di Salvo in (4))

Resistivity versus temperature curves for both 1T and 2H polytypes are shown in Fig. 8. The resistivity anomalies clearly visible all correspond to transitions in the PLD-CDW structure; since not all the Fermi surface is destroyed by PLD-CDW formation they remain metallic at all temperatures, in contrast to KCP and TTF-TCNQ (the rise in the low temperature resisitivity in 1T TaS_2 is the result of falling carrier mobility from impurity induced localisation rather than falling carrier concentration (52).)

1T POLYTYPES

The two 1T polytypes of Ta, TaS_2 and $TaSe_2$ both show strong distortions, with both commensurate and incommensurate structures at high and low temperatures. Fig. 9 shows electron diffraction photographs of 1T TaS_2 at various temperatures. The simple hexagonal diffraction pattern of the undistorted lattice is present at 600K (by which temperature there is an irreversible transition to the 2H polytype). The three phases of superlattice shown in the other three photographs are bounded by the two resistivity transitions in Fig. 8, one at 350K, the other centred at 200K with ∿ 20K hysteresis. Above 350K, in the $1T_1$ phase, an incommensurate structure

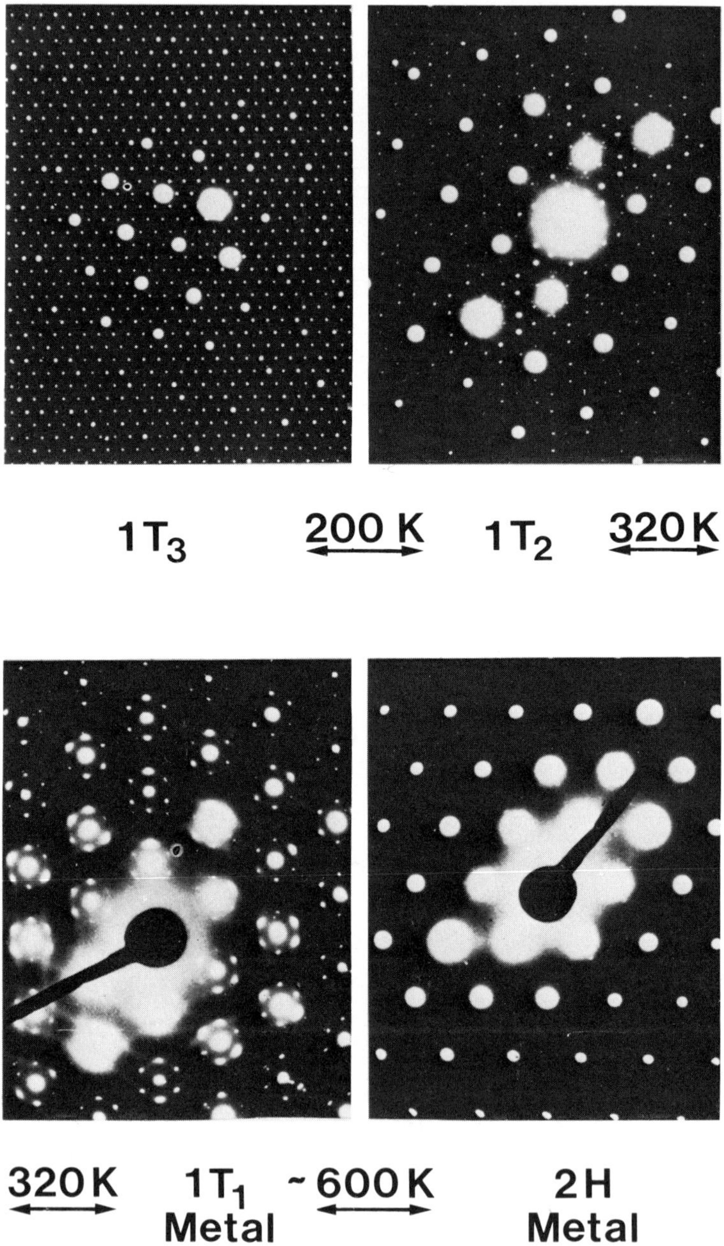

Fig.9 Transmission electron diffraction patterns for the 3 superlattice phases, and undistorted phase of 1T TaS_2 (53)

with superlattice spots in line with the principal spots is seen. As for all the layer compounds, with hexagonal or trigonal symmetry in the layer, a triple PLD-CDW is observed, with three plane wave PLD-CDW's at 120° to one another. Between 350K and 200K, in the $1T_2$ phase, the superlattice reflections have rotated away from the line of the principal reflections. Although it is the tendency to form the commensurate superstructure, the $1T_3$ phase, that must rotate the superlattice in this regime, the superlattice is not commensurate. The commensurate structure is only achieved below 200K, by rotation of the superlattice by $13^o54'$, so that the axes of the hexagonal supercell are $3\underline{a} + \underline{b}$ ($\underline{a}$ and $\underline{b}$ in-plane lattice vectors), or $\sqrt{13}\ |\underline{a}|$. 1T $TaSe_2$ does not exhibit the $1T_2$ phase, and the lock-in from a similar $1T_1$ type phase to the $\sqrt{13}$ superlattice occurs at 475 K (19)

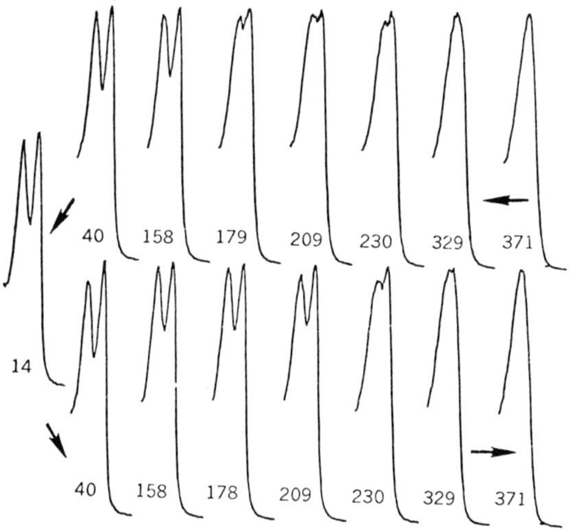

Fig. 10 XPS from Ta $4f_{7/2}$ levels in 1T TaS_2 taken during a cooling and heating cycle, showing hysteresis at the 200K transition. The spitting between the two peaks at the lowest temperatures is ∿ 1eV (56)

The distortions in the 1T polytypes are large, with atomic displacements of ∿ .25 Å (54). One of the more spectacular indications of the changes taking place is X-ray photoemission from the Ta $4f_{5/2}$ and $4f_{7/2}$ levels (55, 56). Changes in the charge density around the Ta sites create splittings in the energy levels of up to 1eV. WERTHEIM et al (55) estimate that such a large splitting corresponds to a charge redistribution between sites of as much as one electron. Fig. 10 shows measurements at various temperatures through the two transitions in 1T TaS_2. In the $1T_3$ phase HUGHES and F ⸤AR (56) argue that the 13 metal sites in the supercell are distributed into 3 inequivalent types , in the ratio 6:6:1, and the two peaks arise from the two more common sites. What is intriguing is that even in the $1T_2$ phase, when the superlattice is incommensurate, the two peaks are still resolved; this must indicate that the superlattice is locally commensurate, and invites a description similar to the discommensurate picture developed by McMILLAN (29) for the 2H polytypes.

2H POLYTYPES

PLD-CDW formation in the 2H polytypes is weaker than in 1T polytypes, and the characteristic commensurate superlattice is $3\underline{a} \times 3\underline{b}$. The two best studied compounds are 2H $TaSe_2$ and 2H $NbSe_2$. The neutron diffraction investigation by MONCTON et al (24) shows that both compounds undergo a distortion to an incommensurate structure with $\underline{q} = 1/3\underline{a}^* (1-\delta)$ and $\delta \sim 2\%$, $TaSe_2$ at 122K and $NbSe_2$ at 33K. Below these temperatures, δ decreases, and at 90K, $TaSe_2$ locks into the commensurate phase. $NbSe_2$ however remains incommensurate down to low temperatures. In the incommensurate regime of $TaSe_2$ MONCTON et al noticed in addition to satellites at $1/3\underline{a}^*(1-\delta)$ weaker satellites at $1/3\underline{a}^*(1+2\delta)$, as is shown in Fig 11.

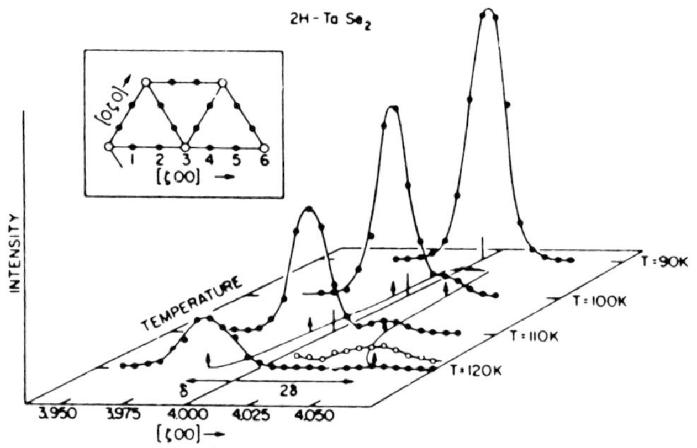

Fig. 11 Elastic neutron scattering measurements of the satellite peak intensity and position in 2H $TaSe_2$ versus temperature (24)

The presence of harmonics in the PLD-CDW is consistent with the discommensuration picture (29), although the treatment by MONCTON et al of the lock-in energy gain is somewhat different.

Nb^{93} is suitable for NMR ($I = 9/2$), and the formation of the PLD-CDW in 2H $NbSe_2$ produces changes in both the Knight shift and the electric field gradient (57). Fig. 12 shows the $(\frac{1}{2},-\frac{1}{2})$ central line at various temperatures. The broadening of the line below the transition at 33K is compared to the calculated lineshape for a triple incommensurate CDW (shown as the solid line). BERTHIER et al (57) consider the general agreement with the experimental line to be good, but the peak in the experimental line at the high field edge is not consistent with the step that is calculated, and may be an indication for a tendency towards local

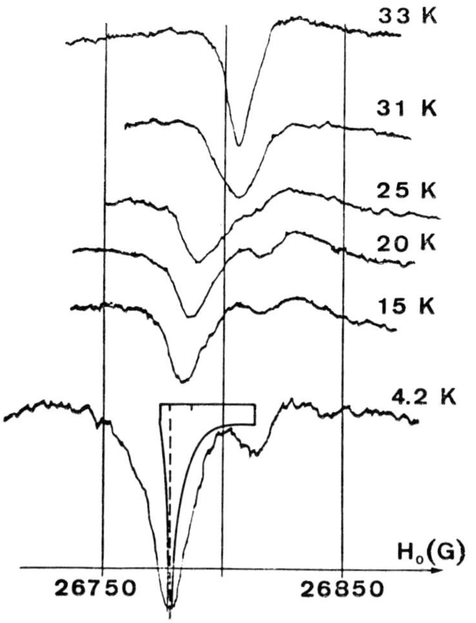

Fig. 12 NMR in $NbSe_2$ showing evolution of the central line $(\frac{1}{2},-\frac{1}{2})$ below 33K. The solid line is the calculated lineshape corresponding to a Knight shift distribution associated with a triple incommensurate CDW (57)

commensurability in this incommensurate phase. BERTHIER et al (57) estimate the change in charge density below the transition to be 10%, a much smaller change than in the 1T polytypes.

I would like to thank D. Jérome and A. D. Yoffe for many helpful discussions.

REFERENCES

1. R.E. Peierls, 'Quantum Theory of Solids', O.U.P. (1955)
2. H. Fröhlich, Proc. Roy. Soc. A223, 296 (1955)
3. 'Low-Dimensional Cooperative Phenomena', Ed. H.J.Keller, Plenum (1975)
4. 'Chemistry & Physics of One Dimensional Metals', Ed. H.J.Keller, Plenum (1977)
5. 'Electron-Phonon Interactions and Phase Transitions', Ed. T. Riste, Plenum (1977)
6. A.D. Yoffe, Chem. Soc. Rev. 5, 51 (1976)
7. A.J. Berlinsky, Contemp. Phys. 17, 331 (1976)
8. G.A. Toombs, Phys. Reports C 40, 181 (1978)
9. M.J. Rice and S. Strässler, Solid State Commun. 13, 125 (1973)
10. J. Friedel in (5)
11. S-K. Chan and V. Heine, J. Phys. F 3, 795 (1973)
12. B. Horovitz, H. Gutfreund and M. Weger, Phys. Rev. B 12, 3174 (1975)
13. B. Mogensen, R.H. Friend, D. Jérome, K Bechgaard and K. Carneiro, Solid State Commun. (1978)
14. N.J. Doran, B. Ricco, M. Schreiber, D. Titterington and G. Wexler, J. Phys C 11, 699 (1978)
15. H.W. Myron, J.Rath and A.J. Freeman, Phys. Rev. B 15, 885 (1977)
16. G. Wexler and A.M. Woolley, J. Phys. C 9, 1185 (1976)
17. H.W. Myron and A.J. Freeman, Phys. Rev. B 11, 2735 (1975)
18. N.J. Doran, G. Wexler, V. Heine and B. Ricco, Nuovo Cimento B 38, 544 (1977)
19. J.A. Wilson, F.J. Di Salvo and S. Mahajan, Adv. Phys. 24, 117 (1975)
20. P.A. Lee, T.M. Rice and P.W. Anderson, Phys. Rev. Lett. 31, 462 (1973)
21. M.J. Rice and S. Strässler, Solid State Commun. 13, 1389 (1973)
22. K. Saub, S. Barisić and J. Friedel, Phys. Lett. 56A, 302 (1976), A. Bjelis and S. Barisić, Lecture Notes in Physics 65, 291 (1977)
23. W.L. McMillan, Phys. Rev. B 16, 643 (1977)
24. D.E. Moncton, J.D. Axe and F.J. Di Salvo, Phys. Rev. Lett. 34 734, (1975), Phys. Rev. B 16 801 (1977)
25. N.P. Ong and P. Monceau, Phys. Rev. B 16, 3443 (1977)
26. K. Tsutsumi, T. Takagaki, M. Yamamoto, Y. Shiozaki, M. Ido, T. Sambongi, K. Yamaya and Y. Abe, Phys. Rev. Lett. 39, 1675 (1977)
27. R.H. Friend, M. Miljak and D. Jérome, Phys. Rev. Lett. 40, 1048 (1978)
28. W.L. McMillan, Phys. Rev. B 12, 1187 (1975)
29. W.L. McMillan, Phys Rev. B 14, 1496, (1976)
30. H.R. Zeller in (3)
31. B. Renker and R. Comès in (3)
32. M. Thielemans, R. Deltour, D. Jérome and J.R. Cooper, Solid State Commun. 19, 21 (1976)
33. A.J. Heeger and A.F. Garito in (3)
34. D. Jérome and M. Weger in (4)
35. G.A. Thomas et al, Phys, Rev. B 13, 5105 (1976)
36. M.J. Cohen, L.B. Coleman, A.F. Garito and A.J. Heeger, Phys. Rev. B 13, 5111 (1976)
37. D.E. Schafer, F. Wudl, G.A. Thomas, J.P. Ferraris and D.O. Cowan, Solid State Commun. 14, 347 (1974)
38. R. Comès in (4)
39. Y. Tomkiewicz, A.R. Taranko and J.B. Torrance, Phys. Rev. Lett. 36 751, (1976), Phys. Rev. B 15 , 1017 (1977)

40. E.F. Rybeczewski, L.S. Smith, A.F. Garito, A.J. Heeger and B.G. Silbernagel, Phys. Rev. B 14, 2746 (1976).
41. P. Bak and V.J. Emery, Phys. Rev. Lett, 36, 978 (1976)
42. F. Herman, D.R. Salahub and R.P. Messmer, Phys. Rev. B 16, 2453 (1977)
43. J. Hubbard, Phys. Rev. B 17, 494 (1978).
44. J.B.Torrance in (4).
45. V.J. Emery, Phys. Rev. Lett. 37, 107 (1976), and in (4).
46. C. Weyl, E.M. Engler,K. Bechgaard, G. Jehanno and S. Etemad, Solid State Commun. 19, 925 (1976).
47. J.P. Pouget, S. Megtert, and R. Comès, Bull. Am. Phys. Soc. 23, 380 (1978).
48 R.F. Frindt, R.B. Murray, G.D. Pitt and A.D. Yoffe, J. Phys. C 5 L154 (1972).
49 J.A. Wilson and A.D. Yoffe, Adv. Phys. 18, 193 (1969).
50. F.J. Di Salvo, D.E. Moncton and J.V. Waszczak, Phys. Rev. B14, 4321 (1976).
51. R.M. White and G. Lucovsky, Nuovo Cimento B 38, 280 (1977)
52. F.J. Di Salvo and J.E. Graebner, Solid State Commun. 23, 825 (1977).
53. P.M. Williams, G.S. Parry and C.B. Scruby, Phil. Mag. 29, 695 (1974) ibid. 31, 255 (1975).
54. R. Brouwer and F. Jellinek, 5th International Conference on Solid Compounds of Transition Elements, Uppsala, Sweden (1976).
55. G.K.Wertheim, F.J. Di Salvo and S. Chiang, Phys. Lett. 54 A, 304 (1975)
56. H.P. Hughes and R.A. Pollak, Phil. Mag. 34, 1025 (1976).
57. C. Berthier, D. Jérome and P. Molinié, J. Phys. C 11, 797 (1978)

Solitons in Incommensurate Systems

Per Bak

Nordita, Blegdamsvej 17, DK-2100 Copenhagen, Denmark

1. Introduction

Many interesting physical systems undergo phase transformations to periodically ordered phases which are incommensurate with the underlying lattices, i.e. the wave vector describing the modulation cannot be formed by simple rational fractions of the reciprocal lattice vectors. The ordered structure may be a static charge density wave (CDW), a modulated lattice distortion, a spin density wave (SDW) or helical magnetic structure, or even a separate atomic lattice or "mass density wave" (MDW).

The periodic potential of the lattice may cause complicated non-linear distortions of the condensed wave. If the coupling is strong enough this potential may even drive a separate phase transition, the commensurate-incommensurate transition, where the period of the modulated phase becomes commensurate with the lattice. The transition takes place as a consequence of the competition between local "Umklapp" terms in the Hamiltonian which favour the commensurate phase, and the remaining "elastic" terms which must favour a periodicity slightly displaced from the commensurate one. This transition is the subject of the present article.

It is rather easy to show that a simple Landau theory, in which the order parameter is assumed to be homogeneous throughout the system, leads to a first order transition. In a very elegant paper [1], however, McMillan demonstrated that by allowing the phase of the order parameter to fluctuate while keeping the amplitude constant one finds in fact a second order transition. According to this theory, the incommensurate phase near the phase transition consists of very large regions which are essentially commensurate with the lattice, separated by relatively narrow domain walls or "discommensurations" where the phase of the order parameter changes rapidly. Bak and Emery [2] subsequently demonstrated that the domain walls in fact appear as the solutions of the 1d sine-Gordon equation; hence the name "solitons". The order parameter of the transition is the soliton density, which in turn is proportional to the displacement of the wave vector from the commensurate one.

Here the theory of the commensurate-incommensurate transition will be reviewed and extended to describe a variety of widely different physical systems. McMillan's theory in its original form applies to a one-dimensional modulated structure, where the modulation is described by a single wave vector, $\vec{q}$. On the other hand, most systems which have been observed to exhibit commensurate-incommensurate (I-C) transitions are in fact two-dimensionally modulated, i.e. the modulation is described by two or three symmetric $\vec{q}$-vectors in a common plane. It will be demonstrated that a second order transition is still possible with a linear temperature dependence of the order parameter ($\beta=1$) in contrast to the logarithmic behaviour ($\beta=0$) in the 1d case. This difference is caused by the topologically different nature of the incommensurate phases in the two cases. This theory applies to the charge density wave (CDW) in the layered metal chalgogenides [3], the I-C transition in the "triple $\vec{q}$" modulated magnetic structure of neodymium [4], and also to the registry-non-registry transitions in a monolayer of rare gas or deuterium atoms adsorbed on graphoil or graphite [5,6]. We shall also consider the coupling to alternative degrees of freedom. It will be shown that a coupling to a macroscopic strain does not change the nature of the transition, and the induced strain will be of a inhomogeneous soliton-like nature.

We start with a brief review and discussion of the nature of the modulated phases in various physical systems.

2. Modulated Structures. 1d and 2d Modulations

The most well known modulated structures are probably the sinusoidal or helical magnetic structures observed in the rare earth metals [7]. Near T_0, where one wave vector is predominant, the magnetic structures may be written on the form

$$M_x(\vec{r}) = \psi_{\vec{q}} \exp(i\vec{q}\cdot\vec{r}) + \text{complex conjugate} \tag{2.1}$$

where $\psi_{\vec{q}}$ is the so-called order parameter. By the term "incommensurate structure" we shall here understand a structure on the more general form

$$M_x(\vec{r}) = M_x(\vec{q}\cdot\vec{r}+\phi) \tag{2.2}$$

where $M_x(\vec{q}\cdot\vec{r})$ is periodic with period 2π. The non-commensurate wave vector $\vec{q}$ is defined by this equation. This structure includes all higher harmonics. A charge density wave, or a modulated lattice distortion may be written on a similar form. Until now, no I-C transitions have been reported in such systems.

There exist, however, a number of modulated systems characterised by two or three symmetric $\vec{q}$-vectors. For example, the CDW in 2H-$TaSe_2$ [3] is given by

$$\rho(\vec{r}) = \sum_{i=1}^{3} \psi_i \exp(i\vec{q}_i\cdot\vec{r}) + \psi_i^* \exp(-i\vec{q}_i\cdot\vec{r}) \tag{2.3}$$

with $|\psi_1|=|\psi_2|=|\psi_3|$ and $\vec{q}_1 = q(1,0,0)$, $\vec{q}_2 = q(-\frac{1}{2}, \sqrt{3/2},0)$, $\vec{q}_3= q(-\frac{1}{2},-\sqrt{3/2},0)$. This is the "triple $\vec{q}$" structure. At T_0=122 K, 2H-$TaSe_2$ undergoes an I-C transition to a commensurate phase where $q = 2\pi/3a$. In the one-dimensional conductor TTF-TCNQ there are three ordered phases, two of which are characterised by single incommensurate wave vectors [2,8]. The third phase includes two commensurate wave vectors [9].

Quite recently it was discovered that the magnetic structure of neodymium is also a "triple $\vec{q}$" structure [4]. Near T_0, the magnetic moments in one of the hexagonal layers are given by

$$M(\vec{r}) = \sum_i \psi_i \hat{q}_i \exp(i\vec{q}_i \cdot \vec{r}) + cc \tag{2.4}$$

This structure is shown in Fig. 1. The magnetic structure is

Fig. 1 "Triple $\vec{q}$" magnetic structure of neodymium (left) and the accompanying lattice distortions (right)

accompanied by a similarly modulated lattice distortion. This distortion is very similar to the one found in the layered metal chalgogenides. The modulation period is temperature dependent, and there are strong indications that the structure locks into a periodicity of 8 lattice units at T_0 = 8 K [10]. The modulated structure forms a hexagonal pattern on top of the basic lattice.

The last group of systems that will be considered seems at first glance to be of fundamentally different nature. A monolayer of rare gases, or deuterium, adsorbed on graphoil [5,6] may undergo transitions from a liquid state to a two-dimensional crystal. The two-dimensional graphoil surface here plays the role of the basic lattice. This solidification may be described as the formation of a "mass density wave"

$$\rho_m(\vec{r}) = \sum_{i=1}^{3} \psi_i \exp(-i\vec{q}_i \cdot \vec{r}) + cc. \qquad (2.5)$$

When the pressure is changed, some of these systems undergo phase transitions to the commensurate "$\sqrt{3}$ structure" with $\vec{q}_{01} = \frac{2\pi}{a}(0,\frac{\sqrt{3}}{2})$ or the two symmetric vectors. The symmetry, therefore is essentially the same as in the preceeding systems.

In general, the incommensurate state of the triple $\vec{q}$ system also includes higher harmonics. We shall assume that these harmonics will not lower the symmetry further by destroying the hexagonal pattern. This requirement can be formulated in a very elegant way using symmetry groups of dimensionality 3+n, where n is the number of wave vectors involved [11].

3. The Commensurate - Incommensurate Transition. McMillan's Theory for 1d-modulated Systems

In this section we shall briefly review McMillan's theory for the I-C transition in a system where the ordering is characterised by a single wave vector. In general, the order parameter has several components corresponding to the equivalent, symmetric, wave vectors. For the system that we consider there are six equivalent vectors $\pm\vec{q}_1$, $\pm\vec{q}_2$, and $\pm\vec{q}_3$. The order parameter therefore has six components $\psi_{\pm 1}$, $\psi_{\pm 2}$ and $\psi_{\pm 3}$. In the Landau theory the free energy is expanded in terms of these components. According to McMillan [1] the appropriate expansion for the commensurate order parameter in $2H\text{-}TaSe_2$ is

$$\begin{aligned} F = \int d^3x \Big[& \tfrac{1}{2} r \sum_{i=1}^{3} |\psi_i|^2 + u\Big(\sum_{i=1}^{3} |\psi_i|^2\Big)^2 \\ & + v \sum_{i=1}^{3} |\psi_i|^4 \\ & + d_1(\psi_1\psi_2\psi_3 + \psi_{-1}\psi_{-2}\psi_{-3}) \qquad (3.1) \\ & + d_2 \sum_{i=1}^{3} (\psi_i^3 + \psi_{-i}^3) \\ & + e' \sum_{i=1}^{3} |(\vec{\nabla} - i\vec{\delta}_i)\psi_i|^2 \\ & + f' \sum_{i=1}^{3} |\hat{q}_i \times \vec{\nabla}\psi_i|^2 \Big] . \end{aligned}$$

Here $\vec{\delta}_i$ is a small vector along the commensurate wave vector q_{0i}. The term with coefficient e' favours an order parameter $\psi_i = A\exp(i\vec{\delta}_i \cdot r)$, i.e. $\rho(\vec{r}) = A\exp[i(\vec{q}_{0i}+\vec{\delta}_i)\cdot\vec{r}]$, according to (2.3). This elastic term thus tends to induce an incommensurate CDW with wave vector $\vec{q}_i = \vec{q}_{0i}+\vec{\delta}_i$. On the other hand, the Umklapp term with coefficient d_2 favours the commensurate wave vector given by $\psi_i = -A$. The phase transition takes place when $|\vec{\delta}_i|$ becomes small enough for the Umklapp term to compensate the elastic term. For the other system that we have in mind the Umklapp terms are of a slightly different form leading to different rational wave vectors. It is a trivial matter to generalize the theory to these cases.

Depending upon the values of the parameters in (3.1) either a single $\vec{q}$ wave, with $|\psi_1| \neq 0$, $|\psi_2| = |\psi_3| = 0$, or a triple $\vec{q}$ wave, with $|\psi_1| \sim |\psi_2| \sim |\psi_3|$ is stable. The basic assumption entering McMillan's theory is that near the I-C transition the amplitude, A, of the order parameter is essentially constant, but the phase, ϕ, is allowed to accomodate to the underlying lattice by fluctuating in space. Inserting

$$\psi_1 = A\exp i\phi(x), \psi_2 = \psi_3 = 0$$

we get the following energy density (apart from a constant term and a constant factor):

$$F(x) = \tfrac{1}{2}\left(\frac{d\phi(x)}{dx} - \delta\right)^2 - v(\cos\{3\phi(x)\}-1) - \tfrac{1}{2}\delta^2 \qquad (3.2)$$

where x is measured along the ordering wave vector and $\delta=|\vec{\delta}_1|$. The actual phase function is the one which minimizes the integral of F(x). The function $\phi(x)\equiv 0$ yields the commensurate phase with energy normalized to zero. The incommensurate phase is thus formed by space-dependent local displacements of the commensurate phase. The functions $\phi(x)$ for which this free energy has an extremum satisfy the sine-Gordon equation,

$$\phi''(x) - 3v\sin\{3\phi(x)\} = 0 \qquad (3.3)$$

with periodic boundary conditions $\phi(x+L) = \phi(x) + \frac{2\pi}{3}$. The average wave vector, $\bar{q}$, which may be measured in a diffraction experiment is

$$\bar{q} = \frac{2\pi}{3L} \qquad (3.4)$$

according to the definition (2.2). The solutions of this equation are the "solitons" [2] or domain walls separating almost commensurate regions with $\phi = \frac{2\pi n}{3}$ and $\phi = \frac{2\pi}{3}(n+1)$, n integer.

Inserting the solutions into (3.2) we find the free energy density

$$F(x) = \frac{4\sqrt{V}}{\pi}\,\bar{q}\left[1 + 4\exp\left(-\frac{2\pi\sqrt{V}}{\bar{q}}\right)\right] - \bar{q}\delta . \qquad (3.5)$$

Minimizing (3.5) with respect to $\bar{q}$ one finds $\bar{q}$ versus δ. The commensurate-incommensurate transition takes place when $4\sqrt{V}/\pi = \delta$. The terms in (3.5) may be interpreted in a simple way. Noting that $\bar{q}$ is proportional to the soliton density we may interpret the coefficient of $\bar{q}$ as the energy required to create a soliton. The remaining term depends exponentially on the distance between solitons and is also proportional to the number of solitons, and may thus be interpreted as a repulsive interaction between solitons.

The equilibrium soliton density results from a balance between the energy gain from forming solitons and the energy loss from the repulsive interaction. In section 5 we shall see that this picture is substantially different for the 2d-modulated case because of a different topological structure of the solitons! At the transition the density of solitons goes smoothly to zero, and the transition is thus of second order. The density of solitons may serve as the order parameter for the transition. The soliton density depends logarithmically on δ and hence also upon the temperature

$$\bar{q} \sim -\ln^{-1} \frac{T-T_0}{T_0}$$

or $\bar{q} \sim -\ln^{-1} \dfrac{P-P_0}{P_0}$

if the pressure is the field variable.

4. Coupling to Strain or "how to count solitons"

Before proceeding to consider the triple $\vec{q}$ case, let us consider the coupling to alternative degrees of freedom. To be specific, we consider the coupling to an elastic strain. It turns out that the transition remains second order; in fact one may argue that the soliton picture and the second order transition are stable against any such local coupling.

The theory presented in this section closely follows the calculation of Timonen and the author [12]. Recently Bruce, Cowley and Murray [13] considered the coupling to a uniform strain η. They argued that such a coupling will necessarily drive the transition first order. Here it will be demonstrated that the induced strain will in fact be of a very inhomogeneous, soliton-like nature, and that the transition remains second order. Cowley, Bruce and Murray showed that the symmetry of the system allows a linear coupling between the macroscopic strain η and the phase gradient. This coupling gives rise to additional terms in the free energy density

$$F_\eta(x) = w\eta\frac{d\phi(x)}{dx} + \tfrac{1}{2}c\eta^2. \tag{4.1}$$

Proceeding as before they found that the free energy density now includes a negative quadratic term $-\frac{w^2}{2c}q^2$ which clearly implies a first order transition.

In the calculation presented in [12], η is allowed to be position dependent. To simplify the calculation without changing the physical picture the cosine term in (3.2) is replaced by the potential (Fig. 2)

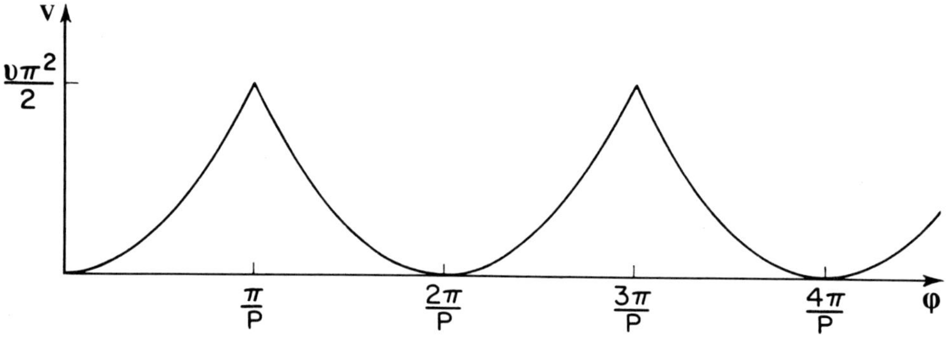

Fig. 2 Potential $V(\phi)$ used in the calculation. [12]

$$V(\phi) = \frac{9}{2}v\phi^2(x) \equiv \tfrac{1}{2}r\phi^2(x), \quad -\frac{\pi}{3}<\phi<\frac{\pi}{3}$$

$$V(\phi) = V(\phi + \frac{2\pi}{3}) \tag{4.2}$$

This is a periodic parabolic potential. In general the actual potential for a real physical system is unknown. The present calculation may thus also serve the purpose of illustrating that the soliton picture does not depend on the particular choice of potential. The complete free energy is

$$F = \frac{1}{L} \int_{-\frac{L}{2}}^{\frac{L}{2}} F(x)dx,$$

$$F(x) = \tfrac{1}{2}(\phi'(x)-\delta)^2 + \tfrac{1}{2}r\phi^2(x)$$

$$+ w\eta(x)\phi'(x) + \tfrac{1}{2}c\eta^2(x) + \tfrac{1}{2}d\eta'(x)^2 \qquad (4.3)$$

where a term quadratic in the strain gradient has been included for completeness. The Euler equations determining the extrema of this functional are

$$\phi''(x) - r\phi(x) + w\eta'(x) = 0 \qquad (4.4)$$

$$d\eta''(x) - c\eta(x) - w\phi'(x) = 0$$

These equations have the complete solutions

$$\phi(x) = A\sinh(\kappa x), \quad -\frac{L}{2} \leq x \leq \frac{L}{2}$$

$$\phi(x+L) = \phi(x) + \frac{2\pi}{3} \qquad (4.5)$$

$$\eta(x) = B\cosh(\kappa x), \quad -\frac{L}{2} \leq x \leq \frac{L}{2}$$

$$\eta(x+L) = \eta(x)$$

Here

$$\kappa^2 = \frac{1}{2d}\{c+rd-w^2 - \left[(c+rd-w^2)^2 - 4rdc\right]^{\frac{1}{2}}\}$$

$$A = \pi/3\sinh(\kappa L/2)$$

$$B = Aw\kappa/(d\kappa^2-c)$$

If the coupling is unphysically strong, $(\sqrt{rd}+w)^2>c$, the phase soliton picture breaks down and this type of solution ceases to exist. The parameter κ gives the width of the solitons. The free energy versus $\bar{q}$ is now

$$F = \frac{\pi u}{6\kappa}\bar{q}\left[1+2\exp(-\frac{2\pi\kappa}{3\bar{q}})\right] - \bar{q}\delta \qquad (4.6)$$

where

$$u = r + d\kappa^2(\frac{B}{A})^2 .$$

This free energy is of exactly the same form as (3.5) and includes no quadratic term in $\bar{q}$. The transition therefore remains second order. The main effect of the coupling to the strain is to modify the soliton width $1/\kappa$, and to change the transition temperature given by $\delta = \frac{\pi u}{6\kappa}$.

The phase solitons and the induced strain are shown in Fig. 3. The strain is highly inhomogeneous, being localized near

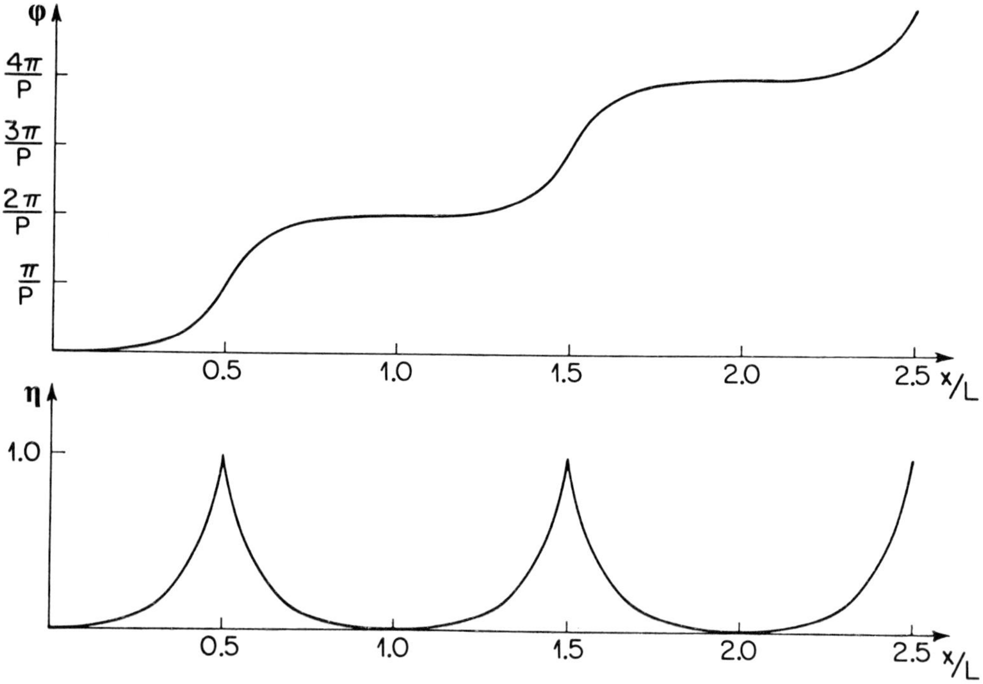

Fig. 3 Phase solitons and induced strain calculated for κ=10 (After [12])

the phase solitons. The cusps originate from our special choice of potential which also has cusps. The macroscopic, measurable strain, $\bar{\eta}$, is given by the integral

$$\bar{\eta} = \frac{1}{L} \int_{-\frac{L}{2}}^{\frac{L}{2}} dx \; \eta(x) = \bar{q} \frac{B}{\kappa A} \tag{4.7}$$

This is an interesting relation. The strain is proportional to the soliton density $\bar{q}$! To count the number of solitons in a crystal one may simply measure its distortion near T_0, i.e.

$$\bar{\eta} \sim \bar{q} \sim -\ln^{-1} |T-T_0| . \tag{4.8}$$

One may argue that any physical quantity coupling to the phase gradient or higher derivatives gives a contribution proportional to the number of solitons, or $\bar{q}$, since these derivatives are small except in the very narrow soliton regions. It is therefore expected that the coupling to such degrees of freedom will not change the soliton picture and drive the transition first order.

Our calculation points out a simple analogy: just as the assumption of a uniform phase ϕ leads to a first order transition which becomes second order of the phase is allowed to fluctuate in space, the assumptions of a constant strain leads to a first order transition which becomes second order if one allows it to be position dependent.

5. Phase Solitons in 2d-modulated Systems

For the physical systems which are known to exhibit I-C transitions, the modulated structure cannot be described by one wave vector but is formed as a superposition of waves with two or three symmetric wave vectors. To be specific we shall study a system characterized by three symmetric vectors $\vec{q}_1$, $\vec{q}_2$, and $\vec{q}_3$. This is the situation for the layered metal chalgogenides, for the magnetic structure of Nd, and for the rare gases adsorbed on substrates.

The starting point will be again the free energy expansion (3.1), but now we assume that the parameters favour the "triple $\vec{q}$" structures. Inspired by the 1d-modulated case we assume the amplitudes of the three waves to be fixed, but allow the phases to fluctuate,

$$\begin{aligned} \psi_{\pm 1}(\vec{r}) &= A\exp(\pm i\phi_1(\vec{r})) \\ \psi_{\pm 2}(\vec{r}) &= A\exp(\pm i\phi_2(\vec{r})) \\ \psi_{\pm 3}(\vec{r}) &= A\exp(\pm i\phi_3(\vec{r})), \end{aligned} \tag{5.1}$$

to obtain the following free energy functional

$$F = \int d^3r \Big[u\cos(\phi_1+\phi_2+\phi_3) + v(\cos3\phi_1 + \cos3\phi_2 + \cos3\phi_3) + e\sum_{i=1}^{3}(\vec{\nabla}\phi_i-\vec{\delta}_i)^2 + f\sum_{i=1}^{3}(\hat{q}_i x\vec{\nabla}\phi_i)^2\Big]. \quad (5.2)$$

Again we have ignored constant terms and factors. The first cosine term favours the phases $\phi_1=\phi_2=\phi_3 = 0$, and the equivalent combinations both in the commensurate and in the incommensurate phases. Just as we expect the amplitude fluctations to be inessential and therefore keep the amplitudes constant, we now make the additional ansatz that the sum of the phases remains zero. It turns out then to be convenient to define new phase variables

$$u_x = \phi_1 - \frac{\phi_2}{2} - \frac{\phi_3}{2}$$

$$u_y = \frac{\sqrt{3}}{2}(\phi_2-\phi_3) \quad (5.3)$$

$$\phi_1 + \phi_2 + \phi_3 = 0.$$

The variables u_x and u_y describe local displacements of the commensurate modulated structure in the perpendicular x- and y-directions in the plane spanned by $\vec{q}_1$, $\vec{q}_2$, and $\vec{q}_3$. This description is quite analogous to the 1d-situation, where the incommensurate phase was described by local displacements of the commensurate structure along $\vec{q}$. In fact, the continuous symmetry which is broken at the I-C transition is the translation of the CDW with respect to the lattice. In the 1d case the translational symmetry is broken in one direction, in the 2d-triple $\vec{q}$ case it is broken in two perpendicular directions. Our free energy functional thus seems to include the appropriate degrees of freedom.

The free energy functional becomes

$$F = \frac{1}{\Omega}\int c_1\Big(\frac{du_x}{dx} + \frac{du_y}{dy}\Big)^2 + c_2\Big[\Big(\frac{du_x}{dx} - \frac{du_y}{dy}\Big)^2 + \Big(\frac{du_x}{dy} + \frac{du_y}{dx}\Big)^2\Big] + c_3\Big(\frac{du_x}{dy} - \frac{du_y}{dx}\Big)^2 - \delta\Big(\frac{du_x}{dx} + \frac{du_y}{dy}\Big) + v\Big[3 - \cos(2u_x) - \cos(\sqrt{3}u_y-u_x) - \cos(\sqrt{3}u_y+u_x)\Big] \quad (5.4)$$

where $c_1=c/3$, $c_2=c/6 + f/6$, $c_3=f/3$. This free energy has a very interesting interpretation which is most clearly understood in terms of the two-dimensional monolayers on graphoil. The two first terms are the elastic energies associated with the two irreducible strains defined for a 2d-crystal. For the CDW these terms give the energies associated with the analogous elastic strains of the wave. The third term with coefficient c_3 is the elastic energy of an infinitesimal rotation of the 2d crystal. This term, of course, is zero for a free crystal. Here it represents an interaction with the lattice. If this term is negative, the equilibrium rare gas lattice or CDW will be rotated relative to the underlying lattice, and we would get the "orientational epitaxi" discussed by Novaco and McTague [15].

In fact, the functional (5.2) is one that one would intuitively apply directly to the rare gas lattice on the hexagonal graphoil surface. Our transformation has thus illustrated in a simple and efficient way the equivalence between this system and the "normal" triple $\vec{q}$ structures in 2H-$TaSe_2$ and Nd.

The actual distortions $u_x(x,y)$ and $u_y(x,y)$ are the ones which minimize the free energy functional. The commensurate phase is given by $u_x(x,y) \equiv 0$, $u_y(x,y) \equiv 0$.

In the 1d-case it was assumed that the solution is periodic, i.e. corresponds to a single $\vec{q}$ vector. In a similar manner it will be assumed here that the functions $u_x(x,y)$ and $u_y(x,y)$ have hexagonal symmetry, i.e. the condensed wave will always have sixfold symmetry, consistent with our general understanding of the incommensurate phase described in section 2. It can be shown that this symmetry will be conserved to any order in perturbation theory. [15]. The diffraction pattern consists of a hexagonal array of satellites situated around each reciprocal lattice point. For mathematical simplicity the cosine potential is again replaced by a truncated parabolic potential:

$$V(u_x,u_y) = Au_y^2 + Bu_x^2, \quad 0 \le u_y \le \pi/\sqrt{3}$$

$$0 \le u_x \le u_y/\sqrt{3} \qquad (5.5)$$

The potential defined in this way within the irreducible part of the hexagonal cell is continued periodically to cover the whole (u_x,u_y) space. (See Fig. 4).

In summary, the free energy density that we shall consider is

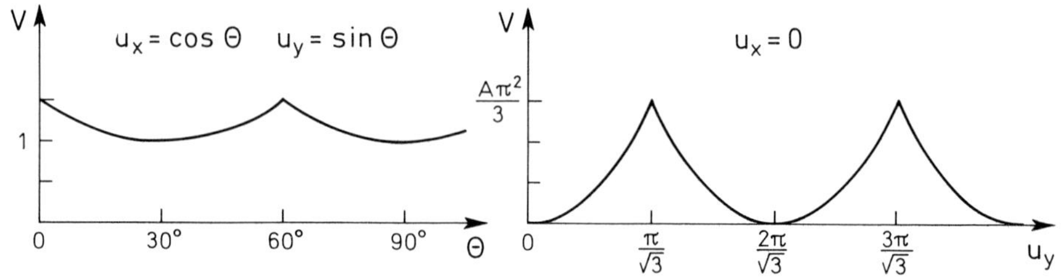

Fig. 4 Potential $V(u_x, u_y)$ used in the calculation of two-dimensional phase solitons

$$F = \frac{8\sqrt{3}}{L^2} \int^{\nabla} c_1 \left(\frac{du_x}{dx} + \frac{du_y}{dy}\right)^2 + c_2 \left[\left(\frac{du_x}{dx} - \frac{du_y}{dy}\right)^2 + \left(\frac{du_y}{dx} + \frac{du_x}{dy}\right)^2\right]$$
$$+ c_3 \left(\frac{du_x}{dy} - \frac{du_y}{dx}\right)^2 + Au_y^{\,2} + Bu_x^{\,2} \qquad (5.6)$$
$$- \delta \left(\frac{du_x}{dx} + \frac{du_y}{dy}\right)$$

where the integration area is described in Fig. 5.

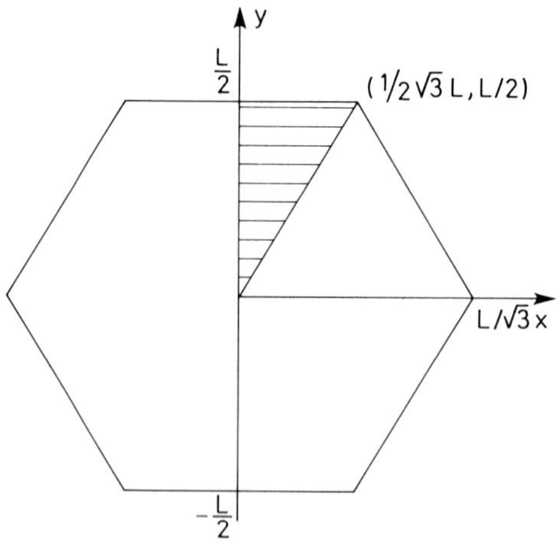

Fig. 5 Integration area used in (5.6)

The Euler equations are

$$(c_1+c_2)\frac{d^2u_x}{dx^2} + (c_2+c_3)\frac{d^2u_x}{dy^2} + (c_1-c_3)\frac{d^2u_y}{dxdy} - Bu_x = 0 \qquad (5.7)$$

$$(c_1+c_2)\frac{d^2u_x}{dy^2} + (c_2+c_3)\frac{d^2u_y}{dx^2} + (c_1-c_3)\frac{d^2u_x}{dxdy} - Au_y = 0$$

with the boundary conditions dictated by the haxagonal symmetry:

$$u_y(x,\tfrac{L}{2}) = \pi/\sqrt{3}$$

$$u_y(y/\sqrt{3},y) = \sqrt{3}u_x(y/\sqrt{3},y)$$

$$u_x(0,y) = 0. \qquad (5.8)$$

These equations can probably not be solved analytically in general, but for B=3A the complete solution can be found. There seems to be little reason to believe that this particular choice will affect the physical picture.

The solution is

$$u_x = \frac{\pi}{3}\,\frac{\sinh(\sqrt{3}\kappa x)}{\sinh(\kappa L/2)}$$

$$u_y = \frac{\pi}{\sqrt{3}}\,\frac{\sinh(\kappa y)}{\sinh(\kappa L/2)} \qquad (5.9)$$

with $\kappa^2(c_1+c_2) = A$. For large L this solution represents a network of solitons (Fig. 6). The incommensurate phase consists of large areas (or volumes) of essentially commensurate phase separated by a hexagonal lattice of solitons where the phases u_x and u_y change rapidly. This is consistent with the numerical result obtained by Nakanishi and Shiba [15]. For the 2d-monolayers these solitons are in fact the "misfit dislocations" introduced by Venables and Schabes-Retchkiman [16].

By feeding the solution back into (5.6) we find

$$F = \frac{4A\pi^2}{3L\kappa}\left[1+2\exp(-\kappa L)\right] - \frac{4\pi\delta}{\sqrt{3}L}$$

$$+ \left(\frac{c_1-c_2}{c_1+c_2}\right)\frac{8A\pi^2}{3\kappa^2L^2}\ . \qquad (5.10)$$

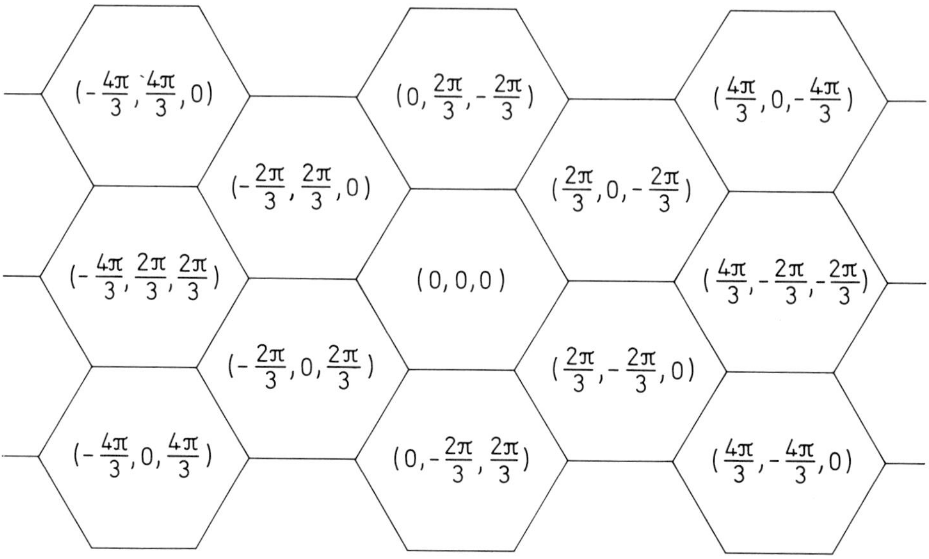

Fig. 6 Hexagonal soliton network near the I-C transition. The numbers in the brackets are the commensurate phases approached in the almost commensurate regions

The average wave vector, $\bar{q} = \frac{2\pi}{\sqrt{3}L}$ is inserted to yield

$$F = \left(\frac{2A\pi}{\sqrt{3}\kappa} - 2\delta\right)\bar{q} + \frac{2A}{\kappa^2}\left(\frac{c_1-c_2}{c_1+c_2}\right)\bar{q}^2 \qquad (5.11)$$

$$+ \frac{4A\pi}{\sqrt{3}\kappa}\bar{q}\exp\left(-\frac{2\pi\kappa}{\sqrt{3}\bar{q}}\right).$$

By minimizing this function with respect to $\bar{q}$ one finds $\bar{q}$ vs δ (or T, or P). The transition is first order for negative coefficient of $\bar{q}^2$, second order for positive coefficient. For the specific model studied by Nakanishi and Shiba the transition seems to be first order, but the present calculation lends support to the conclusion that it may indeed be second order.

The various terms in (5.11) may be interpreted in a simple way. The first terms is the usual soliton formation energy which becomes zero at the transition temperature given by $A\pi=\sqrt{3}\kappa\delta$. The third term is the positive repulsion between solitons. The second term is proportional to the square of the number of solitons, which in turn is proportional to the number of intersections in Fig. 6. This term thus gives the soliton intersection energy, which of course does not exist in the 1d-case. One finds

$$\bar{q} \sim \delta \sim (T_0 - T) \qquad (5.12)$$

in contrast to

$$\bar{q} \sim -\ln^{-1}(T_0 - T)$$

in the 1d-case (Fig. 7).

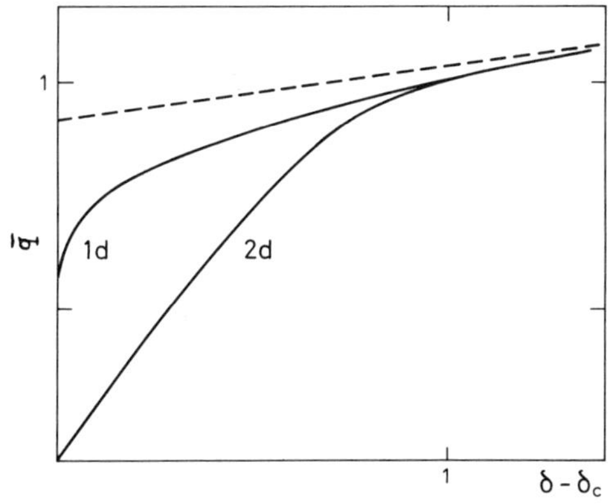

Fig. 7 Temperature dependence of average wave vector $\bar{q}$ near the I-C transition for single $\vec{q}$ and for triple $\vec{q}$ structures, respectively

This fundamental difference has a very simple topological explanation. The structure is stabilized by the competition between soliton formation energy and soliton intersection energy in the 2d-case, and not by the competition between the soliton energy and the repulsion energy as in the 1d-case.

The prediction (5.12) should be tested against experiment. The transition in 2H-$TaSe_2$ is first order [3], which of course is possible within the theory. More interesting in this respect is the LEED experiment performed by Chinn and Fain (Fig. 8) on krypton on graphite. The experiment does indeed show a second order I-C transition when the pressure is varied, and the lattice constant near T_0 seems to be consistent with the linear form (5.12). The neutron scattering experiment by McTague and Nielsen on graphoil was performed at constant coverage and not at constant pressure. The continuous result (Fig. 9) is however consistent with a second order transition. Otherwise a phase separation would have taken place near T_0. Preliminary neutron scattering experiments on Nd indicate a lock-in at a period of eight lattice units at T_0=8 K [10]. It would be interesting to study the temperature dependence of the wave-vector in this material and compare with (5.12).

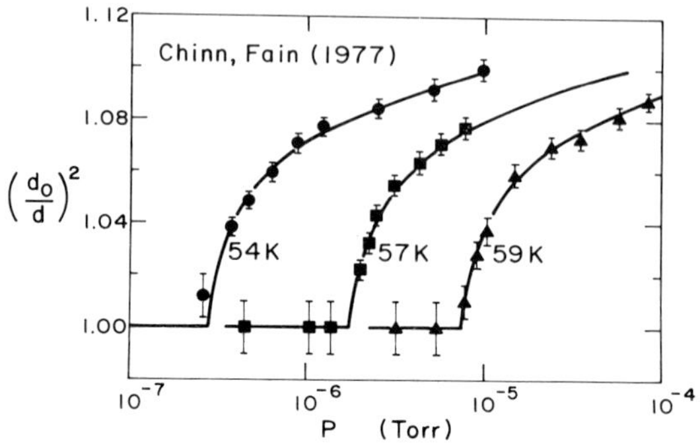

Fig. 8 Lattice parameter measurements for krypton on graphite. [5]

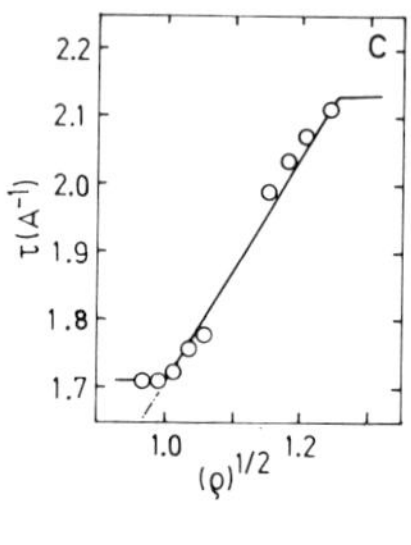

Fig. 9 Neutron scattering measurements of lattice parameter of D_2 on grapoil. [6]

There is one interesting complication. A positive intersection energy is required for a second order transition to take place. One may therefore ask whether an arrangement of parallel solitons might be energetically favourable at not too large δ. To test this, the hexagonal boundary conditions (5.8) are replaced by the conditions

$$u_y(x,\frac{L}{2}) = \pi/\sqrt{3}$$
$$u_x = 0. \tag{5.13}$$

These conditions break the hexagonal symmetry. The free energy becomes

$$F = \frac{1}{2}\left[(\frac{2A\pi}{\sqrt{3}\kappa} - 2\delta)\bar{q} + \frac{4A\pi}{\sqrt{3}\kappa}\bar{q}\exp(-\frac{2\pi\kappa}{\sqrt{3}\bar{q}})\right]. \tag{5.14}$$

For δ small enough, this free energy may be lower than that obtained by minimizing (5.11). The equilibrium soliton density is more than twice that given by (5.11) and goes logarithmically near T_0. At larger δ, the repulsive interaction will compensate the intersection energy and a first order transition to a symmetric state will take place. The picture that emerges thus includes two transitions and an intermediate low symmetry phase. This suggests that the second order transition may take place only through metastable states, and one should investigate this problem for example by looking for hysteresis effects.

Acknowledgements

I would like to thank J. Timonen and B. Lebech for a very pleasant collaboration on much of the research reported here. Also thanks are due to J.D. Axe, J. van Böhm, R.A. Cowley, V.J. Emery J.P. McTague, D. Mukamel, M.Nielsen, A.D. Novaco, V. Prokrovsky, and J.A. Venables for many interesting discussions. I am grateful to S.C. Fain, Jr. and M. Nielsen for their permission to reproduce Figs. 8 and 9, and to E. Riedel and G. Grinstein for a careful reading of the manuscript.

References

1. W.L. McMillan, Phys. Rev. B14, 1496 (1976).
2. P. Bak and V.J. Emery, Phys. Rev. Lett. 36, 978 (1976).
3. D.E. Moncton, J.D. Axe, and F.J. DiSalvo, Phys. Rev. Lett. 34, 734 (1975). Phys. Rev. B16, 801 (1977); J.A. Wilson, F.J. DiSalvo, and S. Mahajan, Advances in Phys. 24, 117 (1975).
4. P. Bak and B. Lebech, Phys. Rev. Lett. 40, 800 (1978).
5. S.C. Fain, Jr. and M.D. Chinn, Journal de Physique, 38, C4-99 (1978), Phys. Rev. Lett. 39, 146 (1977).
6. M. Nielsen, J.P. McTague, and W. Ellenson, Journal de Physique, 38, C4-10 (1978).
7. See, for example, W.C. Koehler in "Magnetic properties at rare earth metals", R.J. Elliott, ed., Plenum, New York, 1972 p. 81.
8. R. Comes, S.M. Shapiro, G. Shirane, A.F. Garito, and A.J. Heeger, Phys. Rev. Lett. 35, 1518 (1975), W.D. Ellenson, R. Comes, S.M. Shapiro, G. Shirane, A.F. Garito, and A.J. Heeger, Solid. State Commun. 30, 53 (1976).
9. P. Bak, Phys. Rev. Lett. 37, 1071 (1976); A. Bjelis and S. Barisic, Phys. Rev. Lett. 37, 1515 (1976).
10. B. Lebech, Private communications.
11. A. Janner and T. Janssen, Phys. Rev. B15, 643 (1977); P. Bak and T. Janssen, Phys. Rev. B17, 436 (1978).
12. P. Bak and J. Timonen, J. Phys. C, to be published.
13. A.D. Bruce, R.A. Cowley, and A.F. Murray; A.D. Bruce and R.A. Cowley, J. Phys. C, to be published.
14. A.D. Novaco and J.P. McTague, J. Physique 10, C4-116 (1977).
15. K. Nakanishi and H. Shiba, J. Phys. Soc. Japan, 44, 1465 (1978).
16. J.A. Venables and P.S. Schabes-Retchkiman, Surface Science 71, 27 (1978); Journal de Physique 38, C4-105 (1978).

Fluctuations and Freezing in a One-Dimensional Liquid: $Hg_{3-\delta}AsF_6$

J.D. Axe

Department of Physics, Brookhaven National Laboratory
Upton, NY 11973, USA

1. Introduction

In this talk we examine some of the properties which may arise in solids composed of two (or more) interpenetrating sublattices with spacings which are incommensurate one with another. The reason to suspect something out of the ordinary is shown by the following simple considerations. Imagine the two sublattices to be perfectly periodic and write their interaction energy as a product of the charge density, $\sigma_A(\vec{r})$, of one times the potential, $\Phi_B(\vec{r})$, of the other,

$$
\begin{aligned}
V_{AB} &= \int \sigma_A(\vec{r})\Phi_B(\vec{r})d\vec{r} \\
&= \sum_{GG'} \sigma_A(\vec{G})\Phi_B(\vec{G}') \int e^{i(\vec{G}-\vec{G}')\cdot\vec{r}}d\vec{r} \qquad (1)\\
&= \sum_{GG'} \sigma_A(\vec{G})\Phi_B(-\vec{G}')\delta_{G,G'}
\end{aligned}
$$

This shows that the two sublattices interact only by virtue of common reciprocal lattice vectors. Suppose that both sublattices can be thought of as two-dimensional arrays of chains arranged on a common rectangular lattice, but with different and incommensurate interatomic spacings along the common chain direction, z. It then follows trivially that the only common reciprocal lattice vectors have $G_z = G_z' = 0$ and the resulting forces, while constraining the chains in the x,y plane, do not fix the relative positions of the two sublattices along z. (In fact, the system can gain additional interaction energy by a mutual modulation of the natural period of one chain type with the period of the other, but this is a small effect and does not change the qualitative conclusion that the forces which act to localize the atoms on their chains are, at best, abnormally weak.)

1.1 $Hg_{3-\delta}AsF_6$. Perhaps the best studied example to date of the type of structure we have in mind is the mercury chain compound $Hg_{3-\delta}AsF_6$. It consists of an ordered body-centered tetragonal (bct) lattice of AsF_6^- anions (the host lattice) through which pass linear chains of polymercury cations arranged in two identical perpendicular nonintersection arrays, one parallel to $\vec{a}_T$, the other to $\vec{b}_T$. See Fig. 1. These will be referred to as the x- and y-arrays, respectively. Room temperature diffraction studies have shown in addition to the expected Bragg reflections, strong diffuse scattering arranged into series of thin sheets in reciprocal space[1-3]. Fig.2

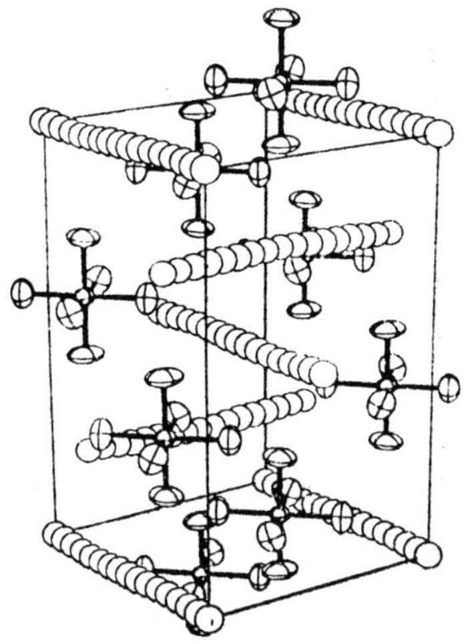

Fig. 1 Structure of $Hg_{3-\delta}AsF_6$. The octahedral AsF_6 groups carry one negative charge. The Hg-atoms on the chains are shown schematically. Above T_C=120 K the average Hg density is uniform along the chains. After A. J. SCHULTZ et al. (Ref. 3)

is a sketch of the (HKO) scattering plane. It is established that the diffuse sheets arise from the Hg-atoms and the narrow width of the sheets shows that the intrachain Hg-Hg distance, d, is well defined, and the nearly uniform distribution of intensity within a sheet shows that there is little or no interference between scattering from different chains [3,4]. Thus positions of the atoms along the chains are virtually uncorrelated from one chain to the next. Finally, from the spacing of the diffuse sheets, the interchain Hg distance d = 2.67 Å, which is incommensurate with a_L = 7.53 Å. This results from a non-stoichiometric composition $Hg_{3-\delta}AsF_6$ with $3-\delta = (a_L/d) = 2.82$. (A puzzling fact is that chemical analyses consistently find $\delta = 0$. Whether this is due to "pools" of excess Hg, to random vacancies on the host lattice, or neither, is at present unresolved.)

Further work by HASTINGS et al.[4] extended the diffraction study to low temperatures and concentrated on the behavior of the diffuse scattering in the m = 1 sheets. Fig. 3 shows that what is essentially a uniform distribution of intensity within a sheet at room temperature evolves into a pronounced modulation at 180 K. The modulation was interpreted as arising from short range correlations between the position of Hg atoms on nearby parallel chains. In the vicinity of T_c = 120 K sharp Bragg peaks grow out of the sheet of diffuse scattering with a temperature dependence typical of a continuous second order transformation (see inset, Fig. 3) and which must be associated with interchain ordering. Very peculiar, however, is the fact that the Bragg peaks do not develop at the positions on the sheets where the modulated diffuse intensity is strongest, but grow instead from regions

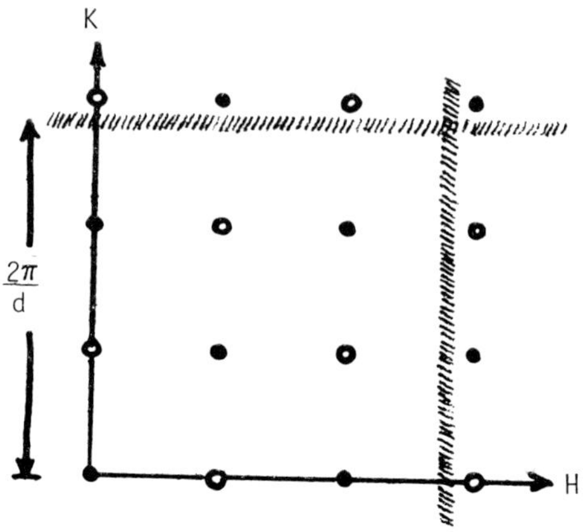

Fig. 2 A schematic representation of the diffraction pattern of $Hg_{3-\delta}AsF_6$ at room temperature. The straight lines represent the intersection of sheets of diffuse scattering lying perpendicular to the figure with the HK0 scattering plane

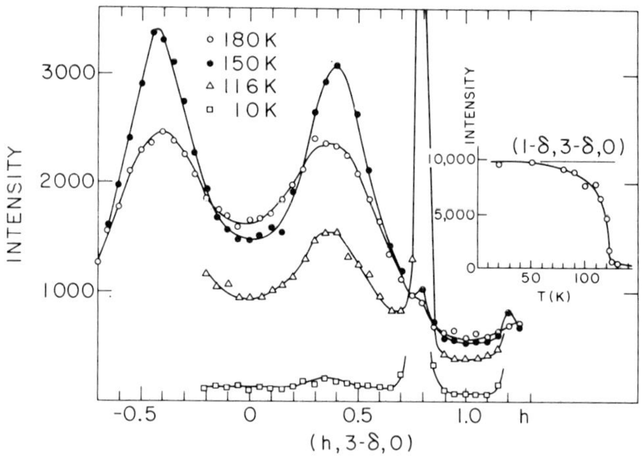

Fig. 3 Temperature dependent evolution of short and long range order as seen by the modulation of the m=1 diffuse sheet. Parallel chain interactions are responsible for the broad peaks at $h \approx \pm 0.4$. The long range order appears in the sharp Bragg peaks at $h = (1-\delta) \approx 0.82$. The inset shows the temperature dependent growth of the Bragg scattering below T_c. After J. M. HASTINGS et al. (Ref. 4)

of low intensity, i.e. the Bragg peaks are preceded by little or no "critical" scattering. The nature of the resulting ordering was deduced by HASTINGS et al. by noting that the positions of the Bragg peaks on the sheets were such that a reciprocal lattice vector from the x-array coincided with one from the y-array (at a point on the intersection of the two m=1 sheets). This fact, in conjunction with the theorem of the first paragraph, strongly implicates interactions between perpendicular chains as the dominant factor in the ordering. Unexplained, however, was the apparent sudden reversal of the relative importance of the parallel chain interactions, responsible for the short range order, and perpendicular chain interactions responsible for the long range order.

At high temperatures HASTINGS et al. also found that emanating from all points of the diffuse scattering sheets are inelastic scattering surfaces with linear dispersion depending only upon the component, Q, of momentum along the chain direction. That is $\omega = \pm v\,|Q-Q_m|$ and $Q_m = 2\pi m/d$ specifies the position of the m'th diffuse sheet. They ascribed this scattering to 1-d longitudinal phonons propagating along the independent Hg chains and found that $v = 4.4 \times 10^5$ cm/sec.

The remainder of this talk is devoted to a discussion of a simple model developed and analyzed by EMERY and AXE[5] for $Hg_{3-\delta}AsF_6$ (although with little modification it should be useful in thinking about other linear incommensurate phases as well.) It incorporates competing parallel and perpendicular chain interactions, predicts correctly the long range order and clarifies the apparent failure of the system to anticipate this ordering in the fluctuations above T_c. In addition, it treats carefully the effects of one-dimensional fluctuations, and predicts that the Hg chains at high temperatures behave as a one-dimensional liquid. The subsequent phase transformation can be thought of as a freezing of the 1-d Hg liquid, and can be discussed in terms of self-consistent solutions of the sine-Gordon Hamiltonian.

2. The Model Hamiltonian

The Hamiltonian is the sum of intra- and inter-chain contributions. $\mathcal{H}_{intra}$ assumes harmonic interactions, $\mathcal{H}_{intra} = \sum_{\ell i} \mathcal{H}^{o}_{\ell i}$ where for example

$$\mathcal{H}^{o}_{\ell x} = \frac{1}{2}\sum_{\alpha=1}^{N} \frac{\Pi^2(\ell x,\alpha)}{m} + K(x(\ell x,\alpha+1) - x(\ell x,\alpha)-d)^2 \tag{2}$$

where $(\Pi(\ell x,\alpha),\ x(\ell x,\alpha))$ are the components of the momentum and position vectors of the α'th particle on the ℓ_x'th chain. (The subscript i = x,y is to be used to specify the x- or y- array of chains.) The effective near-neighbor stiffness constant $K = mv^2/d^2$ is chosen to give the measured 1-d phonon velocity. m is the bare Hg atom mass.

The configuration of the ℓi'th chain is specified by the particle density operators, for example

$$\rho_{\ell x}(x) = \sum_{\alpha} \delta(x-x(\ell x,\alpha)). \tag{3}$$

In the disordered (high temperature) phase the Hg density is uniformly distributed along the chains, i.e. the thermodynamic averages

$\langle\rho_{\ell x}(x)\rangle = \langle\rho_{\ell y}(y)\rangle$= constant. In terms of the Fourier transformed variables, $\langle\rho_{\ell x}(Q)\rangle = \langle\rho_{\ell y}(P)\rangle = 0$ except for P = Q = 0. The quantities ($\langle\rho_{\ell x}(Q_m)\rangle$, $\langle\rho_{\ell y}(P_m)\rangle$ for $(P_m, Q_m) = 2\pi m/d$ can be taken as a complete set of order parameters specifying the chain ordering transformation. We will see that the instability is associated with the primary order parameters ($\langle\rho_{\ell x}(Q_1)\rangle$, $\langle\rho_{\ell y}(P_1)\rangle$). Note that we have retained the notion of a local chain variable by Fourier transforming only the position along the chain direction. Although it is useful in what follows to introduce wave vector components perpendicular to the chain directions as well, it is still important to distinguish between parallel and perpendicular components, as the latter are conjugate to discrete chain positions and can be restricted to the first Brillouin zone, whereas the former is associated with a continuous distribution along the chains and are thus unrestricted.

We introduce local coupling between chains of the form $\mathcal{H}_{inter} = \mathcal{H}_{xx} + \mathcal{H}_{yy} + \mathcal{H}_{xy}$ where

$$\mathcal{H}_{xx} = \frac{1}{2}\sum_{\ell x,\ell x'}\int dx \int dx' v^{\parallel}_{\ell x,\ell x'}(x-x')\rho_{\ell x}(x)\rho_{\ell x}(x') \tag{4a}$$

$$\mathcal{H}_{xy} = \frac{1}{2}\sum_{\ell x,\ell y}\int dx \int dy v^{\perp}_{\ell x,\ell y}(x-x^{o}_{\ell y}, y-y^{o}_{\ell x})\rho_{\ell x}(x)\rho_{\ell y}(y) \tag{4b}$$

These equations can be rewritten in terms of their Fourier transforms, e.g.

$$\rho_{\ell x}(Q) = N^{-1/2}\sum_{\alpha} e^{-iQx(\ell x,\alpha)}$$

and for example

$$\mathcal{H}_{xx} = \frac{1}{2}\sum_{Q} v^{\parallel}_{\ell x,\ell x'}(Q)\rho_{\ell x}(Q)\rho_{\ell}(-Q) \tag{4c}$$

where N is the number of atoms per chain.

2.1 Range of Interactions. We find that only rather near-neighbor coupling is necessary to explain the observed behavior of $Hg_{3-\delta}AsF_6$. The short range of the interchain coupling is understandable. If we associate a charge density, $\sigma_{\ell x}(x) = e^{*}\rho_{\ell x}(x)$ with the atomic density and calculate the Coulomb coupling between two parallel chains ($\ell x, \ell x'$) separated by a distance R, we find

$$v^{\parallel}_{x,x'}(Q) = \frac{2e^{*2}}{d}\int_{-\infty}^{\infty}\frac{\cos Q(x-x')}{[(x-x')^2+R^2]^{1/2}}dx = \frac{2e^{*2}}{d}K_o(QR) \tag{5a}$$

$$\approx \frac{2e^{*2}}{d}\left(\frac{\pi}{2QR}\right)^{1/2} e^{-QR} \quad (QR \gg 1) \tag{5b}$$

where $K_o(z)$ is a Bessel function. This shows that the coupling between charge modulations on parallel chains is exponentially small if the wave vector of the modulation is large compared to the inverse interchain spacing, R^{-1}. The coupling between perpendicular chains shows similar behavior. The important charge fluctuations are at multiples of $Q_1 = (2\pi/d)$ and for near-neighbor parallel chains $Q_1 R = 2\pi(a_L/d) = 2\pi(3-\delta)$. Although neighbor perpendicular chains are closer, we are justified not only in neglecting interactions between widely separated chains, but also in neglecting interactions involving harmonics of the fundamental chain spacing even on nearby chains. That is, the secondary order parameters $<\rho_{\ell x}(Q_m)>$, etc. with $m > 1$ play a vanishingly small role in the interchain coupling.

3. High Temperature Properties ($T > T_c$)

We discuss the thermodynamics using a generalized mean field theory in which the interchain coupling is approximated by a mean field but the resulting one-dimensional chain problem is solved exactly [6]. At low temperatures, where the full nonlinear response of the chains is important, this formulation leads to a sine-Gordon Hamiltonian, and thus is of most direct relevance for this conference. It is worthwhile, however, to sketch some results for $T > T_c$ since they display several unusual features of this system and establish much of the necessary justification for the model itself. For $T > T_c$ we need only the linear response, χ°, of the harmonic chain, so that

$$<\rho_x(\vec{q})> = \chi^o(\vec{q})h^{eff}(\vec{q}) \tag{6a}$$

$$h^{eff}(\vec{q}) = h^o(\vec{q}) - v^{\parallel}(\vec{q})<\rho_x(\vec{q})> - \sum_p v^{\perp}(-\vec{q},\vec{p})\Delta(\vec{p}-\vec{q})<\rho_y(\vec{p})> \tag{6b}$$

where we have now introduced Fourier components perpendicular to the chain directions, so that for the x-array $\vec{q} \equiv (Q,q_y,q_z)$ and for the y-array $\vec{p} \equiv (p_x,P,p_z)$. The notation emphasizes the mixed nature of the momentum variables, with the components represented by lower case symbols being defined modulo a reciprocal lattice vector and thus reducible to the first Brillouin zone. This mixed momentum representation is also in evidence through the function

$$\Delta(\vec{p}-\vec{q}) \equiv 1 \text{ if } p_x = Q(\text{mod}\vec{G});q_y = P(\text{mod}\vec{G}'); \; p_z = q_z;$$
$$\equiv 0 \text{ otherwise.}$$

where $\vec{G}(\vec{G}')$ is a reciprocal lattice vector of the x(y)-array.

Eq.(6), together with a similar set defining $<\rho_y(p)>$ are to be solved for the coupled response $\chi(\vec{q}) \equiv <\rho_x(\vec{q})>/h^o(\vec{q})$, or equivalently the pair correlation functions $<\rho_x(\vec{q})\rho_x(-\vec{q})> = kT\chi(\vec{q})$. (We will justify shortly the use of the classical form of the fluctuation-dissipation theorem.) Because of the umklapp momentum terms, the solutions can only be developed perturbatively. Their character depends upon the relationship of the momenta components along the two chain directions.

3.1 Uncoupled Solutions In regions of reciprocal space such that P and Q are not approximately equal the two chain arrays are effectively decoupled, and for the x-array

$$<\rho_x(\vec{q})\rho_x(-\vec{q})> = \frac{S^o(Q)}{1+\beta v''(\vec{q})S^o(Q)} \tag{7}$$

where $\beta \equiv (kT)^{-1}$ and a similar expression holds for the y-array. $S^o(Q) = kT\chi^o(Q)$ is the pair correlation function for an independent one-dimensional harmonic chain. It is given by

$$S^o(Q) = \sum_\alpha e^{iQ(x_\alpha^o - x_o^o)} <e^{iQu_\alpha} e^{-iQu_o}> = \sum_\alpha e^{iQd\alpha} e^{-\frac{1}{2}Q^2<(u_\alpha - u_o)^2>} \tag{8}$$

and $<(u_\alpha - u_o)^2>$ may be evaluated as an ensemble average over the single chain Hamiltonian, $\mathcal{H}_o$,

$$<(u_\alpha - u_o)^2> = \frac{d^2kT}{4\pi mv^2} \int_{-\pi/a}^{\pi/a} dq \frac{(1-\cos qd)}{\sin^2(\frac{qd}{2})} \equiv |\alpha|\sigma^2 \tag{9}$$

where $\sigma^2 = \frac{kT}{mv^2} d^2$ is the mean square fluctuation in nearest neighbor distance. As is well known, even though there is a well-defined average spacing, αd, for α'th neighbors, the harmonic 1-d chain lacks long range order since the mean square fluctuation about αd increases linearly with $|\alpha|$. Substituting (9) into (8) yields a geometric series which can easily be summed to give

$$S^o(Q) = \frac{\sinh(\frac{1}{2}\sigma^2Q^2)}{\cosh(\frac{1}{2}\sigma^2Q^2) - \cos(Qd)} \tag{10}$$

This is a typical liquid-like scattering function (see Fig. 4). Using the measured phonon velocity, we find for $Hg_{3-\delta}AsF_6$, $(\sigma/\alpha)^2 = 6.4 \times 10^{-4}$ at room temperature, which justifies the use of the harmonic approximation within the chain.

In the high temperature limit (somewhat above room temperature for $Hg_{3-\delta}AsF_6$) that we may set the denominator of (7) to unity and we recover the independent chain limit. For $Qd >> (\sigma/d)^2$ which is easily fulfilled in this case, $S^o(Q)$ consists of a series of nearly Lorentzian peaks (the sheets of scattering) centered at $Q_m = 2\pi m/d$ with a half width at half maximum, κ_m, given by $\kappa_m d = 2\pi^2(\sigma/d)^2m^2$.

The predicted value of κ_1 is consistent with a resolution limited width for the m = 1 sheet, but it should be possible to measure the widths of the higher order sheets and to verify whether or not they are proportional to m^2T [7].

As the temperature is lowered, the form of (7) and (8) shows that the effect of parallel chain interaction is first evidenced near $Q = Q_1$ since successive maxima in $S^o(Q_m)$ are weaker, $S^o(Q_m) = S^o(Q_1)/m^2$. (This explains the failure to observe modulation on the m = 2 sheet at temperatures where such modulation was pronounced at m = 1.) The modulation along the sheet is determined by $v''(\vec{q})$ and the existing data can be fit semiquantitatively with contributions from near neighbor and next near neighbor chains only, with $v_{nnn} \approx -2v_{nn} \approx 0.14$ K. (The interaction seems other than direct

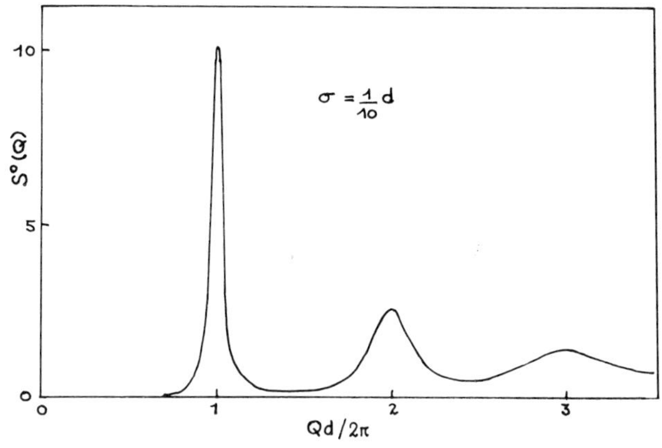

Fig. 4 The scattering function, $S^o(Q)$, for a 1-d harmonic model (see Eq.(10) shows a typical liquid-like pattern. For this case, $\sigma/d = 1/10$, the correlations are weak

Coulomb as v_{nn} is the wrong sign and both are $\sim$ 50 too large.) Although the interactions are weak, they are sufficiently enhanced by the long coherence length within a chain as to tend toward an ordered state only a few degrees below T_c = 120 K.

3.2 Coupled Solutions. The character of the solutions of (6), together with the corresponding ones for $\rho_y(\vec{p})$ are of a different character if the momenta along the two chains are nearly equal, P = Q. For these momenta, the x- and y-arrays are strongly coupled, giving rise to new fluctuation modes, $\rho^{\pm}(\vec{q}) = [\rho_x(\vec{q}) \pm \rho_y(\vec{q})]$ and the fluctuation scattering is proportional to

$$\sum_{i,j} \langle\rho_i(\vec{q})\rho_j(-\vec{q})\rangle = \frac{S^o(Q)}{1+\beta(v^{\parallel}(\vec{q})+v^{\perp}(\vec{q}))S^o(Q)} \tag{11}$$

which for reasons discussed above is enhanced for $Q = Q_m$, that is along the line of intersection of the m'th sheets (a reflection of the simple physics of (1)) and most enhanced for m = 1. Whether an instability first arises on the m = 1 sheet at (Q_1,Q_1,q_z) due to perpendicular chain coupling or at a more general position (Q_1,q_y,q_z) due to parallel chain coupling depends upon whether the denominator is smaller in (11) or (7); the former is the case for $Hg_{3-\delta}AsF_6$. We believe that the apparent failure to observe critical scattering above T_c is the result of the fact that the region of enhanced scattering is restricted to a linear dimension of order $2\kappa_1$ in the (a_T,b_T) plane. Since this width is below the existing experimental resolution, the basal plane scans should have the appearance of weak Bragg scattering persisting above T_c. Just such scattering has been observed, and it should be possible to establish its true character by determining whether the scattering is broad or narrow in the z direction, perpendicular to both chain arrays. In both sense (repulsive) and magnitude v seems roughly consistent with Coulombic interactions.

4. Long Range Order

As usual, we associate the order parameter with the mode giving rise to the divergent fluctuations (i.e. with the coupled mode solutions discussed above) and thus define a complex order parameter, $\eta_1 e^{i\psi} = \langle\rho_x(\vec{q}_c)\rangle = \pm\langle\rho_y(\vec{q}_c)\rangle$. The arbitrary phase factor $e^{i\psi}$ plays no role in determining the energetics of the system and is associated with a zero energy "sliding mode," familiar in incommensurate systems. For convenience, we set $\psi = 0$. η_1 specifies the amplitude of the sinusoidal modulation of the mean atomic density on a chain, e.g. $\langle\rho_{\ell x}(Q_1)\rangle = \eta_1 e^{i\phi^o_{\ell x}}$, where $\phi^o_{\ell x}$ is a phase associated with the perpendicular components of q_c and can be made to vanish by an appropriate choice of origin for each chain. Using this convention the mean field potential $\bar{v}$, obtained by replacing one of the density operators in (4) by its mean value, is identical for each chain. This allows us in what follows to suppress the chain index, ℓ_i, and we are left with the problem of a 1-d harmonic chain in a (commensurate) staggered field,

$$\bar{v}(\eta_1) = \sum_\alpha \left\{ \frac{K}{2}(u_{\alpha+1}-u_\alpha)^2 + \eta_1 h\cos Q_1 u_\alpha \right\} \tag{12}$$

where $h = 2(v^{\parallel}(\vec{q}_c) + v^{\perp}(\vec{q}_c))$. To discuss the evolution of the low temperature phase we must calculate the growth of all of the Fourier components of the atomic density on a chain. This can be done classically using transfer matrix techniques [8] since except at very low temperatures the effect of zero point fluctuations are negligible. The long intrachain coherence length, $\kappa_1^{-1} \approx 200$ d, allows us to pass to the continuum limit $((u_{\alpha+1} - u_\alpha) \to d(\partial u(x)/\partial x))$ and (12) reduces to the classical sine-Gordon potential and we must calculate

$$\eta_m = \langle\sum_\alpha \cos Q_m u_\alpha\rangle = \sum_\alpha \int du_1 \ldots du_N \cos(Q_m u_\alpha) e^{-\beta\bar{v}} \tag{13a}$$

$$= \langle\Phi_o|\cos Q_m u|\Phi_o\rangle / \langle\Phi_o|\Phi_o\rangle \tag{13b}$$

where $\Phi_o \equiv ce_o(q,v)$ is the lowest eigen vector of the transfer matrix and satisfies the Mathieu equation.

$$\left[\frac{d^2}{dv^2} + (a_o - 2q\cos 2v)\right]\Phi_o(v) = 0 \tag{14a}$$

$$q = \frac{4K\beta^2 h\eta_1}{Q_1^2} = -2\left(\frac{\beta}{\beta_c}\right)^2 \eta_1 \tag{14b}$$

where $2v = Q_1 u$. The transformation temperature $T_c = (k\beta_c)^{-1}$ is obtained by setting the denominator of (9) to zero for $\vec{q} = \vec{q}_c$.

Eq. (13) can be readily evaluated by developing $\Phi_0(v)$ in a Fourier series. When m = 1 (13) must be solved self-consistently with (14b). The temperature dependence of the first three Fourier components of the atomic density are shown in Fig. 5. For small η_1 ($T \approx T_c$)

$$\eta_1^2 \approx \frac{16}{7} [1 - \frac{T}{T_c}] \tag{15}$$

and η_m is proportional to $\eta_1{}^m$, while near T = 0

$$\eta_m = \eta_1{}^{m^2} \approx 1 - \frac{m^2}{\sqrt{8}} \left(\frac{T}{T_c}\right) \quad .$$

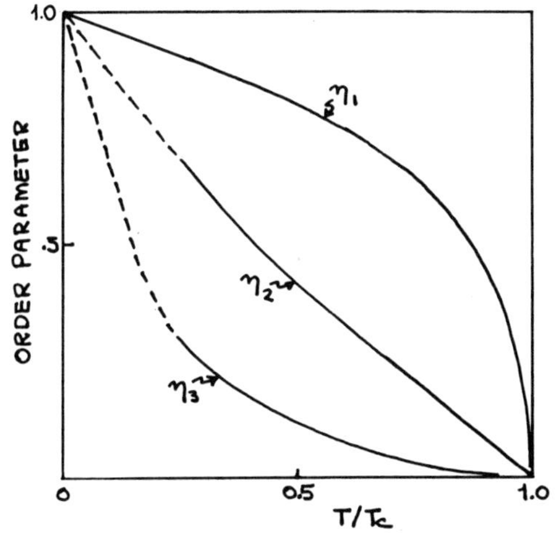

Fig. 5 Self-consistent solutions of the sine-Gordon Hamiltonian for the order parameter, η_m, associated with the first three density wave components

As $T \to 0$, the density distribution on the chain approaches that of a sum of Gaussians centered at x_α = nd with a mean square fluctuation $\langle (x-x_\alpha)^2 \rangle = (T/\sqrt{2}\, Q_1^2 T_c)$.

The temperature dependence of the Bragg scattering associated with Q_1 as studied by HASTINGS et al (see inset Fig. 3), rises much more quickly than predicted by (15). We believe that the reason for this is that the fluctuations associated with v'' may not be neglected for $T \approx T_c$ because they are divergent at $T = T_1$ a few degrees below T_c. The coupling of the two types of order parameters not only has the effect of promoting a more rapid growth in $\eta_1(T)$ but also in suppressing the v'' fluctuations below T_c, a feature which is very noticeable in Fig. 3. As this aspect of the theory is specialized to $Hg_{3-\delta}AsF_6$ we will not pursue it further here.

5. Dynamics

We conclude with a brief discussion of the dynamical properties that are to be expected in a system of loosely coupled harmonic chains. The dynamics are readily susceptible to calculation and contain several novel features which one can compare with neutron scattering experiments in progress.

In the high temperature limit, it is possible to redo the calculations summarized in (8) and (9) for the time dependent pair correlations

$$S^{o}(Q,\omega) = (2\pi)^{-1}\int dt e^{i\omega t}\sum_{\alpha} \langle e^{iQx_{\alpha}(t)} e^{iQx_{o}(0)}\rangle \equiv kT\chi^{o}(Q,\omega)$$

The result, in the vicinity of the m'th diffuse sheet (i.e. $|\Delta Q_m| = |Q-Q_m| << d^{-1}$) is

$$S^{o}(\Delta Q_m,\omega) = \frac{4}{\pi dv}\left\{\frac{\kappa_m}{(\Delta Q_m-\frac{\omega}{v})^2+\kappa_m^2}\right\}\left\{\frac{\kappa_m}{(\Delta Q_m+\frac{\omega}{v})^2+\kappa_m^2}\right\} \tag{16}$$

Eq.(16) deserves several comments.

1) The unusual product-of-Lorentzian form is characteristic of correlation functions of one-dimensional problems [9].

2) In deriving (16) one cannot proceed through the familiar separation into a product of a time dependent and time independent parts, as the latter (Debye-Waller) term vanishes while the time dependent fluctuations diverge. Similarly, there is no separation into one- and multiphonon terms. Eq. (16) represents the total density response.

3) When (16) is integrated over frequency one recovers (10), and as with a 3-d liquid, there is no truly elastic scattering (i.e. no term proportional to $\delta(\omega)$).

It is possible to extend the above results to include interchain coupling in the random phase approximation. The dynamical analogs of (7) and (11) are obtained by replacing $\langle\rho_i(\vec{q})\rho_j(\vec{q})\rangle$ by

$$S_{ij}(\vec{q},\omega) \equiv (2\pi)^{-1}\int e^{i\omega t}\langle\rho_i(\vec{q},t)\rho_j(-\vec{q}0)\rangle dt$$

and $S^o(Q)$ by $S^o(Q,\omega)$ in those expressions.

Below T_c the dynamics can be discussed in terms of weakly coupled sine-Gordon systems. For an individual chain there are two sorts of excitations to consider [10]. The first are free solitons for which

$$\omega_s^2 = \Delta_s^2 + v^2Q^2;\ \Delta_s^2 = \frac{16mv^2h\eta_1}{(\pi\hbar)^2} .$$

At low temperatures, the minimum energy necessary to create a soliton is so large ($\hbar\Delta_s/k \sim 700$ K) that these are not important thermal excitations, but the gap vanishes as $\eta_1{}^{1/2}$ near T_c. It may be possible to observe these excitations directly by neutron spectroscopy. The second kind of excitations can be described as bound soliton-antisoliton pairs or doublets for which

$$\omega_\nu^2 = \Delta_\nu^2 + v^2Q^2;\ \Delta_\nu^2 = 4\Delta_s^2\sin^2(\frac{\pi\theta\nu}{2})$$

where $\nu = 1,2,\ldots\theta^{-1}$ and $\theta = (\pi\hbar/2mvd)$. The maximum value of ν is the boundary of stability for breakup into a free soliton-antisoliton pair, whereas for small ν, $\omega_\nu \approx \pi\left(\frac{4h\eta_1}{md^2}\right)^{1/2}\nu$ and the excitations can be thought of as ordinary phonons near the bottom of the sinusoidal potential. These single-chain excitations form the basis for coupled collective modes which satisfy the lattice translational symmetry. Of course at low temperatures, the results may be obtained by making an harmonic approximation to the full Hamiltonian, including the interchain coupling, but near T_C the interactions are highly anharmonic.

6. Acknowledgments

It is a pleasure to acknowledge the collaboration of V. J. EMERY, who brought to this endeavor not only many of the ideas presented here, but a considerable sophistication which is missing in this account of it. We both profited from discussions of the experiments with J. M. HASTINGS, I. U. HEILMANN, J. P. POUGET, and G. SHIRANE. Research at Brookhaven was supported by the Division of Basic Energy Sciences, U. S. Department of Energy, under Contract No. EY-76-C-02-0016.

7. References

1. I. D. Brown, B. D. Cutforth, C. G. Davies, R. J. Gillespie, P. R. Ireland, and J. E. Verkris, Can. J. Chem. 52, 791 (1974).
2. C. K. Chiang, R. Spal, A. Denenstein, A. J. Heeger, N. D. Miro, and A. G. MacDiarmid, Solid State Commun. 22,293 (1977).
3. A. J. Schultz, J. M. Williams, N. D. Miro, A. G. MacDiarmid, and A. J. Heeger, Inorg. Chem., March, 1978.
4. J. M. Hastings, J. P. Pouget, G. Shirane, A. J. Heeger, N. D. Miro, and A. G. MacDiarmid, Phys. Rev. Lett. 39, 1484 (1977).
5. V. J. Emery and J. D. Axe, Phys. Rev. Lett. 40, 1507 (1978).
6. This approximation was first described in a systematic way by D. J. Scalapino, Y. Imry, and P. Pincus, Phys. Rev. B 11, 2042 (1975).
7. Neutron scattering measurements have verified these predictions. (Private communication, I. U. Heilmann, G. Shirane, and J. D. Axe).
8. S. F. Edwards and A. Lenard, J. Math. Phys. 3, 778 (1962); N. Gupta and B. Sutherland, Phys. Rev. A 14, 790 (1976).
9. See, for example, S. A. Brazovskii and I. E. Dzyaloshinskii, Zh. Eksp. Teor. Fiz. 71, 2338 (1976). (English transl., Sov. Phys. JETP 44, 1233 (1977)); A. Luther and I. Peschel, Phys. Rev. B 9, 2911 (1974).
10. R. F. Dashen, B. Hasslacher, and A. Neveu, Phys. Rev. D 11, 3424 (1975); A. Luther, Phys. Rev. B 14, 2153 (1976).

Charge Density Wave Systems: The ϕ-Particle Model

M.J. Rice

Xerox Webster Research Center, Webster, NY 14580, USA

1. Introduction

In this lecture I would like to introduce the subject of quasi-one-dimensional (1-d) charge density wave (CDW) condensates [1]; and to discuss the ϕ-particle model [2] which is a model of a charged non-linear excitation of a weakly-pinned CDW condensate. Such CDW condensates exist at low temperature in a number of well-known organic [1] and inorganic [3] linear chain conductors which at room temperature exhibit quasi-one-dimensional metallic behavior. It is intended that this lecture will provide useful background material for the following lectures of Drs. Luther, Friend and Horovitz and also provide the non-specialist with a simple introduction to the physical ideas basic to CDW phenomena in quasi-one-dimensional systems.

2. The Idea of Peierls (1955)

In 1955 Peierls [4] pointed out that a hypothetical 1-d metal would probably never exhibit metallic behavior because at absolute zero it would be energetically more favorable for the linear lattice to develop a periodic distortion with wavevector q_0 equal to twice the Fermi wavevector k_F of the conduction electrons. Such a distortion leads to an insulating state at absolute zero.

The argument is quite simple. In the uniform (metallic) linear chain system at absolute zero (T=0) all the conduction electron states with wavevectors $|k| \leq k_F$ are occupied, while those with $|k| > k_F$ are unoccupied. The density of states at the Fermi energy is finite. Now suppose the uniform chain is subject to a periodic distortion in which the j th atom or molecule is displaced from its equilibrium position R_j by an amount u_j given by $u_j = u \cos(q_0 R_j)$. Then, because of the electron-ion interaction, the conduction electrons will be subject to a new periodic potential with wavevector $2k_F$. Such a periodic potential will open up an energy gap of magnitude 2Δ in the electron spectrum precisely at the wavevector k_F. The density of states at the Fermi energy will be zero. Since the electronic states below k_F will now have been pushed lower, it follows immediately that the electronic energy of the distorted chain will be lower than that of the undistorted chain. If, for the distorted chain, the increase in elastic distortion energy δE_L is smaller than this decrease in the electronic energy δE_{el}, then the distorted linear chain will be the stable structure. For the linear chain this will always be the case at absolute zero since δE_L will be quadratic in the amplitude u of the periodic lattice distortion (PLD) whereas δE_{el} will be of the form $\delta E_{el} = b\, u^2 \log(1/|u|)$ where b is a constant. The logarithmic term is peculiar to one-dimension.

3. The Ideas of Fröhlich (1954)

Independent studies of the 1-d metal were made by Fröhlich [5]. He stressed the electronic response to the PLD and pointed out a series of interesting consequences. The main points are as follows:

(a) The PLD is balanced by a periodic modulation in the macroscopic conduction electron density n of the form $n(x) = n_0 + \delta n(x)$ where n_0 is the uniform density of the metallic state and $\delta n(x) = n_0 A \cos q_0 x$. $\delta n(x)$ describes a CDW with wavevector q_0 and its amplitude A may be related to the amplitude u of the PLD by use of force balance considerations [5,6]. The CDW and the PLD sustain each other and may be termed the condensate. In particular, it is to be borne in mind that the CDW is tied to the PLD and not to the equilibrium positions of the underlying atoms or molecules.

(b) The condensate will have a phase ϕ specifying its position relative to an external laboratory frame, so that in general the PLD and CDW are described by the equations

$$u_j = u \cos (q_o x + \phi)$$

$$\delta n(x) = n_o A \cos (q_o x + \phi)$$

A variation $\delta\phi$ in the phase translates the entire condensate — and hence n_0 electrons per unit length — over the distance $\delta x = \delta\phi/q_0$.

(c) For a translationally invariant system, such as was considered by Fröhlich, the condensate energy E_c is independent of the phase ϕ, $E_c \neq E_c(\phi)$: consequently the condensate may be set into a state of uniform motion with velocity v. This corresponds to the time-dependent phase $\phi(t) = \phi_0 + vt$ and a furnishes the collective current

$$i = n_o e \dot{\phi}/q_o$$

In the presence of an electric field E the equation of motion of the condensate may be written

$$M^* \ddot{\phi}/q_o = e^* E$$

where M* and e* are the total inertial mass and charge of the condensate, respectively. Consequently Fröhlich proposed the translationally invariant CDW condensate of the 1-d metal as a mathematical model for the then unexplained phenomenon of superconductivity.

4. The Ideas of Lee, Rice & Anderson (1974)

Now I should like to come to the ideas of Lee, Rice and Anderson (LRA) [7] who with the advent of the experimental realization of quasi-1-d metals, pointed out that, of course, in real systems translational invariance will be broken by a number of factors. The principle factors will be:

(a) The discreteness of the lattice (commensurability).
(b) Interchain interaction (presence of CDW's on neighbouring chains).
(c) Lattice defects — particularly charged impurities.

The condensate energy $E_c = E(A,\phi)$ will now depend upon the phase ϕ as well as on the amplitude A of the CDW. Consequently the condensate will now be subject to a finite restoring force

$$F_R = -\frac{\partial E_c}{\partial x} = -q_o\ \partial E_c/\partial\phi.$$

Thus a preferred position, denoted by $\phi = \phi_o$, may be expected for the condensate corresponding to $F_R(\phi_o) = 0$. The condensate may be said to be "pinned". For small displacements $\delta\phi$ from ϕ_o, we may assume the harmonic form $F_R = K\delta\phi$, and writing the spring constant K as $K = M^*\omega_F^2$, where ω_F has the dimensions of frequency, the equation of motion of the pinned condensate will be

$$M^*\ \delta\ddot{\phi}/q_o = e^*E - (\omega_F^2 M^*/q_o)\delta\phi$$

This equation describes simple harmonic motion of the phase with frequency ω_F and leads to a frequency dependent conductivity of the form

$$\sigma(\omega) = \frac{1}{L}\frac{e^{*2}}{i\omega M^*}\frac{\omega^2}{\omega_F^2-\omega^2}$$

and a static dielectric constant ε_s equal to

$$\varepsilon_s \simeq \frac{4\pi e^{*2}}{LM^*}\frac{1}{\omega_F^2}$$

which diverges as $\omega_F \to 0$. Here L denotes the length of the system. Weak pinning, that is small ω_F, leads to a low frequency IR absorption and to a large static dielectric constant. For the quasi-1-d Pt-complex compound KCP [3], ω_F is observed to be of the order of 2 meV and $\varepsilon_s \sim 3000$.

5. Pinning Potentials

I would now like to discuss briefly the form of the various pinning pontentials [7].

(a) Commensurability potential: this arises when the ratio of the linear chain reciprocal lattice vector G and q_o is equal to an integer M. There will then arise in $E_c(A,\phi)$, a term of the form $V(\phi) \propto (\Delta/W)^M (1-\cos(M\phi))$ where W denotes the electronic bandwidth. The CDW thus moves in a periodic potential with phase period $2\pi/M$.

(b) Interchain interaction: The Coulomb interaction V_{12} between CDW's on adjacent parallel chains will be a function of the relative phases $\phi_1-\phi_2$ of the two CDW's:

$$V_{12} \propto \Delta_1\Delta_2 \cos(\phi_1-\phi_2)$$

This interaction leads to ordering of the relative phases at low-temperatures and gives rise to an effective periodic pinning potential for a single chain $V(\phi) \propto \cos\phi$ with phase period 2π.

(c) Defect pinning: this is more complicated and in general leads to a pinning potential which is not a single periodic function of ϕ. E.g., for a random distribution of charged impurities $V(\phi)$ takes the form

$V(\phi) = \sum_i V_i \cos(q_0 x_i + \phi(x_i))$ where x_i describe the locations of the impurities and V_i the magnitude of the impurity potential. As discussed by Fukuyama and Lee [8] this leads to the phase of the CDW becoming pinned locally and consequently to a CDW ground state with an inhomogeneous phase!

6. Order Parameter and its Behavior

I should now like to make some general remarks concerning the order parameter of the CDW state and the determination of its behavior.

Clearly, the order parameter is complex and we may define it to be $\Psi = A\,e^{i\phi}$, where A denotes the amplitude and ϕ the phase of the CDW. In general, both A and ϕ may be functions of space and time, so that

$$\delta\rho(x,t) = A(x,t)\cos(q_0 x + \phi(x,t)) \qquad (\delta\rho \equiv \delta n/n_0)$$

The behavior of Ψ can be obtained if $E_c[\Psi]$ is known. A microscopic theory is required to provide the latter. Unfortunately microscopic theory is available only for simple models [1]. Thus one must proceed phenomenologically.

One way to proceed is to construct a Lagrangian density $L(x,t) = L[\Psi(x,t)]$ which is argued to have a physically resonable form. Then minimization of the classical action

$$\delta \int_{t_1}^{t_2} dt \int dx\, L[\Psi] = 0$$

yields the equations of motion for A and ϕ. These provide the ground state amplitude and phase profiles, and of course, also the excited state profiles. Also, for fields with one space dimension, the thermodynamic properties of the Lagrangian field may be rigorously investigated by functional integral techniques such, as for example, those followed by Scalapino, Sears and Ferrell [9].

A physically resonable form for $L[\Psi]$ is

$$L[\Psi] = c|\dot{\Psi}|^2 + \frac{a}{2}|\Psi|^2 - \frac{b}{4}|\Psi|^4 - d|\nabla_x\Psi|^2 - V_{pin}[\Psi]$$

where a, b, c and d are positive constants. In the absence of space and time dependences and the pinning energy this Lagrangian generates a double-well potential as a function of the amplitude A and is a form well-known in other order-parameter contexts [10].

In general, however, the present Lagrangian describes two-coupled non-linear fields [11]. Nevertheless if the magnitude of the pinning potential is very small by comparison to the depth of the double-well potential for A, the low lying excitations of the condensate will be dominated by deformations in the phase ϕ. These may be investigated by substituting into the Lagrangian the form $\psi(x,t) \simeq A_0\exp(i\phi(x,t))$ where A_0 is the uniform ground state amplitude. In this case the Lagrangian becomes a function of the phase ϕ only. This brings us to the ϕ-particle model [2] which is a model of non-linear phase deformations for cases involving such weak pinning potentials.

7. ϕ-Particle Model

In the ϕ-particle model, introduced by Rice, Bishop, Krumhansl and Trullinger [2], the Lagrangian is constructed as follows. n_s electrons per unit length are assumed to have condensed to form the CDW, each with a mass m^*. A ground state with a perfectly uniform phase ϕ_o is assumed and in excited states the local phase $\phi(x,t)$ is measured relative to ϕ_o. The kinetic energy density is then equal to $\frac{1}{2} m^* n_s \dot{\phi}(x,t)^2/q_o^2$. Spacial variation in the phase leads to a strain potential energy density and this is taken to be PE = $n_s \frac{1}{2} K(q_o^{-1}\nabla_x\phi)^2$. Re-expressing the spring constant K by $K = m^* c_o^2$, where c_o has dimensions of velocity, we have $PE = \frac{1}{2} m^* n_s q_o^{-2} c_o^2|\nabla_x\phi|^2$. The pinning potential is written as $n_s V_p(\phi)$ where $V_p(\phi)$ is an even function of ϕ and periodic in ϕ with period 2π. Without loss of generality, $V_p(\phi)$ is expressed in the reduced form $V_p(\phi) = m^* q_o^{-2} \omega_F^2 V(\phi)$. For small ϕ, $V(\phi)$ is assumed to possess the harmonic form $V(\phi) = \frac{1}{2} \phi^2$. Combining these three energy densities we obtain the Lagrangian density

$$L[\phi] = L(\phi_o) + n_s m^* q_o^{-2} \left\{\frac{1}{2} \dot{\phi}^2 - \frac{c_o^2}{2} (\nabla_x\phi)^2 - \omega_F^2 V(\phi)\right\}$$

The Lagrangian is valid for small gradients in the phase but arbitrary large variations in the magnitude of the phase.

It is useful to supplement this Lagrangian with the following physical information [2]. The time derivative of the local phase generates the local current density $j_s(x,t)$ according to $j_s(x,t) = n_s\, e\, q_o^{-1}\dot{\phi}(x,t)$. Also, a spacial gradient in the local phase leads to a modulation δn_s in the local condensed electron density according to $\delta n_s(x,t) = -n_s\, q_o^{-1}\nabla_x\phi(x,t)$. This important consideration follows from the physical definition of the local phase. It is evident that j_s and δn_s satisfy the conservation equation $e\, \delta\dot{n}_s(x,t) + \nabla_x j_s(x,t) = 0$.

It follows from the Lagrangian that the equation of motion for the local phase is $\ddot{\phi} - c_o^2 \nabla_x^2\phi + \omega_F^2 (dV/d\phi) = 0$. This is a non-linear wave equation of the Sine-Gordon type. Small amplitude extended wave solutions of this equation correspond to the linear phase phonons obtained by LRA [7]. These have the dispersion relation $\omega_q^2 = \omega_F^2 + c_o^2 q^2$. Here, of course, we shall be interested in the solitary wave solutions.

8. Solitary Wave Solutions: ϕ-Particles

The solitary wave solutions $\phi_\pm(x-vt)$ are given by

$$\pm(x-vt)/(1-v^2/c_o^2)^{1/2} = (\ell/\sqrt{2}) \int_{\pm\pi}^{\phi_\pm} d\phi V(\phi)^{-1/2}$$

in which the length $\ell = c_o/\omega_F$. These describe a local shift in phase of $\pm 2\pi$ propagating at uniform velocity v ($|v| < c_o$). The distance D over which the phase variation takes place is $D \simeq 2\ell \sqrt{1-v^2/c_o^2}$. Evidently the solutions describe propagating domain walls which separate segments of the condensate having common uniform phase. However, in view of the relation between the local condensed electron density and phase gradient, they also imply propagating localized compressions or rarefactions in the local condensed electron density $n_s(x,t)$. The solitary wave solutions can thus be regarded as describing pseudo relativistic particles of (rest) mass M and charge $\pm e^*$; these may be termed ϕ and anti-ϕ particles.

The excitation energy of the ϕ-particles may be computed from the Hamiltonian density to be $E_\pm(v) = M c_0^2/\sqrt{1-v^2/c_0^2}$ where the rest mass M is given by

$$M = 8n_s \, \omega_F \, M^*G/q_0^2 \, c_0$$

in which G is a numerical factor depending on the shape of the periodic pinning potential (G = 1 for a SG potential):

$$G = \frac{1}{\sqrt{8}} \int_0^\pi d\phi V(\phi)^{1/2}$$

Note that $M \to 0$ as $\omega_F \to 0$.

The ϕ-particle charge e^* is obtained by an integration over the phase gradient

$$e^* = -en_s \, q_0^{-1} \int_{-\infty}^{\infty} dx \, \nabla_x \phi_\pm$$

yielding $e^* = \mp 2e(n_s/n)$. At absolute zero, for a non-gapless condensate, $n_s = n$ and $e^* = \mp 2e$.

Since the total number of condensed electrons is a conserved quantity, the spacial integral over δn_s must be zero at all times. Thus, ϕ-particles are created in pairs of positively and negatively charged particles.

9. ϕ-Particles as Elementary Excitations

Can the non-linear ϕ-particles be regarded as elementary excitations of the condensate? The answer appears [2] to be yes for k_BT small compared to the rest energy $E_0 = M c_0^2$. The functional integral technique [9] may be employed to compute the classical free energy F. In the limit $T \to 0$ one finds a term in the free energy equal to $F_t = -2k_BT \, (L/2\ell) \exp(-E_0/k_BT)$. This is just what we expect for a "lattice gas" of $N(T) = (L/2\ell) \exp(-E_0/k_BT)$ ϕ-particle pairs distributed at random over $L/2\ell$ ϕ-particle sites and $(L/2\ell)$ anti-ϕ-particle sites.

It is of interest to know the magnitudes of the ϕ-particle parameters for assumed typical values of the phenomenological constants. The results of a representative calculation are shown in the Table below. There we have assumed $c_0 \sim 10^5$ cm sec^{-1}, $m^*/m \sim 10^2$, $\omega_F \sim 1.5$ meV, $q_0 \sim 5 \times 10^7$ cm^{-1}, $n_s = n_0$ and $G \sim 1$.

Table 1 Representative values of the ϕ-particle constants

e^*	2ℓ	$M c_0^2$	M/m
$\mp 2e$	88Å	220°K	11.5

10. Conductivity

It is interesting that although the condensate is pinned, and therefore unable to contribute a collective Fröhlich conductivity [5], the charged ϕ-particles now render it conducting. Indeed, for weak pinning the ϕ-particle excitation energy $E_0 = M c_0^2$ may be anticipated to be much smaller than the single-electron energy gap, in which case ϕ-particles will be the dominant current carriers at low-temperatures. In this context we mention that prior to the formulation of the ϕ-particle model the possibility that non-linear phase deformations could contribute a conductivity was investigated variationally by Pietronero, Strässler and Toombs (1975). If we introduce a phenomenological transport lifetime τ for the ϕ-particles the conductivity due to them will be given by [2]

$$\sigma_F = \frac{\pi^2}{2G}\frac{n_s e\tau}{m^*}\exp(-E_0/k_B T)$$

This describes a thermally activated conductivity with activation energy equal to E_0.

In recent work Larkin and Lee [13] have shown that if charged impurities are present ahd pin the CDW locally the transport lifetime will be of the form $\tau = \tau_0 \exp(-b\sqrt{m^*/m})$ where b is a constant. The exponential factor represents the probability amplitude for tunneling of a ϕ-particle through the impurity-pinned phase and severely reduces the ϕ-particle transport time. In other recent work Maki [14] has shown that at absolute zero an applied electric field F can lead to ϕ anti-ϕ pair-production. This leads to an interesting non-linear conductivity of the form $\sigma(F) \propto F_0/F \exp(-F_0/F)$ where F_0 is a characteristic (de-pinning) field.

The low-temperature, thermally activated, conductivities of the various organic quasi-1-dimensional metals have been tentatively [15] interpreted by the University of Pennsylvania group as arising from ϕ-particle transport. The experimental investigations in this direction, however, are still in an early phase of development.

11. Analogy with Schottky Defects

In conclusion, I would like to point out that the creation of ϕ-particles in the CDW condensate is analogous to the creation of Schottky defects in an ionic crystal. This is illustrated in the figures drawn below which depict the condensed electron density n_s as a function of x. At T = 0, n_s is everywhere uniform and corresponds to the perfectly formed CDW (Figure 1).

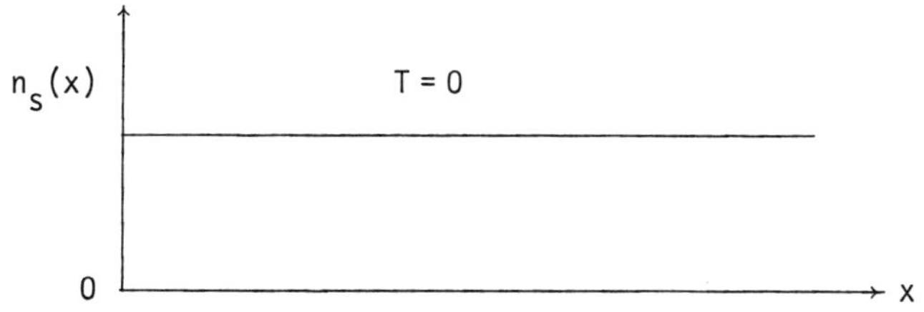

Fig.1 The condensate at absolute zero

At small but finite T thermodynamic considerations guarantee that there will always be a finite number of ϕ-particle pairs, $N(T) \neq 0$. These disorder the otherwise perfect CDW (Fig. 2).

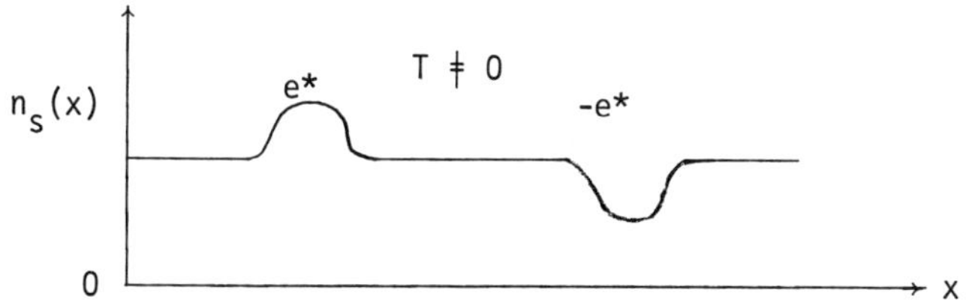

Fig.2 Disordered condensate at $T \neq 0$

In this respect we may regard the ϕ-particles as the inevitable Schottky defects of the perfect "ionic crystal" of which the perfectly formed CDW is representative.

References

1. For a recent general review see G. A. Toombs: Phys. Reports 40C, 181 (1978)
2. M. J. Rice, A. R. Bishop, J. A. Krumhansl, S. E. Trullinger: Phys. Rev. Lett. 36, 432 (1976)
3. P. Brüesch, S. Strässler, H. R. Zeller: Phys. Rev. B12, 219 (1975).
4. R. E. Peierls: Quantum Theory of Solids (O.U.P., London 1955) p. 108
5. H. Fröhlich: Proc. Roy. Soc. A223, 296 (1954)
6. M. J. Rice, S. Strässler, W. R. Schneider: Lecture Notes in Physics 34, 282 (1975)
7. P. A. Lee, T. M. Rice, P. W. Anderson: Solid State Comm. 14, 703 (1974).
8. H. Fukuyama, P. A. Lee: Phys. Rev. B17, 535 (1978)
9. D. J. Scalapino, M. Sears, R. S. Ferrell: Phys. Rev. B6, 3409 (1972).
10. For example, see J. A. Krumhansl, J. R. Schriefer: Phys. Rev. B11, 3535 (1975)
11. For example, see S. Sarker, S. E. Trullinger, A. R. Bishop: Phys. Letts. 59A, 255 (1976)
12. L. Pietronero, S. Strässler, G. A. Toombs: Phys. Rev. B12, 5213 (1975)
13. A. I. Larkin, P. A. Lee: Phys. Rev. B17, 1596 (1978)
14. K. Maki: Phys. Rev. Lett. 39, 46 (1977)
15. W. J. Gunning, S. K. Khanna, A. F. Garito, A. J. Heeger: Solid State Comm. 21, 765 (1977)

The Soliton Lattice: Application to the ω Phase

Baruch Horovitz

Laboratory of Atomic and Solid State Physics and Materials Science Center
Cornell University, Ithaca, NY 14853, USA

1. Introduction

In the present work we study the connection among three subjects. 1) Dislocation defects in crystals. 2) The physics and thermodynamics of solitons. 3) Displacive phase transitions and the quasi-elastic scattering in the disordered phase (the "central peak"). The idea that connects these subjects is that a displacive phase transition also involves a change in the lattice constant, i.e. a volume change. If a domain of the ordered phase appears within the disordered phase, it will suffer a strain due to the discrepancy in lattice constants. The interaction of these incommensurate structures in a one dimensional (1D) model leads to the solitons as described in some of the previous contributions in this volume, (see A.D. BRUCE; P. BAK; J.D. AXE).

An excellent case for the study of this problem is the β-ω transformation in Zr-Nb and similar alloys [1-5]. These alloys transform from a bcc (β) into a hcp structure below ~610°C. However, when alloys with 5-30 wt% Nb are quenched from their high temperature bcc phase into room temperature, regions of the so called "ω-phase" appear, imbedded in a bcc matrix.

The ω phase can be described by considering the [1,1,1] planes of the bcc lattice, which form a sequence of the type ABCABCA... (Fig.1). The A-A separation is the nearest neighbor distance in the bcc-3a, where a is the nearest [1,1,1] plane distance in the bcc. The ω_1 subvariant [6] is formed by collapsing the B-C planes towards their midplane while the A planes remain unshifted. The ω_2 and ω_3 subvariants are formed by retaining the B or C planes respectively and collapsing the other pairs of planes.

The ω structure corresponds to a longitudinal displacement modulation in the [1,1,1] direction of the form $u \sin(2\pi n/3 + \varphi)$, where n indexes the [1,1,1] planes and $\varphi = 0, 2\pi/3, 4\pi/3$ for the three possible subvariants. In the ideal ω phase $u = u_m \equiv a\sqrt{3}$ and pairs of planes are fully collapsed to form a hexagonal structure. For a partial collapse $u<u_m$ and the structure is trigonal.

The ω-structure has been studied extensively by neutron [1], electron [2,5] x-ray [3] and Mossbauer [4] techniques. The interesting features of

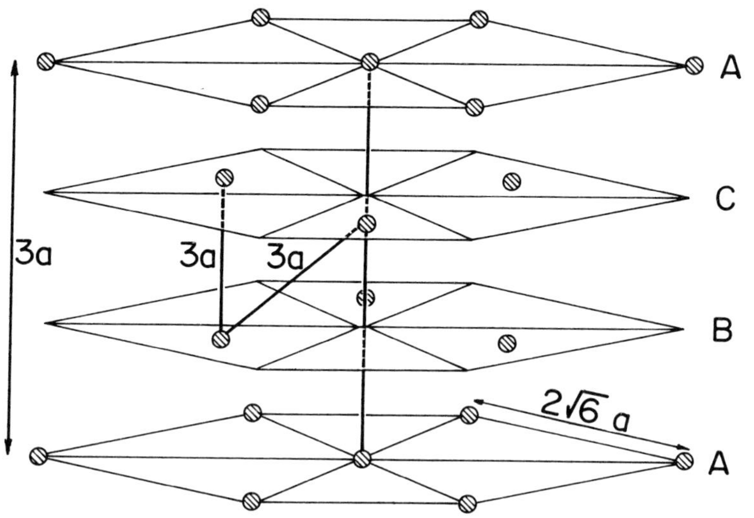

Fig. 1 The bcc structure viewed along its [1,1,1] direction. For convenience the [1,1,1] axis is expanded by a factor of 3 relative to the perpendicular axes. The nearest neighbor distance is 3a. The ω phase is obtained by collapsing planes B, C, while planes A are unshifted

the experimental data are the following: a) As Nb concentration increases the ω phase reflections become more diffuse and tend to elongate perpendicular to one of the [1,1,1] directions. Intensity ratios show [6] that u is close to u_m for the 8% alloy, decreases to $\sim \frac{1}{2} u_m$ for the 20% alloy, and remains $\sim \frac{1}{2} u_m$ for the 30% alloy. b) In the 5-15% wt Nb alloys the ω reflections appear at (or very close to) the expected positions of the ω structure, while in the 20-30% wt Nb the diffuse peaks are shifted away from these positions corresponding to a larger modulation wavevector. This suggests some kind of transition at about 17% wt Nb. c) Preliminary results [7] show that as temperature is lowered to $\sim 5^{\circ}$K the peak position of the 30% alloy is hardly affected, while the peak shift in the 20% alloy becomes somewhat smaller but it is definitely present. d) The diffuse peaks are elastic within 3.10^{-9} eV [4] (at room temperature).

The shape of the ω phase reflections (feature a) shows that the ω phase appears in rod shaped clusters [2,3] along the $[1,1,1]_\beta$ direction. Thus we assume that the ω clusters are one-dimensional objects which interact with the surrounding β matrix. We do not use a microscopic theory for the β-ω transition. Instead we assume that the ω phase prefers a smaller lattice parameter than that of the β phase. It will turn out that this information is enough to construct an effective Hamiltonian for an appropriate phase variable in terms of which the scattering can be explained. This effective Hamiltonian leads to the well known non-linear soliton solutions which we call here stacking solitons. We then show that stacking solitons are stabilized if the discrepancy δ in the lattice constant is larger than some δ_c and then the ground state is an array of solitons. Thus for $\delta<\delta_c$ the β and ω phase are commensurate, while for $\delta>\delta_c$ they are incommensurate. We suggest that δ changes with Nb concentration, and that a commensurate-incommensurate transition occurs at $\sim$17% Nb concentration.

In the next section we define the stacking solitions, following [8]. In the third section we study the thermodynamics of solitons, compare with experiment and suggest other experiments.

2. Stacking Solitons

Let us first examine the defect proposed by BORIE, SASS and ANDREASSEN (BSA) [6]. They found that the correct shifts in the diffuse scattering can be reproduced if a very particular sequence of subvariants, ω_1, ω_3, ω_2, ω_1, ... exists along the ω cluster. (Fig.2a). If the positions of the [1,1,1] planes are described by

$$x_n = na + u \sin(q_o na + \varphi_n) \tag{1}$$

where $q_o = 2\pi/3a$, then as n increases the BSA defect sequence implies that φ_n jumps by + $2\pi/3$ from one subvariant to the next. Since subvariants are degenerate ground states this defect may be thought of as a soliton [9]. An antisoliton, which would vary locally from 0 to $-2\pi/3$ is not compatible with the BSA sequence. In fact, the form (1) implies immediately that a monotonically increasing function φ_n leads to a local wavevector which is larger than q_o, and this leads to the observed shifts in the diffuse peak positions.

The BSA sequence was developed as an ad-hoc explanation for the shifts in the diffuse peaks. This sequence, viewed as a soliton, is a proper excitation mode, however it is not acceptable here for the following reasons: a) individual solitons can be thermally excited but are not a property of of the ground state. However, experimentally the peak shifts are present even at low temperatures and the defects should be described by the ground state. b) There is no reason for the system to prefer solitons or antisolitons; if present in equal numbers the peaks would not be shifted.

It was suggested [10] that the bcc instability is associated with a phonon mode whose wavevector q_m differs from q_o. Such a modulation q_m is incommensurate with the lattice and solitons could result [9]. However, the Mossbauer spectroscopy [3,4] shows that the inelastic scattering is centered at the exact ω positions, although the elastic portion is shifted from these positions. Therefore the phonon intensity is centered at q_o and not at q_m.

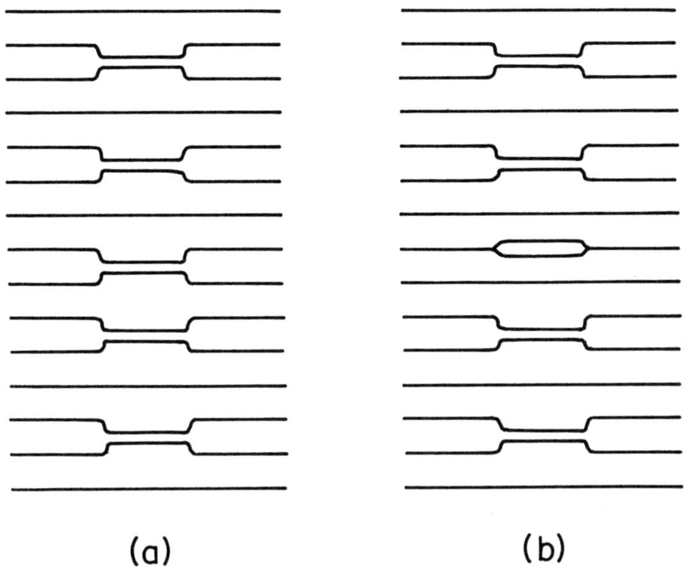

Fig.2 Defects in an ω phase cluster embedded in the bcc [1,1,1] planes. a) The BSA structure [6] representing a defect in the modulation pattern. b) The stacking soliton - the ω modulation pattern is maintained but there is a local density defect. Note that away from the defects the structures (a) and (b) are in exact registry. Hence the structure factors of (a) and (b) are very similar

In the present theory we concentrate on static properties; the Nb atoms are smaller than the Zr atoms and act as a pressure source. This favours the appearance of ω clusters with a natural lattice parameter $a - \delta$ embedded in a matrix with a lattice parameter a.

The average positions of the atoms in the chain are specificed by a sequence $\langle x_n \rangle$ from which the displacement modulation is defined. The values $\langle x_n \rangle$ are not necessarily separated by $a - \delta$, since the matrix strains the chain and prefers values of $\langle x_n \rangle$ separated by a. The actual positions of the atoms are $x_n = \langle x_n \rangle + u_n \sin(q_o \langle x_n \rangle + \varphi_n)$, and in general u_n and φ_n are position dependent amplitude and phase of the order parameter. Here we consider the simpler case $u_n = \text{const.} = \bar{u}$. [8]

The important physics is that variations in φ_n are coupled to variations in the density. This is derived from the identification of the local wave-vector with the local lattice constant by the relationship

$$q_o a = q_{loc} \cdot a_{loc} \tag{2}$$

where $a_{loc} = \langle x_{n+1} \rangle - \langle x_n \rangle$ and

$$q_{loc} = q_o + (\varphi_{n+1} - \varphi_n)/a_{loc} \tag{3}$$

Note that the modulation pattern is defined by $r = q_o a/2\pi$ (for the ω phase $r = 1/3$), and is independent of $\langle x_n \rangle$. Eq. (2) means that this pattern on the chain is indeed locked into its own lattice, in spite of possible shifts in the positions $\langle x_n \rangle$. Eqs. (2,3) lead to

$$\langle x_n \rangle = na - \varphi_n/q_o \tag{4}$$

$$a_{loc} = a - (\varphi_{n+1} - \varphi_n)/q_o \tag{5}$$

The relation (5) between variations in the phase and the local density in a continuum limit is well known for charge density wave systems [11,12]. From Eq. (4) we obtain the following equivalent relations:

$$x_n = na + \bar{u}\sin(q_o\langle x_n \rangle + \varphi_n) - \varphi_n/q_o \tag{6}$$

$$x_n = na + \bar{u}\sin(2\pi rn) - \varphi_n/q_o \tag{7}$$

These results are different from the form (1). In particular, a solution for φ_n which varies from o to $q_o a$ leads to the BSA defect in (1) while in (7) there is no defect in the modulation pattern; it is the density which is modified locally. (Fig.2). This demonstrates the physics behind (2): The phase is locked to its own lattice i.e. the chain, but may be unlocked relative to the surrounding matrix. Therefore the BSA type solitons are not allowed in this scheme, but instead we can obtain "stacking solitons", that is, solitons in the phase of the chain relative to the surrounding matrix.

The locking energy V_{lock} is a periodic function of the shift $\langle x_n \rangle - na$ with periodicity a. A simple form is $\sim\cos[2\pi(\langle x_n \rangle - na)/a]$ so that

$$V_{lock} = \omega_o^2 [1 - \cos(\varphi_n/r)] \tag{8}$$

The competing phase dependent energy is the elastic energy of the form $\frac{1}{2}C[\langle x_{n+1} \rangle - \langle x_n \rangle - (a-\delta)]^2/a^2$. Thus we are led to the following phase dependent Hamiltonian

$$\mathcal{H} = \sum_n \left\{ \frac{C}{8\pi^2}(\psi_{n+1} - \psi_n)^2 + \omega_o^2(1 - \cos\psi_n) \right\} - \frac{C}{2\pi}\frac{\delta}{a}(\psi_N - \psi_{-N}) \tag{9}$$

where $\psi_n = \varphi_n/r$. The effects of time dependence are discussed later on.

In the last term of (9) $\psi_N - \psi_{-N}$ is the phase difference of the first and last atom in the chain and measures the length change (Eq.4). Therefore δ is identified as the pressure on the chain, and positive pressure ($\delta>0$)

favors increasing solutions ψ_n and length contraction.

The Hamiltonian (9) can be derived directly from Eq. (4) without identifying φ_n as a phase. This actually has been done [13] and should be classified as an epitaxy problem. [14]. Our approach shows that changes in the lattice constant are coupled intrinsically to the phase of a complex order parameter and therefore the solutions of (9) appear in the nucleation process of a phase transition.

The equation of motion for (9) is the sine Gordon equation. The soliton solution, when inserted in Eq. (6), leads to the defect in Fig.2b which we call a stacking soliton. On the other hand a soliton solution in (1) leads to the BSA defect of Fig.2a. Far from the defect centers the two structures are in exact registry; thus for a narrow soliton the x-ray scattering from the two types of defects are very similar to each other and to the experimental data [6,8]. In the next section we show why the stacking solitons should be the proper type of defect- as they explain all the other experimental features.

3. The Soliton Lattice

Denote the soliton mass of (9) (for δ=o) by E_s. (For explicit expressions see [8,14].) The last term of (9) reduces this mass by $\mu = C\,\delta/a$, and if $\mu > E_s$ the ground state will have a finite density ρ of solitons. This is described by the solutions of the soliton lattice [13-16]. Thus we obtain a transition in the ground state properties of the chain. For $|\delta| < \delta_c (\delta_c = aE_s/C)$ there no solitons and the phase is locked $\psi(X)$ = const., while for $\delta > \delta_c$ a soliton lattice appears. As δ grows the soliton density increases, until for $\delta/\delta_c \to \infty$ the solitons merge into a line $\psi(x) = 2\pi\delta x/a2$ which is the free chain with its lattice constant a-δ. (Clearly, if $\delta < \delta_c$ an array of anti-solitons will appear.)

The behaviour of the density ρ as function of temperature T can be studied in the continuum limit using the transfer integral technique. [15]. In this limit the mass of an isolated soliton is $E_s = 4\omega_o\sqrt{C}/\pi$ and its width is $\ell = a\sqrt{C}/(2\pi\omega_o)$. As $T \to \infty$ the locking potential is less effective and $\rho \to \delta/a^2$ as for the free chain. For a finite T one can use various expansions. For $T \ll E_s - \mu$, [17]

$$\rho = (2E_s/\pi\, T)\ \exp\ [-(E_s - \mu)/T] \tag{10}$$

Eq. (126) of [15] is a high T expansion; the same method can be used to obtain an expansion in $1/\mu$ valid for all T. The result is

$$\rho = \frac{2\mu}{\pi^2\, \ell E_s}\, [1-\left(\frac{\pi}{8}\right)^4 \frac{8E_s^4}{(\mu^2 + \pi^2\, T^2)^2} + 0\ (1/\mu^8)] \tag{11}$$

The function $\rho(\mu)$ is shown in Fig.3 at various temperatures. There are three qualitatively different regions: 1) For $\mu \ll E_s$ the ground state has $\rho = 0$, while at low temperatures (compared with $E_s-\mu$ and μ) solitons are thermally excited [17] as described by the leading term of $\ln \rho$ in (10). 2) At $\mu \sim E_s$ temperature has the most dramatic effect. At T = 0 there is a sharp phase transition - the commensurate-incommensurate transition at $\mu = E_s$. It is a continuous transition, but "almost first order" since $\rho \sim [\ln(\mu - E_s)]^{-1}$ near the critical μ. For $T \neq 0$ the transition is smeared, as expected in a one dimensional system. Therefore ρ is a sensitive function of T when $\mu \sim E_s$. 3) In the third region $\mu >> E_s$ temperature has a small effect on ρ- see (11). The ground state has a finite ρ which is close to the high T value.

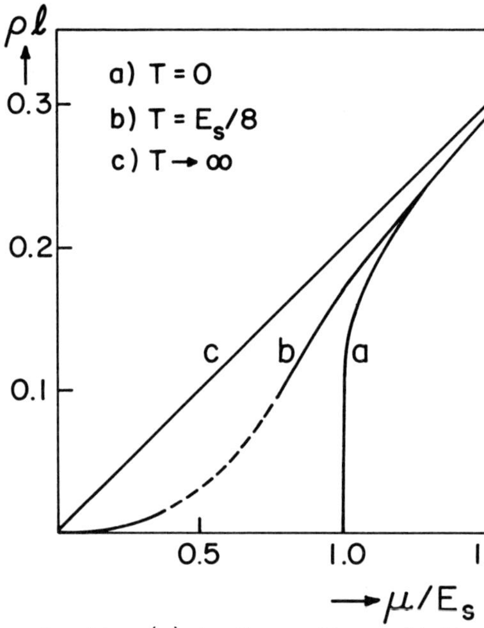

Fig.3 Soliton density $\rho(\mu)$ in the continuum limit at various temperatuares. Curve (b) is obtained from (10) and (11) which are valid at the lower and upper regions respectively; the dashed portion is an extrapolation between these regions. The mass and width of an isolated soliton are E_s and ℓ respectively

In a discrete model the exact solution $\rho(\mu,T)$ is different from Fig.3 [14], however the qualitative features should be the same. These features can explain the experimental data on the ω phase systems. This can be done if we assume: a) the ω cluster is long enough compared with the lattice constant and the soliton width. Therefore the effects of the end atoms on the amplitude u and the soliton solution can be neglected. From the fit to the x-ray data on the 20% and 30% alloys [8] an ω cluster has $\sim$40 planes, the soliton width is $\sim$a and each cluster has $\sim$1 soliton. b) Each plane within the ω cluster consists of a small and equal number of atoms so that the locking potential is constant within this plane and has the periodicity a.

This is consistent with the observed rod shaped ω cluster [3,5].

We proceed now to show that the ω phase systems are a laboratory test for the features of Fig.3. As Nb concentration increases the discrepancy $\delta(\sim\mu)$ in lattice constants increases; therefore the μ axis in Fig.3 is a measure of the Nb concentration. The stacking solitons shift the scattering intensity away from the commensurate wavevector q_o; therefor the ρ axis in Fig. 3 measures the magnitude of the peak shift. The commensurate-incommensurate transition is at $\sim$17% Nb; in higher %Nb alloys stacking solitons appear in the ground state, partially unlocking the ω cluster from the β matrix.

The temperature dependence is consistent with Fig.3, even if the ampli- $\bar{u}$ (or the coefficients in (9)) do not change with temperature [7,8]; the temperature dependence comes from the 1/T factor in the partition function. The 30% alloy correspond to $\delta >> \delta_c$ ($\mu >> E_s$) and its peak shift should depend weakly on temperature. The 20% alloy has δ closer to δ_c and its peak shift should depend more strongly on temperature. These results were indeed observed in a preliminary experiment [7]. We predict that the alloy with $\sim$17% Nb is the most sensitive to temperature change since then $\delta \sim \delta_c$ ($\mu \sim E_s$) and that the slight peak shift of the 15% alloy disappears at low temperature ($\mu < E_s$).

We also suggest the following experiments: Apply high pressure on the quenched Zr-Nb alloys at room temperature and then release the pressure. As in the pure Zr case [18], the alloy should transform into a 3D ω phase, and this phase should be retained after the pressure release. We predict that the resulting ω phase will have a smaller lattice constant in its [0,0,1] direction than the previous bcc lattice in its [1,1,1] direction. The change δ in the lattice constant should increase with Nb concentration and the critical δ_c at $\sim$17% Nb can be determined.

In a discrete system the solitons are free to move only if their width is large compared with the lattice constant [14]. Otherwise the solitons are locked and the vibration spectrum is above some finite frequency. In the ω clusters the solitons are narrow and therefor they form a static configuration. This is consistent with the extreme narrowness of the quasi-elastic peak. [4].

Finally we suggest that stacking solitons are relevant to the general problem of the central peak. Evidently we look for transitions which involve a change in the lattice constant, but this is probably true for all displacive transitions. Also one should study the appearance of stacking solitons on a surface of 3D clusters [19].

Models for the central peak [20,21] consider 1D systems, and it is still not understood why domains in 3D systems are so stable and quasi-static. These domains can either move around (phase fluctuations) or disappear (amplitude fluctuations). Both phenomena can be seen by computer simulations [22], and they tend to broaden the scattering intensity $S(q_0,\omega)$. Consider now the effect of stacking solitons: If the soliton is not too wide compared with the lattice spacing, translation requires finite energy and phase fluctuations are suppressed; this can happen in a discrete model also without stacking solitons. In addition however, the stacking soliton suppresses also amplitude fluctuations, since by taking the amplitude to zero the density defect is not removed (Fig.2b). Thus there is an energy barrier against motion or decay of domains, and then the central peak becomes narrower. Experimentally stacking solitons are seen by a shift of the peak in $S(q_0,\omega)$ away from q_0, and we suggest that this shift correlates with increasing narrowness of the peak around $\omega = 0$.

A well known example are the A15 compounds Nb_3Sn and V_3Si [23,24] which transform from cubic to tetragonal at 45°K and 21°K respectively. Evidently, this transition involves changes in the lattice constant and stacking solitons are favored. In fact, the central peak was found to exist also away $q = 0$ which is analogous to the shift in the ω reflection peaks. Various properties of this transition were explained by postulating the existence of defects [25]. Stacking solitons are natural and intrinsic structures which can represent these defects.

Acknowledgments

A large part of this work was carried out with J.A. Krumhansl and J.L. Murray. I am very grateful to D.H. Bilderback and B.W. Batterman for presenting me their unpublished data. I also wish to thank S. Aubry, T.S. Kuan and S.L. Sass for very useful discussions. This work has been supported by the Cornell University Materials Science Center, Grant #DMR-76-81083, technical report #3042.

References

1. D.T. Keating and S.J. LaPlaca, J. Phys. Chem. Solids 35, 879 (1973).
2. S.L. Sass, J. of Less Common Metals 28, 157 (1972); C.W. Dawson and S.L. Sass, Metall. Trans. 1, 2225 (1970).
3. W. Lin, H. Spalt and B.W. Batterman, Phys. Rev. B13, 5158 (1976).
4. S.K. Andersen and B.W. Batterman, (to be published).
5. T.S. Kuan and S.L. Sass, Phil. Mag. 36, 1473 (1977).
6. B. Borie, S.L. Sass and A. Andreassen, Acta Crystallographica A29, 594 (1973).
7. D.H. Bilderback and B.W. Batterman, (private communication).
8. B. Horovitz, J.L. Murray and J.A. Krumhansl, Phys. Rev. B, (to be published).
9. This has also been noticed by R. Pynn, J. Phys. F8, 1 (1978).
10. H.E. Cook, Acta Metal. 23, 1041 (1975); J.M. Sanchez and D. DeFontaine, J. Appl. Cryst. 10, 220 (1977).
11. T.M. Rice, Solid State Commun. 17, 1055 (1975).
12. B. Horovitz and J.A. Krumhansl, Solid State Commun. 26, 81 (1978).

13. F.C. Frank and J.H. van der Merwe, Proc. Roy. Soc. A198, 205 (1949).
14. S. Aubry, J. Math. Phys. (to be published). See also in this volume.
15. N. Gupta and B. Sutherland, Phys. Rev. A14, 1790 (1976).
16. W.L. McMillan, Phys. Rev. B14, 1496 (1976); P. Bak and V.J. Emery, Phys. Rev. Lett. 36, 978 (1976).
17. J.F. Currie, M.B. Fogel and F.L. Palmer, Phys. Rev. A16, 796 (1977).
18. J.C. Jamieson, Science 140, 72 (1963).
19. B. Horovitz, J.A. Krumhansl and E. Domany, Phys. Rev. Lett. 38, 778 (1977).
20. J.A. Krumhansl and J.R. Schrieffer, Phys. Rev. B11, 3535 (1975).
21. S. Aubry, J. Chem. Phys. 62, 3217 (1975); 64, 3392 (1976).
22. T. Schneider and E. Stoll, Phys. Rev. B13, 1216 (1976). See also T. Schneider in this volume.
23 S.M. Shapiro, J.D. Axe, G. Shirane and T. Riste, Phys. Rev. B6, 4332 (1972).
24. J.D. Axe and G. Shirane, Phys. Rev. B8, 1965 (1973).
25. C.M. Varma, J.C. Phillips and S.T. Chui, Phys. Rev. Lett. 33, 1223 (1974); J.C. Phillips, Solid State Commun. 18, 831 (1976); K.L. Nagi and T.L.Reinecke, Phys. Rev. B16, 1077 (1977).

The New Concept of Transitions by Breaking of Analyticity in a Crystallographic Model

Serge Aubry*

Laboratoire Léon Brillouin, Orme de Merisiers, CEN-SACLAY, B.P. 2
F-91190 Gif-sur-Yvette, France

1. Introduction

Phase transitions such as for example, incommensurate to commensurate modulated cristal [1] metal-insulator in Peierls quasi-one dimensional metals [2] , non registered to registered in adsorbed (or intersticial) atomic monolayers [3] exhibit new and unexpected features. Despite other problems like the zero-width central peak phenomena [4] , the Anderson-Mott metal-insulator transition [5] , the spin-glass transitions [6] look very different, we believe that the ideas that we develop in this paper, should be relevant to suggest a new approach to them. The main topic of this paper is the description of the phase transitions versus parameters of a very simple one-dimensional model for epitaxy at zero degree K. Most results are rigorous but their proofs which are too long, are omitted. We try to emphasize the role of two coupled concepts of defectibility and of frustration in the origin of this new kind of phase transitions.

These ones are not usual symmetry breaking but involve analytical properties of some characteristic functions. Despite mathematical objects such as discontinuous or non differentiable functions, Cantor set [7] have not very much been considered by physicists until now, it is suggested that they play a fundamental role in the conceptual description of many physical phenomena, such as those mentioned above. In addition, it is shown that they have indirect observable consequences.

2. The model.

We choose to study a very old model used to describe many different phenomena such as the dislocation motion in crystals [8.a] or adsorbed atomic monolayer [8.b] . It is likely the simplest one for our purpose. Its potential energy at 0 K is

$$\phi(\{u_i\}) = \sum_i L_i \tag{1.a}$$

with

$$L_i = \lambda V(u_i) + W(u_i - u_{i-1}) - P\times(u_i - u_{i-1}) \tag{1.b}$$

where u_i is the atomic position of the ith atom of the chain. The potential $V(u)$ with amplitude λ is choosen analytical and periodic with period 2a : $V(u+2a) = V(u)$ and symmetric $V(-u) = V(u)$. The potential $W(u) - P\times u$ is analytical and convex: its second derivative versus u: $W''(u)$ is

*Chargé de Récherche au CNRS.

strictly positive. P is a pressure term choosen such that the minimum of W(u) is obtained at u = 0 (or equivalently at a multiple 2na of the period of V).

Conditions on V and W are not only convenient to get rigorous proofs but avoid the occurrence of more complicated features, which should be considered in further studies.

Without periodic potential $\lambda = 0$, the consecutive atoms, in the classical ground-state, would be at a constant distance ℓ_0 defined by

$$W'(\ell_0) = P \tag{2}$$

This distance is generally different and incommensurate with the period 2a of the substrate potential λ V. When λ is not zero, a conflict arises between these two energy terms W(u) - Pu and $\lambda V(u)$ which we call by definition frustration, by analogy with spin-glasses. Here the frustration is determined by two parameters : the pressure P which controle as we will see, the creation of defects, and the amplitude λ which determines the relative force of the two conflicting terms and by this way, the interaction between defects.

We study the classical ground state of this model at fixed volume $\langle u_{i+1} - u_i \rangle = \ell$ and at fixed pressured P. We discuss first the existence and the properties of defects in (1) and show that they are essentially a consequence of stochastic features produced by the strong non-linearities of (1). Defectibility makes non trivial the frustration effects.

3. Deflectibility and stochasticity

The stationary (time-independent) configuration family of classical crystallographic model may contain other configurations than the ground state if defects exist (i.e.) if the system is defectible.

3.1. The formalism

Stationary configuration of (1) are given by

$$\frac{\partial \phi}{\partial u_i} = V'(u_i) + W'(u_i - u_{i-1}) - W'(u_{i+1} - u_i) = 0 \tag{3}$$

In (1), ϕ has been written as an action with the Lagrangian L_i with a discrete time i ($\sum_i$ replaces the usual time integral $\int dt$).

A conjugate variable p_i to u_i can be defined as

$$p_i = \frac{\partial L}{\partial u_i} = W'(u_i - u_{i-1}) \tag{4}$$

We rewrite (3) and (4) using an operator T_λ defined as

$$\begin{pmatrix} p_{i+1} \\ u_{i+1} \end{pmatrix} = T_\lambda \begin{pmatrix} p_i \\ u_i \end{pmatrix} = \begin{pmatrix} p_i + V'(u_i) \\ W'^{-1}(p_{i+1}) + u_i \end{pmatrix} \tag{5}$$

which is the evolution operator of the canonical system associated with the Lagrangian (1.b) (for details, see ref. [9]). This operator T_λ is inversible and preserve the measure $dp \wedge du$ in the phase space (p,u). All the solutions of equation (3) are generated by the successive applications of the operators T_λ and T_λ^{-1} to all the initial points (u_0,p_0) and are then represented by some trajectory in the phase space (u,p). (This formalism, generalisable to most crystallographic models is the transfer matrix method at 0 K).

u_i is generally unbounded. For convenience and since $V(u_i)$ is periodic, we set $\tilde{u}_i = u_i$ (mod 2a) and $\tilde{T}_\lambda$

$$\begin{pmatrix} p_{i+1} \\ \tilde{u}_{i+1} \end{pmatrix} = \tilde{T}_\lambda \begin{pmatrix} p_i \\ u_i \end{pmatrix} = \begin{pmatrix} p_i + \lambda V'(\tilde{u}_i) \\ (W'^{-1}(p_{i+1}) + \tilde{u}_i) \bmod 2a \end{pmatrix} \tag{6}$$

The analyticity of V and W implies that $\tilde{T}_\lambda$ which applies the cylinder $(p_i,\tilde{u}_i)$ onto itself is analytical.

3.2. Trajectory properties

Figure (1) shows several examples of mapping generated by the trajectories of T_λ (we choose $\lambda = 1$, $V(u) = (E_0/2)(1-\cos(\pi u/a))$ $W(u) = 1/2\ C\ u^2$, and plotted more conveniently, the sequence of points $P_i = ((u_i/a),(u_{i-1}/a))$ mod 2 for i = 1 to 1000 instead of the sequence $(p_i = (u_i - u_{i-1})/C\ ,\ u_i \bmod 2a)$.

We observe :

1. q^{th} order fixed point F (called q^{th} order cycle) such that $\tilde{T}^q(F) = F$. For example, the first order fixed points $H = \tilde{T}_\lambda(H)$ and $F_0 = \tilde{T}_\lambda(F_0)$ and the the second order fixed point set $F_2' = \tilde{T}_\lambda(F_1')$, $F_1' = \tilde{T}^2(F_1')$. Such a trajectory $\{u_i\}$ satisfies for any i

$$u_{i+q} = u_i + 2\,p\,a \tag{7}$$

where p is some integer independent on i. The corresponding stationnary configuration has a unit cell of q atoms and of length 2pa. They are periodic commensurate configurations of model (1).

2. Trajectories dense on q q^{th} order smooth and regular orbits which are mapped for example, by the iterated sequence for $1 \leq n \leq 1\,000$: $\tilde{T}_\lambda^n(M_4)$ (see the closed curve around M_0 which is first order) or the sequence $\tilde{T}_\lambda^n(M_2)$ (see the two closed curves around F_1' and F_2' which are second order) The existence of such kind of $\tilde{T}_\lambda$-invariant curve is proved by the Kolmogorov-Arnold Moser theorem (called K.A.M.) (see ref. [9a] theorem 21.11 or ref.[12] theorem 2.11). In addition on these curves $\tilde{T}_\lambda^q$ (where q is the order of the curve) reduces to a simple rotation with an angle q k incommensurate with 2π . (dependent on the considered orbit).

The corresponding configuration can be considered as a periodic commensurate configuration (7) modulated by a surimposed periodic modulation with a wave vector k/a incommensurate with $2\pi/a$ (see Ref. [10])

A specially important kind of smooth trajectory (which does not exist in the case $Ca^2/E_o = 4$ but which exists for larger values of this parameter) map closed orbit on the two dimensional cylinder $(p_i, \tilde{u}_i)$ which are not homotopic to zero. On these trajectories u_i can be written as

$$u_i = f(i\ell + \alpha) = i\ell + \alpha + g(i\ell + \alpha) \qquad (8)$$

where f and g are analytical functions, f is monotonous increasing and g is periodic with period 2a, ℓ is incommensurate with 2a and α is an arbitrary phase.

These configurations are periodic and called incommensurate with period ℓ. (see section 4).

3. Erratic trajectories which are mapped for example by the sequences $\tilde{T}^n_\lambda(M_1)$ or $\tilde{T}^n_\lambda(M_2)$. The resulting clouds of points are almost not distinguishable on figure 1. The closure of these sets of point are not analytical curves, but are complicated shape set, called in mathematics Cantor sets. The existence of these $\tilde{T}_\lambda$ invariant non analytical sets is proved when some conditions are satisfied which, as we shall see in the next, is the localized defect existence. The corresponding configurations are (generally) disordered without any defined period

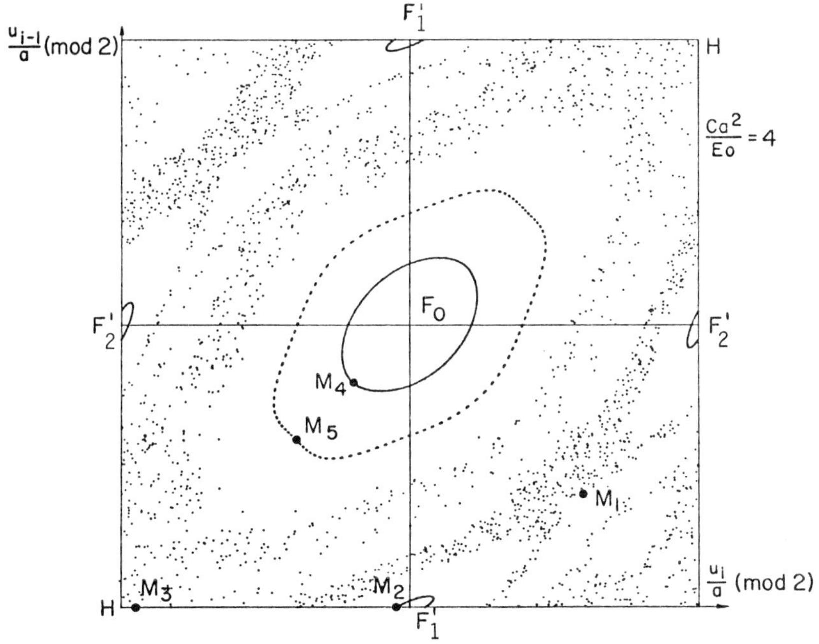

Figure 1

Such a mapping behavior can be considered as the characteristic consequence of strong non-linearities of model (1). For example, the T_λ mapping behavior becomes gradient-like and smooth when V(u) is replaced by a non-linear but single well potential

3.3. T_λ -stability of trajectories

The stochastic property of each trajectory can be simply related to its linear stability. Giving a $\tilde{T}_\lambda$ trajectory $\bar{M}_i = \tilde{T}_\lambda^i(\bar{M}_0) = (p_i, u_i)$ which is not a set of fixed points, a small perturbation $\bar{\varepsilon}_0$ on $\bar{M}_0$ is multiplied by the linearized part $\bar{\bar{J}}_\lambda$ of the operator $\tilde{T}_\lambda$ in $\bar{M}_0, \bar{M}_1, \bar{M}_2, \ldots, \bar{M}_n$ (called Lyapounov matrices). Because T_λ is measure-preserving, their determinant are unity. Then we set at the point $\bar{M}_0$, the matrix definition:

$$\bar{\bar{J}}_n(\bar{M}_0) = \bar{\bar{J}}(\bar{M}_{n-1}) \times \bar{\bar{J}}(\bar{M}_{n-2}) \times \ldots \bar{\bar{J}}(\bar{M}_0) \qquad (9a)$$

with

$$\bar{\bar{J}}(\bar{M}_i) = \begin{pmatrix} 1 & \lambda V''(u_i) \\ \dfrac{1}{W''(u_{i+1} - u_i)} & 1 + \dfrac{\lambda V''(u_i)}{W''(u_{i+1} - u_i)} \end{pmatrix} \qquad (9b)$$

If this matrices product (9.a) diverges for n going either to $+\infty$ or to $-\infty$ this trajectory $\bar{M}_i$ is linearly $\tilde{T}_\lambda$ unstable. In fact, it is unstable even taking into account the non-linearized transformation and this trajectory turns to be generally erratic.

If the matrices product (9.a) is bounded for any n, the trajectory $\bar{M}_n$ is linearly stable and in fact turns to be generally stable by the non-linearized transformation $\tilde{T}_\lambda$. Then its closure is a set of q closed analytical curves.

When the trajectory M_i is stable and represented by a set of q fixed points, these fixed points are called elliptic if the eigen-values of the Jacobian matrix $\bar{\bar{J}}_q$ of $\tilde{T}_\lambda^q$ are complex conjugate with modulus unity $e^{\pm i\theta}$. Then $\bar{\bar{J}}_q$ is equivalent to a rotation on an ellipse. The case where $\theta = 0$, is marginal and the fixed point is then parabolic.

If the trajectory is unstable, the qth order fixed point is hyperbolic. The corresponding eigen-values are real as well the eigen-directions. The eigen-direction which corresponds the eigen-value larger than 1, is dilating and the second one is contracting.

3.4. Homoclinic, heteroclinic point trajectories and defects

Let us consider now a qth order hyperbolic fixed point H_1. Its two real eigen-directions are parallel in H_1 to the dilating and contracting sheets of H_1. The first one is defined as the set of points P such that the sequence $\tilde{T}_\lambda^{qn}(P)$ converges to H_1 when goes to $+\infty$. The second one has the same definition for n going to $-\infty$ (see ref. 10 for example).

When these two lines intersect transversally at some point h called a homoclinic point, then it is proved that erratic trajectories are generated (see for example in ref. [12] theorem 3.7 and figure 17). Their closure is a Cantor set. Such a situation does exist close to most elliptic fixed point of $\tilde{T}_\lambda$ (ref. [12] theorem 3.9). By definition, the trajectory generated by h is identical to the trajectory of H_1 for large i, i going to $+\infty$ or $-\infty$. This trajectory represents physically nothing else than a localized defect configuration of the periodic and commensurate represented by H_1. More generally a localized defect is referred to two periodic configurations which are necessarily both represented by unstable $\tilde{T}_\lambda$ trajectories. These ones cannot be erratic and then are both represented by hyperbolic fixed points H_1 and H_2, with some orders equal to their periodicity at $+\infty$ and $-\infty$. This localized defect must be represented by the trajectory of a point belonging to the intersection of the dilating sheet of H_1 and of the contracting sheet of H_2. (see Ref. [10]). These points are called heteroclinic points. The configuration represented by H_2 may be the same as the one represented by H_1 apart from a phase shift. Then the obtained configuration is by definition a phase defect of a periodic and commensurate configuration.

In continuous models (where the index i is choosen continuous), for example the well known Sine-Gordon equation , these phase defects are easily found as moving kinks or better as solitons.

In fact, in discrete lattice, solitons or moving kinks are exceptional. In general they are locked.

We shew in ref. [10] that this is the fundamental property which explains physically the stochasticity which occur in solutions of equations similar to (3). A defect is submitted to a locking force from the lattice and to interacting forces from the other defects. If the distance between neighboring defects is sufficient, the locking force is larger than the interacting forces. Then to a random distribution on the lattice of defects, corresponds a stationary configuration. It is represented,by its definition, by an erratic trajectory. This is the physical translation of the theorem of (ref.[12] theorem 3.7) on homoclinic points.

For larger concentrations of defects, the resulting interacting forces may be stronger than the locking forces. The defects cannot be maintained at the chosen position in the lattice. If no other lattice effect intervenes, (commensurability of the defect mean distance with 2a) after rearrangement the defect must be regularly spaced. Then the lattice locking forces vanish and the interacting forces balance exactly. The resulting trajectory (representing this configuration) is dense on a set of analytical closed curves. This is the physical interpretation of the K.A.M. theorem.

More generally, the (stationary) defect properties of any classical model at O.K. are exhibited by the stochastic features of the transformation (5) of this model. (This remark is generalizable to continuous models, but since there is no lattice, stochasticity describe only the interactions between the defects).

3.5. Gap in the stationary configurations spectrum and dynamical stability

Assuming a mass m for each atom of the chain the time-Fourier transform of the linearized dynamical equation, of the small motions $\{\varepsilon_i(\omega)\}$ around a stationary configuration $\{u_i\}$ is

$$\begin{aligned}(\lambda V''(u_i) + W''(u_{i+1} - u_i) + W''(u_i - u_{i-1}) - m\,\omega^2)\,\varepsilon_i \\ - W''(u_{i+1} - u_i)\,\varepsilon_{i+1} - W''(u_i - u_{i-1})\,\varepsilon_{i-1} = 0 \qquad (10)\end{aligned}$$

A solution of (10) can be obtained for any ω , by choosing an arbitrary initial vector $(\varepsilon_1(\omega)\ ,\ \varepsilon_0(\omega))$. ε_{i+1} (or ε_{i-1}) is then calculated recursively, as a linear combination, given by (10) of ε_i and ε_{i-1} (or ε_{i+1})

for positive and negative i. However, since ε_i is a perturbation, a solution is physically acceptable if and only if $\varepsilon_i(\omega)$ is uniformly bounded for any i. For $\omega = 0$, omitting some calculations for brevity, this condition is equivalent to the $\tilde{T}_\lambda$ -stability condition of 3.3.

Then we can conclude : The stationary configurations with a zero frequency mode (gapless configurations) are represented by $\tilde{T}_\lambda$-stable trajectories (which generally are dense on closed analytical curves). This stability implies that no other configuration can be asymptotic to one of them at $i = \pm\infty$ and locally different. Such configurations are called undefectible.

If no zero-frequency mode exists, the stationary configuration has a finite frequency gap (except in marginal cases) and is represented by a $\tilde{T}_\lambda$ unstable trajectory. These ones are either a set of hyperbolic fixed points or erratic and dense inside a $\tilde{T}_\lambda$ invariant Cantor set. With the above definition,this configuration is defectible when its corresponding trajectory belongs to such a Cantor set (but is not necessarily dense in it).

To have a dynamically stable configuration (such that all the frequencies ω determined by equation (10) are real) we found that a necessary and sufficient condition is that for $\omega = 0$, there is at most one change of sign in the sequence of ε_i generated by equation (10) from an arbitrary $(\varepsilon_1\ \varepsilon_0)$

Applying this criteria to the different kinds of configurations, we found that

1. Configurations represented by q^{th} order elliptic fixed points and

2. Configurations represented by q^{th} order analytical curves, which are homotopic to zero are both dynamically unstable (see another proof in ref. [10] appendice A and B).

As a result, the dynamically stable configurations of model (1) are necessarily represented either by an hyperbolic fixed points set, an erratic trajectory or an analytical orbit which is non-homotopic to zero as described by (8). This last configuration only is gapless. However, these conditions are not sufficient.

4. Transition by breaking of analyticity versus the amplitude of the periodic potential

Fixing the boundary conditions as

$$u_N - u_{N'} = (N - N')\,\ell \tag{11}$$

for $N \to +\infty$ and $N' \to -\infty$ the atomic mean distance along the chain is fixed to an arbitrary value ℓ . We study in this section, the classical ground-state of model (1) with condition (11).

First, we examine the case where ℓ is commensurate with the period 2a of V(u) $\ell = 2a\,(p/q)$ where p and q are two irreducible integers. We proved (see ref. (16) that the ground state $\{u_i\}$ satisfies (7) and is then represented by a set of hyperbolic fixed points. In addition,we prove [15] the existence of elementary phase defects. These are defect configurations obtained by boundary conditions, generally such :

$$u_N - u_{N'} = (N - N')\,2a\,\frac{p}{q} \pm \frac{2a}{q} \tag{12}$$

and which has the minimum energy (1). They are represented by heteroclinic points (see 3.4).

4.1. Proof of a phase transition versus λ for incommensurate ℓ

We proved [16] that when ℓ is incommensurate with 2a

$$u_i = f(i\ell+\alpha) = i\ell + \alpha + g(i\ell + \alpha) \tag{13}$$

where f is a monotonous increasing function and g is periodic and antisymmetric with period 2a. These functions depend on ℓ , λ V and W but are independent on the arbitrary phase α .

Moreover, we proved [16] the existence of a constant λ_1 independent on ℓ (for ℓ in some interval $\ell_1 < \ell < \ell_2$) such that for

$$\lambda > \lambda_1 \tag{14}$$

functions f or g are necessarily discontinuous. The discontinuity point set A is countable but dense on the real axis. With the same condition (14) we succeeded to prove that the derivative of f is zero except on the set A and that the variation of f is the sum of its only discontinuity jumps. As a result f takes its values inside a Cantor set with zero measure which proves partially the conjecture of ref. [16]. Then the closure of the representing trajectory of (13) is a Cantor set. (In addition, this Cantor set is an invariant minimal closed set of $\tilde{T}_\lambda$ which means that it contains no other closed set which is invariant by $\tilde{T}_\lambda$).

For small values of ℓ when ℓ is sufficiently incommensurate with 2a, the K. A. M. theorem can be applied (ref. [9.a] theorem 21.11 or ref. [12] theorem 2.11).

(The condition on ℓ means that $\ell/2a$ is not a Liouville number, in other words there exists a real positive number β such that the rational numbers set which satisfy

$$\left| \frac{\ell}{2a} - \frac{p}{q} \right| < \frac{1}{q^{2+\beta}} \tag{15}$$

is finite. The measure of the Liouville numbers is zero which means that almost any real number is not a Liouville number. This condition is useful to know the behavior of series, the terms of which contains denominators like $1 - \cos(2\pi n\ell/2a)$. (small denominators). There exists similar small denominator problems in other physical situations e.g. ref. [5]).

When the parameter λ is zero, the canonical system defined by (6) is integrable, since p_i is not dependent on i. It is a motion invariant for $\tilde{T}_\lambda$

$$u_{i+1} - u_i = \ell \quad \text{for any } i,$$

$\tilde{T}_o$ is a rotation with angle ℓ on the analytical circle, defined by $p_i = p_o$. Since ℓ is incommensurate with 2a, the K.A.M. theorem states that for small enough perturbation λ V on this integrable system, defined by some $\lambda_2(\ell)$ and

$$0 < \lambda < \lambda_2(\ell) \tag{16}$$

there exists an invariant analytical curve close to this circle, on which

$\tilde{T}_\lambda$ is conjugated to a rotation with mean angle ℓ. The corresponding configurations are described by (8)

Next, we prove that if for a given ℓ such a trajectory exists, it is the only stationary configuration with atomic mean distance ℓ (apart phase shifts). Therefore it represents the ground state with boundary conditions (11).

These two results show that when λ varies between $\lambda_2(\ell)$ and λ_1, some transition by breaking of analyticity of the hull function f must occur.

Numerical test [19] on the ground state configurations with condition (11) shows a well-defined second order transition at some $\lambda = \lambda_c(\ell)$. For $\lambda > \lambda_c(\ell)$ we clearly see the existence of discontinuities on the function f and a finite gap in the phonon spectrum frequencies which both disappear at $\lambda = \lambda_c(\ell)$.

4.2. Physical interpretation of the analyticity breaking transition

Fixing ℓ is equivalent to fix the number of phase excitations referred to the state $u_i \equiv 0$ (epitaxy dislocations). Increasing λ is equivalent to increase the locking force on the excitations while the interacting force which depends mainly on W change very few.

When $\lambda < \lambda_c(\ell)$ the ground-state perturbation spectrum is gapless (see 3.5) Equation (10) exhibits the translation mode ε_i (at $\omega = 0$) defined as

$$\varepsilon_i = f'(i\ell + \alpha) = 1 + g'(i\ell + \alpha) \qquad (17)$$

ε_i does exist because f is analytical and therefore differentiable.

The atomic monolayer is unlocked on the substrate periodic potential V(u) From 3.5, this ground-state is undefectible which mean that there is no stable rearrangement of the epitaxy dislocations building this ground-state.

When $\lambda > \lambda_c(\ell)$, f' as conjectured in ref. [16] is zero and the gap is numerically observable. The atomic monolayer is locked. This state is defectible which means that there is many possible rearrangements of the epitaxy dislocations. The atoms of the chain, are trapped close to the minima of the periodic potential V(u) and requires, to be moved, to jump its potential barriers.

These features suggest an analogy with those of a metal insulator transition. The metallic phase is gapless, conducting and does not exhibit any charge defect. The insulating phase possesses a gap, is locked and is defectible by exitons.

More precisely, we believe that this model is not too bad to interpret the metal-insulator transition of Peierls systems [2] . A substrate potential λV with wave vector $2k_F$ and amplitude λ , could represent the electronic potential on the atoms, the parameter λ being non zero and increasing for decreasing temperatures below T_p. However more sophisticated models, considering also the quantum motion of the electron in the quasi-periodic lattice potential [19] , could exhibit simultaneously a transformation of the density of state into a purely singular measure. (Such measure has a zero measure support as for example $\psi''(\ell)$ in the following). Simultaneously the electron wave function should localize. There is other arguments suggesting that analyticity breakings and "Cantorization" are involved in metal-insulator transitions of various types [5],[17],[18]. (see [22])

5. Continuous non-differentiable transformations versus pressure

The mean energy per atom of model (1) at $P = 0$ with the atomic mean distance ℓ in $\psi(\ell)$. Finding the ground state of (1) at a given pressure P requires to find the minimum of $\psi(\ell) - P\ell$. The solution of the equation

$$\psi'(\ell) - P = 0 \tag{18}$$

determines $\ell(P)$.

5.1. Differentiability of $\psi(\ell)$ and defect energy

We proved first that $\psi(\ell)$ is a convex function versus ℓ . Its left derivative $\psi'_-(\ell)$ and its right derivative $\psi'_+(\ell)$ are both defined. Although, $\psi'_+(\ell) \geq \psi'_-(\ell)$ they are generally unequal when ℓ is commensurate with 2a.

Let us assume that $\ell_r = (p/q)\,2a$ with p and q two irreducible integers. We saw that the elementary phase defect (12) of this configuration corresponds generally to a phase shift which is $\pm\,(2a/q)$. Changing ℓ_r in $\ell_r + \delta\ell\,(\delta\ell > 0)$ is equivalent to add $q(\delta\ell/2a)$ phase defects per atom. The change of energy per atom is $\psi(\ell_r + \delta\ell) - \psi(\ell_r)$. For $\delta\ell \to 0_+$, the interacting energy between phase defect goes exponentially to zero. and is negligible. The required energy to add a single phase defect is

$$e_+\left(2a\,\frac{p}{q}\right) = \frac{2a}{q}\;\psi'_+\left(2a\,\frac{p}{q}\right) \tag{19a}$$

and identically the energy to suppress a phase defect is

$$e_-\left(2a\,\frac{p}{q}\right) = \frac{2a}{q}\;\psi'_-\left(2a\,\frac{p}{q}\right)$$

We prove [15] that generally $e_+(\ell_r) \neq e_-(\ell_r)$.

Then, the condition (18) shows that $\ell(P)$ is constant in some interval when $\ell = \ell_r$.

The ground-state with mean distance ℓ_r becomes unstable when the energy to add an elementary phase excitation

$$e_+(\ell_r) - \frac{2a}{q}\,P$$

becomes negative (or similarly to suppress an elementary phase excitation). This condition is also given by (18).

5.2. The devil's stair transition and the infinitely many phases diagram

Function $\psi'(\ell)$ which is monotonous increasing, has then infinitely many jumps $(\psi'_+(\ell_r) - \psi'_-(\ell_r))$ for any $\ell = \ell_r$ commensurate with 2a. The inverse function $\ell(P)$ defined by (18) has infinitely many steps and is called a devil's stair [7].

We proved [15] that when the condition (14) is satisfied for $\ell_1 < \ell < \ell_2$, the second derivative $\psi''(\ell)$ is zero for almost any ℓ incommensurate with 2a. (This function must be understood as generalized [20]). Since $\psi''(\ell)$ is positive it is a singular measure.(Its support has a zero usual measure. This implies that $\ell(P)$ is entirely composed of steps. The devil's stair is called complete and then the ground state is commensurate for almost any pressure.

For smaller values of $\lambda < \lambda_1$, we conjecture that the devil's stair $\ell(P)$ is not complete. In other word it is possible to find incommensurate ground states for a finite measure set of pressure. Equivalently, this means that $\psi''(\ell)$ is not purely singular. It contains in its decomposition, an absolutely continuous component, and a singular component. The support of the continuous part has however a very complicated shape since it has a finite usual measure and is included inside the set of ℓ incommensurate with 2a. When λ goes to zero, the singular part of $\psi''(\ell)$ goes to zero. In addition at fixed pressure, the incommensurate ground-states are always unlocked. (See section 3.5).

We did not yet prove this conjecture, but reliable arguments based on measure estimations and also on the conditions of the proof of the singularity of $\psi''(\ell)$ for large $\lambda > \lambda_1$, are obtained.

5.3. An exactly calculable model

All the above results are valid when choosing V and W as analytical functions.(They are still likely qualitatively exact if V and W are at least three time differentiable) [15] . Choosing in model (1)

$$\lambda\ V(u) = \frac{1}{2}\ \omega_0^2(u - 2m\ a)^2 \tag{20.a}$$

with

$$m = \text{integer part of } \left(\frac{u+a}{2a}\right) \tag{20.b}$$

and

$$W(u) - Pu = \frac{1}{2}\ (u - b)^2 \tag{20.c}$$

Potential V is not everywhere differentiable and then, this model does not exhibit all the features mentioned above. Particularly, it is easy to show that all the stationary configurations of this model are locked. From reference [16] theorem 1,

$$m_i = \text{Int}\ \left(\frac{u_i + a}{2a}\right) = \text{Int}\ \left(\frac{i\ell + \alpha}{2a}\right).$$

Equation (3) provides u_i as a linear combination of m_j, and omitting some calculations, we get readily :

$$\psi(\ell) = \frac{2a^2\omega_0^2}{\sqrt{1+\frac{4C}{\omega_0^2}}} \sum_{k=0}^{\infty} \mu^k\ \psi_k(\ell) \tag{21.a}$$

with

$$\mu = 1 + \frac{\omega_0^2}{2C} - \sqrt{\frac{\omega_0^2}{C} + \frac{\omega_0^4}{4C^2}} \tag{21.b}$$

and

$$\psi_k(\ell) = \left(1 + \text{Int}\ \frac{k\ell}{2a}\right)\left(\frac{k\ell}{a} - \text{Int}\ \frac{k\ell}{2a}\right) \tag{21.c}$$

$\psi k(\ell)$ is continuous and linear by parts. When ℓ is incommensurate with 2a $\psi''(\ell) = 0$. The ground-state is commensurate for almost any pressure and model parameters. This is consistent with 5.2 because incommensurate unlocked configuration does not exist and then, cannot be represented in the phase diagram.

Figure (2) shows an idea of the exact phase diagram of this model. The point $(1/\sqrt{C/\omega_0^2}, b/2a)$ lies in the region where the commensurability $\ell/2a$ of the ground state is indicated. The plane should be completely filled by such commensurate phase regions, without any first order line (compare with ref. [8b]). The grey area corresponds to the high order commensurate phases.

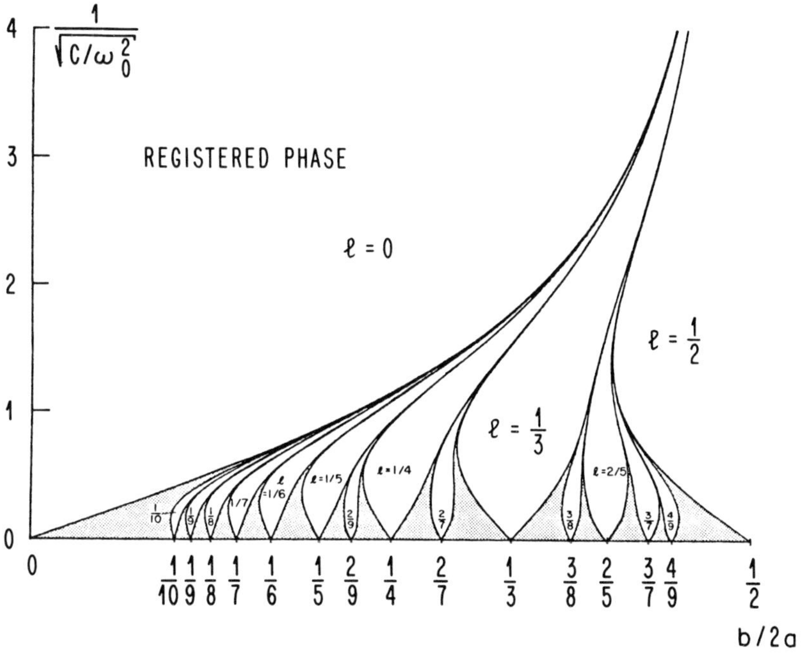

Figure 2

The next figure shows a picture of some steps of the devil's stair transformation $\ell(P)$ in this model [20] .

For analytical models (1), the phase diagram looks qualitatively the same (the pressure P is on the x-axis and $\sqrt{\lambda}$ on the y-axis). However in the grey area of figure 2 there exists close to the x-axis, a finite measure but infinitely disconnected region in-between the commensurate regions, where the ground-state is incommensurate, and gapless

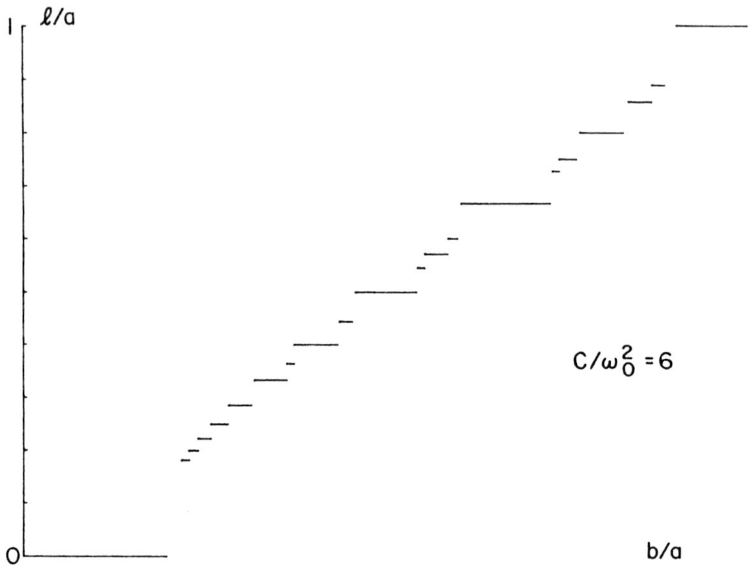

Figure 3

5.4. Physical interpretation of the devil's stair transformation

Of course, it is impossible to distinguish experimentally the details of the curve $\ell(P)$. However, there is observable effects. If the devil's stair is complete, all the ground-states $\ell(P)$ are locked. The transformation of these configurations requires that the atoms or more precisely the defects, jumps the potential barrier of $\lambda V(u)$. An estimation of the effective barrier shows that when λ is large enough (i.e. $\lambda > \lambda_1$) this barrier is weakly dependent on ℓ. There is a global hysteresis schematized by the formula

$$\ell^{\uparrow}_{obs.}(P) \simeq \ell(P-\delta) \tag{22a}$$

and

$$\ell^{\downarrow}_{obs.}(P) \simeq \ell(P+\delta) \tag{22b}$$

The observed value of ℓ : $\ell_{obs.}$, should depend on the history of the system. It is $\ell^{\uparrow}_{obs.}$ if P has been increased before the experiment and $\ell^{\downarrow}_{obs.}$ if it has been decreased.

Such features have been observed in several experiments for apparently continuous transformations (see for example [21]).

For smaller value of λ , the devil's stair becomes incomplete ; δ in formula (22) becomes more dependent on ℓ and vanish for unlocked ground-states, more precisely for incommensurate one such that $\psi''(\ell) \neq 0$.

When λ is small enough, δ is practically zero and the transformation $\ell(P)$ is second order, without almost any hysteresis. At the opposite limit : λ large, $\ell(P)$ varies sharply only on very small pressure intervals with an important hysteresis. Physically this transformation establishes the continuity between a second and a first order transition which may be interpreted also as blurred first order transition, often observed in various experiments.

Acknwoledgments.

I thank for their encouragement and their hospitality J.A. Krumhansl at Cornell University and V. Emery and M. Blume at the Brookhaven National Laboratory where parts of this work have been done. I thank also Y. Pomeau for many useful discussions.

References

[1] M. Izumi, J.D. Axe and G. Shirane, K. Shimaoka, Phys. Rev. B 15, 4392, (1967).
[2] R.E. Peierls, Quantum Theory of Solids (Clarendon, Oxford 1955, p. 108). R. Comes, M. Lambert, M. Launois and H.R. Zeller, Phys. Rev. B 8, 571 (1973).
[3] M.D. Chinn and S.C. Fain, Phys. Rev. Letters 39, 146 (1977); A. Novaco and J.P. Mc Tague, Phys. Rev. Letters 38, 1286 (1977).
[4] T. Riste, E.J. Samuelsen, K. Otnes and J. Feder, Solid State Communication 9, 1455 (1971). S.M. Shapiro, J.D. Axe, G. Shirane and T. Riste, Phys. Rev. B6, 4332 (1972).
[5] D.J. Thouless, Physics Reports 13, 93 (1974).
[6] K. Binder, Advance in Solid State Physics, XVII 55 (1977). G. Toulouse , Commun. Phys. 2, 115 (1977). J. Villain, J. Phys. C 10, 4793 (1977).
[7] Benoit B. Mandelbrot, Form, chance and dimension, W.H. Freeman and Company, San Francisco (1977).
[8a] F.C. Frank and J.H. Van der Merve, Proc. Royal Soc. (London) A198, 205 (1949).
[8b] S.C. Ying, Phys. Rev. B 3, 4160 (1971).
[9a] V.I. Arnold and A. Avez, Ergodic problems of classical mechanics, W.A. Benjamin Inc. (1968).
[9b] L. Landau and E. Lifchitz, Mechanics p. 143, Pergamon (1960).
[10] S. Aubry, preprint (to be published in J. of Mathematical Physics).
[11] M. Henon, Quat. Appl. Math. 27, 291 (1970).
[12] J. Moser, Stable and Random motions in Dynamical Systems, Princeton University Press, N.J. (1973).
[13] D. Ruelle, preprint (1977).
[14] G. Benettin, L. Galgani, J.M. Strelcyn, Phys. Rev. A 14, 2338 (1976).
[15] S. Aubry,in preparation
[16] S. Aubry , preprint.
[17] E.I. Dinaburg and Y.G. Sinaï, Functional analysis 9, 279 (1976).
[18] D.R. Hofstadter, Phys. Rev. B 14, 2239 (1976).
[19] G. André and S. Aubry, in preparation.
[20] I.M. Gel'Fand and G.E. Shilov, Generalized functions, Academic Press (1964).
[21] W.D. Ellenson, S.M. Shapiro, G. Shirane and A.F. Garito, Phys. Rev. B 16, 3244 (1977).
[22] W. Rudin, Real and Complex Analysis, Mc Graw Hill (1970).

Textures in Superfluid ^{3}He

Kazumi Maki

Department of Physics, University of Southern California
Los Angeles, CA 90007, USA

1. Introduction

Superfluid ^{3}He is a new phase of liquid ^{3}He at extremely low temperatures (below 3 mK). It consists of two distinct phases; ^{3}He-A and ^{3}He-B (although the third phase ^{3}He-A_1 appears in the presence of magnetic fields). The phase diagram is shown in Fig.1. Superfluid ^{3}He arises from the pair con-

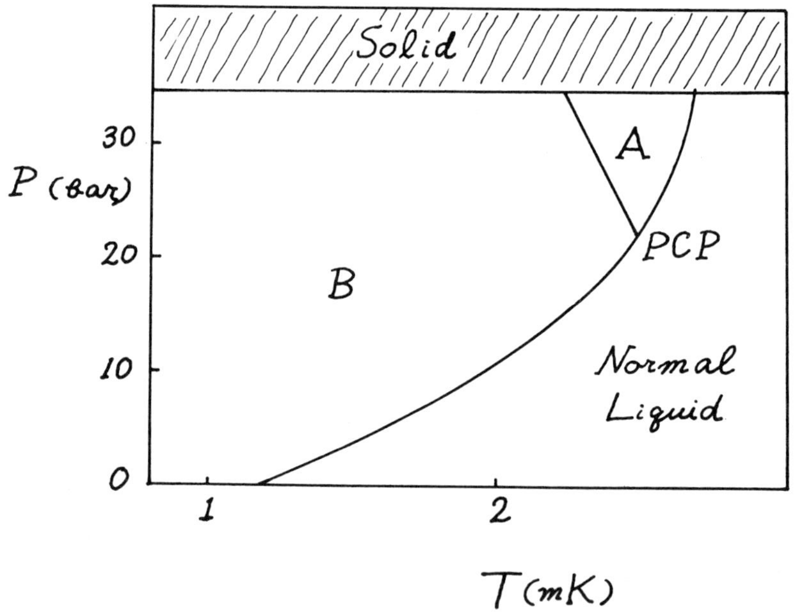

Fig.1 The phase diagram of liquid ^{3}He in the absence of magnetic field

densation of ^{3}He-atoms just like the superconductivity in metals is due to the electronic pair condensation. However, unlike the case of a superconductor, the pairing takes place in the spin triplet P wave state. This indicates that the condensate is characterized by a nine-component complex order parameter $A_{\mu i}$ (μ, i = 1,2,3) where μ and i are spin and orbital indices. This large degree of freedom associated with the order parameter implies high degeneracy in the ground state of superfluid ^{3}He in the absence of external perturbation. In particular superfluid ^{3}He possesses three distinct classes of Goldstone bosons; zero sound, spin wave and orbital wave. Furthermore, superfluid ^{3}He exhibits rich phenomena associated with textures like liquid crystals.

Before going into classification of textures, it is necessary to specify the order parameters for ^{3}He-A and ^{3}He-B and then to look into energies, which control formation of textures. The order parameters of ^{3}He-A and ^{3}He-B are given by [1]

$$A_{\mu i} = \begin{cases} \frac{1}{\sqrt{2}} \Delta_0(T)\ \hat{d}_\mu\ (\hat{\delta}_1 + i\hat{\delta}_2)_i \\ \frac{1}{\sqrt{3}} \Delta_0(T)\ e^{i\Psi}\ R_{\mu i}(\vec{n}, \theta) \end{cases} \tag{1}$$

respectively, where $\Delta_0(T)$ is the temperature dependent amplitude. Here $\hat{d}$ is a unit vector specifying the spin configuration of the ^{3}He-A condensate, $\hat{\delta}_1$, $\hat{\delta}_2$ and $\hat{\ell}$ ($\equiv \hat{\delta}_1 \times \hat{\delta}_2$) are a triad of unit vectors describing the orbital configuration. In particular $\hat{\ell}$ designates the direction of the symmetry axis of the quasi-particle energy gap as shown in Fig.2. The condensate of ^{3}He-B is characterized by a phase function Ψ and a rotation matrix $R_{\mu i}(\vec{n}, \theta)$;

$$R_{\mu i}(\vec{n}, \theta) = \cos\theta\ \delta_{\mu i} + (1 - \cos\theta)\ n_\mu\ n_i + \sin\theta\ \varepsilon_{\mu i k}\ n_k \tag{2}$$

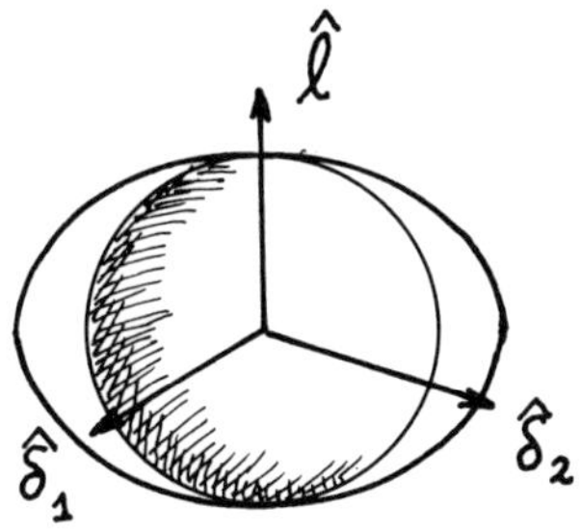

Fig.2 The orbital triad ($\hat{\delta}_1$, $\hat{\delta}_2$, $\hat{\ell}$) and anisotropic energy gap in ^{3}He-A are shown

where $\vec{n}$ is the axis of rotation and θ the rotation angle [2].

In general we mean by textures [3] the spatial conformation of $\hat{d}$ and $\hat{\ell}$ for ^{3}He-A and that of $\vec{n}$ for ^{3}He-B. Textures are controlled by a variety of energies. Of most importance is the kinetic energy associated with spatial distortion of textures, the nuclear dipole energy and the magnetic anisotropy energy [1]. In the Ginzburg-Landau regime the kinetic energy is given by [1]

$$F_{kin} = \frac{1}{2} K \int d^3r \, \{\partial_i A_{\mu i} \partial_j A^{\star}_{\mu j} + \partial_i A_{\mu j} \partial_j A^{\star}_{\mu i} + \partial_i A_{\mu j} \partial_i A^{\star}_{\mu j}\}$$

where

$$K = \frac{6}{5} \zeta(3) \, (8m^{\star})^{-1} \, N \, (2\pi k_B T_c)^{-2} \tag{3}$$

and $m^{\star}$ is the effective mass of the quasi-particle. The nuclear dipole energy E_D and the magnetic anisotropy energy E_H (which appears only in the presence of magnetic fields) are given by

$$E_D = \begin{cases} -\frac{1}{2} \chi_N \, \Omega_A^2 \int d^3r \, (\hat{\ell} \cdot \hat{d})^2 \\ \frac{8}{15} \chi_B \, \Omega_B^2 \int d^3r \, (\cos\theta + \frac{1}{4})^2 \end{cases} \tag{4}$$

and

$$E_H = \begin{cases} \frac{1}{2} (\Delta\chi_A) \, \omega_0^2 \int d^3r \, (\hat{d} \cdot \hat{H})^2 \\ -\frac{1}{2} \chi_B \, \Omega_H^2 \int d^3r \, (\vec{n} \cdot \hat{H})^2 \end{cases} \tag{5}$$

for ^{3}He-A and ^{3}He-B respectively, where χ_N, χ_B, and $\Delta\chi_A$ are the spin susceptibility of the normal liquid, of ^{3}He-B and anisotropic part of χ in ^{3}He-A, Ω_A and Ω_B are Legett frequencies for ^{3}He-A and ^{3}He-B, ω_0 ($\equiv \gamma_0 H$) is the Larmor frequency, Ω_H ($\sim (\omega_0/k_B T_c) \, \Omega_B$) is a small frequency, and $\hat{H}$ is the unit vector parallel to $\vec{H}$.

Comparing F_{kin} with E_D, we obtain a characteristic length $\xi (\equiv C_\perp/\Omega_A \sim 10\mu)$ which we shall call the dipolar coherence length. Here $C_\perp$ is the spin wave velocity in ^{3}He-A. In ^{3}He-B we have also a similar length $\xi = C_\perp/\Omega_B$, where $C_\perp$ is now the spin wave velocity in ^{3}He-B. These lengths play an extremely important role in the following consideration. Furthermore, in ^{3}He-A $E_D \simeq E_H$ yields $H = H_0$ ($\simeq 20$ O_e).

Finally, in the presence of confining walls, $\hat{\ell}$ in ^{3}He-A has to be oriented normal to the wall at the wall surface [4]. For ^{3}He-B, on the other hand, there are two superweak surface energies F_S and F_{SH} [2];

$$F_s = - a \int d^2r \, (\vec{n} \cdot \hat{s})^2 \, (1 - g \, (\vec{n} \cdot \hat{s})^2)$$

$$F_{sH} = - b \int d^2r \left[\hat{s} \cdot \overleftrightarrow{R} \, (\vec{n} \cdot \theta_0) \cdot \vec{H}\right]^2 \tag{6}$$

where the two-dimensional integrals have to be done on the wall surface. Here $a \simeq E_D \, \xi_0(T)$, $b \simeq \gamma_0^2 \, (\chi_N - \chi_B) \, \xi_0(T)$ and $\xi_0(T) (\sim 2 \times 10^{-6} (1-T/T_c)^{-\frac{1}{2}}$ cm) is the microscopic coherence distance. Putting these terms together, we have the total free energy

$$F = F_{kin} + E_D + E_H + (F_s + F_{sH}) \quad (7)$$

Stable textures in superfluid ^{3}He are determined as local minimum of F.

2. Classification

In classifying a variety of textures the homotopy theory is extremely useful. In superfluid ^{3}He most of the analysis has been done by TOULOUSE and KLÉMAN [5] and by VOLOVIK and MINEEV [6].

a. ^{3}He-A

Textures in ^{3}He-A are first classified depending on the importance of the dipole energy into three categories; 1) dipole-locked textures ($\hat{\ell} \parallel \pm \hat{d}$ all over space), 2) dipole-generated textures (planar solitons where $\hat{\ell} = \pm \hat{d}$ is broken locally), and 3) dipole-delocked textures. The first two textures are found in an open geometry ($D > \xi$, where D is the linear dimension of the system), while the last texture is found in a confined geometry ($D < \xi$), where $\hat{d}$ is practically constant all over space. The internal spaces of cases 1 and 3 are the same as SU(2). The homotopy theory tells us then that there are two classes of textures with different dimensionality; linear textures and point-like objects, although the latter appears to be dynamically unstable due to the orbital viscosity [7]. Linear textures consist of singularities like vortex lines and disgyrations with Frank index $\pm$ 1 and linear solitons (i.e., coreless vortices of MERMIN and HO [8] and of ANDERSON and TOULOUSE [9]). Some of the disgyrations are shown in Fig.3.

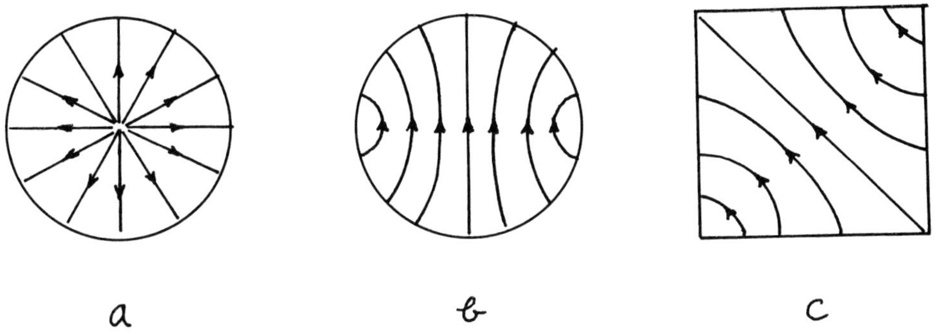

Fig.3 Typical textures with disgyration in ^{3}He-A are shown. Solid lines indicate direction of $\hat{\ell}$

Radial and circular disgyrations have Frank index 1, while hyperbolic disgyration -1. In a long tube with circular cross-section, if $\hat{\ell}$ lies in a plane normal to the axis of the tube, sum of Frank indices of all disgyrations in the tube has to be 1. A coreless vortex of ANDERSON and TOULOUSE [9] is shown schematically in Fig.4. If one circles around the periphery

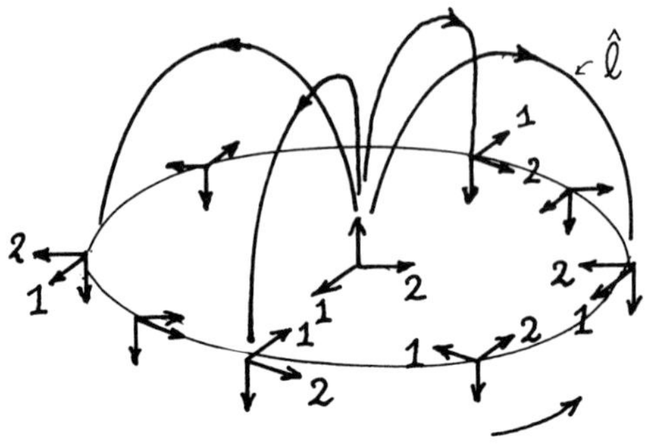

Fig.4 The coreless vortices of Anderson and Toulouse

of the vortex, the orbital triad completes two full rotations. The coreless vortex carries a topological charge

$$N_2 = \frac{1}{2\pi} \int_S dx_i \, dx_j \, \hat{\ell} \, (\partial_i \hat{\ell} \times \partial_i \hat{\ell}) \qquad (8)$$

where the integral is performed over an area, which encloses the total cross-section of the linear soliton as shown in Fig.5. For a MERMIN-HO vortex

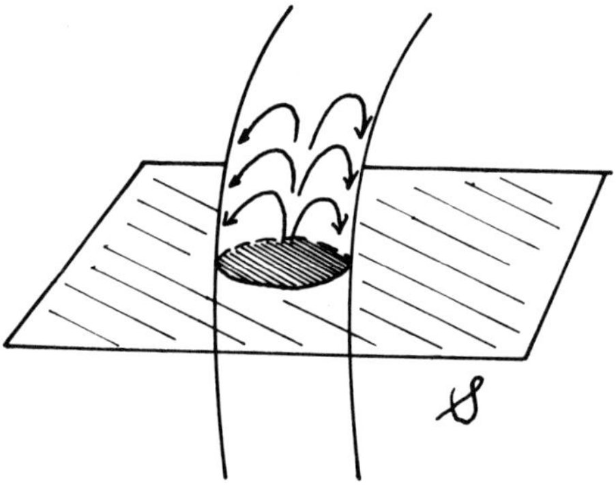

Fig.5 The surface over which the integral (8) must be done

$N_2 = \pm 1$, while for an ANDERSON-TOULOUSE vortex $N_2 = \pm 2$. Indeed N_2 is proportional to the strength of vorticity associated with the linear soliton, as is seen from the MERMIN-HO relation [8]

$$\partial_i v_{sj} - \partial_j v_{si} = \frac{\hbar}{2m} \{\hat{\ell} \cdot (\partial_i \hat{\ell} \times \partial_j \hat{\ell})\} \qquad (9)$$

where $\vec{v}_s$ is the superfluid velocity.

When a coreless vortex terminates at a point on the surface of the wall, this termination point is called boojum [10], which can be either the sink or the source of vorticity. Booja may play an important role in the stability of superflow in ^{3}He-A.

All the dipole-generated textures are planar solitons, which are mobile domain walls. There are three distinct planar solitons; $\hat{d}$-solitons, $\hat{\ell}$-solitons, and composite-solitons, which involve both $\hat{d}$ and $\hat{\ell}$-fields. In an open system the composite solitons are the only stable solitons. However, $\hat{\ell}$-soliton can exist trapped at the wall of the container in the presence of a magnetic field. In particular in the presence of a magnetic field $\vec{H}$, the composite solitons are quite stable, as both $\hat{\ell}$ and $\hat{d}$ are confined in a plane normal to $\vec{H}$ |11|. This stability is guaranteed by a topological charge:

$$N_1 = \frac{1}{\pi} \int_P^Q dx_i\ \hat{H} \cdot \left[\hat{\ell} \times \partial_i \hat{\ell} - \hat{d} \times \partial_i \hat{d}\right] \qquad (10)$$

where the integral is done along a path connecting two points P and Q lying on opposite sides of the soliton as shown in Fig.6. The planar solitons

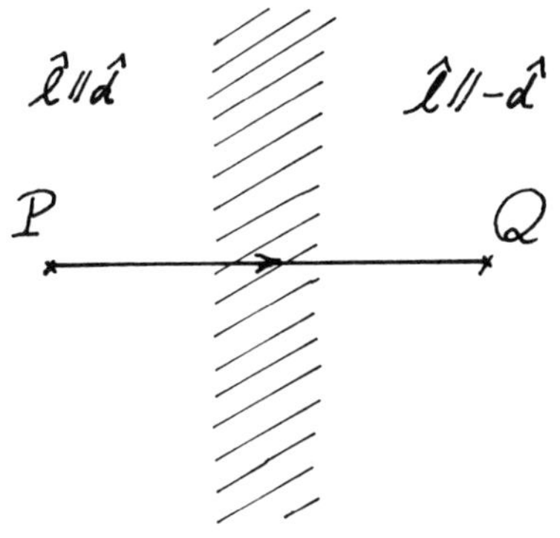

Fig.6 The integral path in (10) is schematically shown

have $N_1 = \pm 1$. These solitons are most abundant in ^{3}He-A.

b. ^{3}He-B

In ^{3}He-B textures are classified into two categories; θ-solitons and $\vec{n}$-solitons. θ-solitons are domain walls generated by the dipole interaction, whereas $\vec{n}$-solitons are generated by the superweak interaction. There are two types of θ-solitons; type I and type II [12], although type II θ-soliton is unstable and decays into $\vec{n}$-soliton [11]. $\vec{n}$-solitons can be, on the other hand, planar, linear, or point-like, although only planar solitons are studied in great detail [13, 14].

3. Nuclear Magnetic Resonance

Perhaps nuclear magnetic resonance provides the most sensitive means to study textures in superfluid ^{3}He, since it can explore small regions of dimension of the order of ξ. In particular spin waves are trapped by textures [15,16] and inhomogeneities [17] in superfluid ^{3}He. These spin wave bound states can be seen as satellites in the nmr experiments since the composite solitons in ^{3}He-A, and $\vec{n}$-solitons in ^{3}He-B have a well-defined set of bound states, nmr is extremely useful in identifying these textures. On the other hand, to explore the dipole-locked textures in ^{3}He-A, nmr is almost useless and one has to rely on other techniques like ultrasonic attenuation.

a. ^{3}He-A

The linear spin dynamics in the presence of textures is most readily formulated in terms of Lagrangian [18];

$$L = T - V$$

$$T = \frac{1}{2} \chi_N \int d^3r \, (\vec{\omega}^2 - 2\vec{\omega}_0 \cdot \vec{\omega})$$

$$V = \delta F \tag{11}$$

where

$$\omega_x = - \sin\alpha \; \beta_t + \cos\alpha \; \sin\beta \; \gamma_t$$

$$\omega_y = \cos\alpha \; \beta_t + \sin\alpha \; \sin\beta \; \gamma_t$$

$$\omega_z = \alpha_t + \cos\beta \; \gamma_t$$

$$\vec{\omega}_0 = \gamma_0 \vec{H}$$

$$(\vec{M} = - \gamma_0 \chi \; \vec{\omega}) \tag{12}$$

and α, β, and γ are Euclerian angles describing the rotation of $\hat{d}$ and γ_0 is the gyromagnetic ratio of the ^{3}He nucleus. Note that the kinetic energy is the same as that for a spherical top. δF is the spin fluctuation free energy, that is obtained from (7) by substituting $\hat{d} = \hat{d}_0 + \delta\hat{d}$ and keeping only the quadratic terms in $\delta\hat{d}$, where $\hat{d}_0$ describes the equilibrium $\hat{d}$ configuration in the texture. Furthermore, $\delta\hat{d}$ can be expressed in terms of Euclerian angles as

$$\hat{d}(t) \; (\equiv \hat{d}(0) + \delta\hat{d}) = \overleftrightarrow{R} \; (\alpha, \beta, \gamma) \; \hat{d}(0) \tag{13}$$

In particular for a composite soliton [16, 19], we can parameterize $\hat{d}$ as

$$\hat{d} = \cos g \; \{\sin(\psi + f) \; \hat{x} + \cos(\psi + f) \; \hat{y}\} + \sin g \; \hat{z}$$

$$\hat{d}_0 = \sin\psi \; \hat{x} + \cos\psi \; \hat{y}$$

$$\hat{\ell}_0 = \sin\chi \; \hat{x} + \cos\chi \; \hat{y} \tag{14}$$

The resulting Euler-Lagrange equation is solved in two steps; first to determine the eigenvalues

$$\lambda_f f = - \xi^2 \frac{d}{ds} \left[(1 - \frac{1}{2} a^2) \ f_s \right] + (1 - 2\sin^2 v) f$$

$$\lambda_g g = - \xi^2 \frac{d}{ds} \left[(1 - \frac{1}{2} a^2) \ g_s \right] + (1 - \sin^2 v - (1 - \frac{1}{2} a^2) \ \xi^2 \ \psi_s^2) \ g \qquad (15)$$

where

$$s = \hat{k} \cdot \vec{x}, \quad a = k_1 \cos\chi + k_2 \sin\chi, \quad \text{and } v = \chi - \psi \qquad (16)$$

Then the satellite frequencies are given in terms of the above eigenvalues as

$$\omega_\ell = (\lambda_f)^{\frac{1}{2}} \ \Omega_A \quad \text{and} \quad \omega_t = \left[\omega_0^2 + \lambda_g \ \Omega_A^2 \right]^{\frac{1}{2}} \qquad (17)$$

for longitudinal and transverse mode respectively. In particular for the twist composite soliton [16], we have

$$\lambda_f f = - \xi^2 \ f_{ss} + (1 - 2 \ \mathrm{sech}^2 \ (\sqrt{5} \ s/2\xi)) f$$

$$\lambda_g g = - \xi^2 \ g_{ss} + (1 - \frac{6}{5} \ \mathrm{sech}^2 \ (\sqrt{5} \ s/2\xi)) g \qquad (18)$$

where we made use of the relation

$$\chi + 4\psi = \text{const}$$

$$v = 2 \tan^{-1} \left[\exp \ (\sqrt{5} \ s/\xi) \right]$$

Eq.(14) yields

$$\lambda_f = \frac{1}{2} \ (\sqrt{65} - 7) \qquad f \propto \left\{ \mathrm{sech} \left[\sqrt{5} \ s/2\xi \right] \right\}^{\frac{1}{2}(\frac{13}{\sqrt{5}} - 1)}$$

$$\lambda_g = \frac{4}{5} \qquad g \propto \left\{ \mathrm{sech} \left[\sqrt{5} \ s/2\xi \right] \right\}^{1/5} \qquad (19)$$

More generally when the magnetic field makes an angle θ to $\hat{k}$ the normal direction of the domain wall as shown in Fig.7, the composite soliton changes continuously from pure twist to splay like as the angle θ increases from 0 to $\pi/2$. The corresponding eigenvalues λ_f and λ_g are calculated numerically and shown in Fig.8 [19]. Here plotted are $R_\ell = (\lambda_f)^{\frac{1}{2}}$ and $R_t = (\lambda_g)^{\frac{1}{2}}$ instead of λ_f and λ_g. The longitudinal satellite in ^{3}He-A was first observed by AVENEL et al., [20]. The satellite frequency was reported to be

$$\omega_\ell / \Omega_A \sim \frac{1}{\sqrt{2}}$$

More recently an extensive study of these satellites in both the longitudinal and the transverse case has been carried out by GOULD and LEE [21]. Their results are summarized as

$$R_\ell = 0.74 - 0.35\,(1 - T/Tc)$$

and

$$R_t = 0.835 \qquad (20)$$

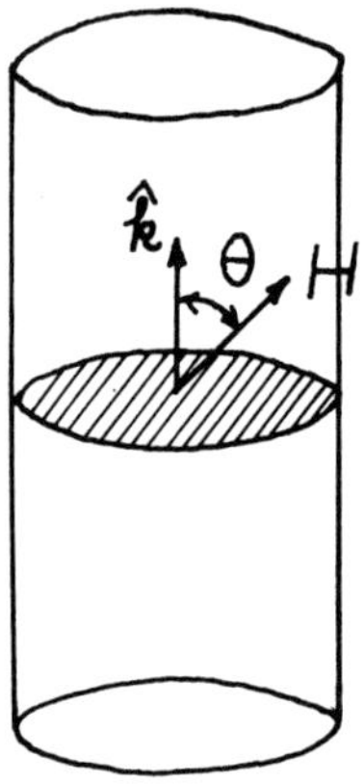

Fig.7 The general configuration is shown

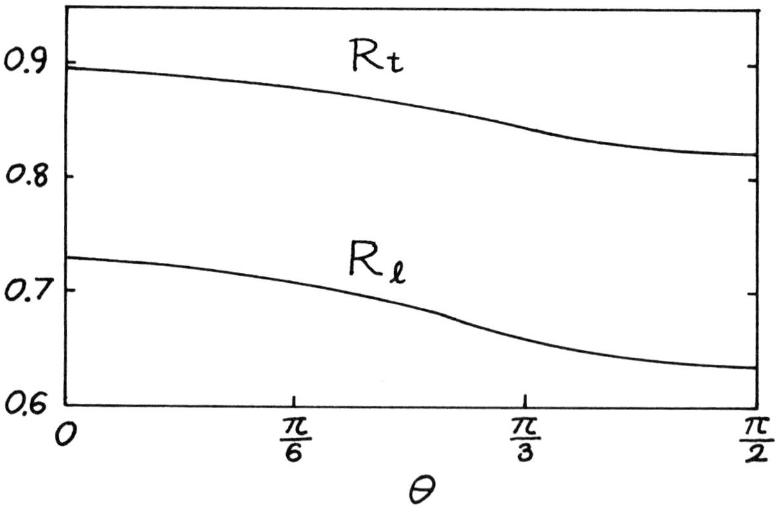

Fig.8 R_t and R_ℓ which appear in the satellite frequencies are shown

These as well as the earlier Orsay-Saclay result [20], are in excellent agreement with the theoretical prediction, if we discard for a moment a small temperature dependent term and if we assume that their longitudinal satellite is due to the twist soliton ($\theta = 0$) whereas their transverse satellite is due to the splay soliton ($\theta = \pi/2$). Indeed in the Cornell experiment, [21] they rotated the static magnetic field by 90^0 rather than the rf field in order to measure both the transverse and the longitudinal resonance, which justified the assumption that longitudianl and transverse satellites arise from related but different textures. Furthermore, the observed temperature dependence in R_ℓ is interpreted to be due to deviation from the Ginzburg-Landau regime [22], which allows us to express $F_1{}^a$ the Fermi liquid coefficient in terms of the coefficient of the temperature dependent term in (20). This yields $F_1{}^a = -1.03$, which is remarkably close to the value deduced by OSHEROFF et al., [15] from the analysis of their spin wave data in superfluid ^{3}He-B.

b. ^{3}He-B

In superfluid ^{3}He-B the most interesting textures are $\vec{n}$-solitons in open geometry [13, 14]. In the presence of a magnetic field the $\vec{n}$-vector has two preferred directions;

$$\vec{n} \parallel \pm \vec{H}$$

The $\vec{n}$-soliton is the domain-wall associated with these two equilibrium $\vec{n}$ configurations. In particular in the limit $\omega_0 >> \Omega_B$ the spin wave spectrum associated with a $\vec{n}$-soliton is given by

$$(\omega^2 \mp \omega\omega_0)\, f^{\pm} = - C_\perp^2 \left(1 - \frac{1}{4} k_\perp^2\right) f^{\pm}_{ss} + \frac{1}{2} \Omega_B^2 (n_{0\perp})^2 f^{\pm}$$

$$\omega^2 f^z = - C_\perp^2 \left(1 - \frac{1}{2} k_z^2\right) f^z_{ss} + \Omega_B^2 (n_{0z})^2 f^z \qquad (21)$$

where

$$s = \hat{k} \cdot \vec{x}, \quad M_\pm \propto f^\pm \text{ and } M_z \propto f^z$$

and $\vec{n}_0$ is the $\vec{n}$ vector at the equilibrium configuration. For a $\vec{n}$-soliton n_{0z} is well approximated by $n_{0z} = \tanh(\eta s)$, where

$$\eta \sim \xi_H^{-1} \; (\equiv \Omega_H / C_\perp) \text{ and } \xi_H \; (\sim 1 \text{ mm})$$

is the magnetic coherence length in ^{3}He-B. Then (21) tells us that a $\vec{n}$-soliton provides a repulsive potential to the transverse mode and an attractive potential to the longitudinal mode. Indeed, the longitudinal mode has a series of bound states, which can be solved exactly in terms of the hypergeometric function [14]. In particular the spin wave spectrum is given by

$$(\omega_\ell)_n = \Omega_B K^{\frac{1}{2}} \left\{(2n+1)\left[1 + \frac{1}{4} K^2\right]^{\frac{1}{2}} - \frac{1}{4} K \left[(2n+1)^2 + 1\right]\right\}^{\frac{1}{2}}$$

$$K = (\eta\xi)\left(1 - \frac{1}{2} k_z^2\right) \qquad (22)$$

and n integer.

The homogeneous rf field couples only with even n mode. The intensities of these modes are given by

$$I_0 = 2\sqrt{\pi}\, K^{\frac{1}{2}} \quad \text{and} \quad I_{2n}/I_0 = \pi^{-1}\, B(2n + \tfrac{1}{2}, \tfrac{1}{2}) \tag{23}$$

where $B(x, y)$ is the beta function and we normalized the intensities by the width of the domain wall η^{-1}. In the usual circumstance $K \sim 10^{-2}$ we expect about 10 satellites associated with $\vec{n}$-soliton. As seen from (21), there is no bound state for the transverse mode. However, in the presence of a field gradient (this effect is included by replacing ω_0 by $\omega_0 + Gs$ in (21)) a series of bound states do appear in the transverse mode as first shown by ENGELSBERG et al., [23]. The energy spectrum is given by [14]

$$(\omega_t)_n = \omega_0 + (2\eta)^{-1}\, G \left[\ln\,(4\eta\, \Omega_B^2/\omega_0 G) + 1\right]$$
$$+ C_\perp\, (G\eta/2\omega_0\, (1 - \tfrac{1}{4}\, k_\perp^2))^{\frac{1}{2}}\, (2n + 1) \tag{24}$$

where again only even n-mode couples with rf field. This provides the most convincing interpretation of one of the unusual nmr's in ^{3}He-B observed by OSHEROFF [24].

4. Creation of Planar Solitons

So far we have described properties of solitons which are well understood. We shall now embark on problems of which further clarification is clearly desirable. It is now well established that planar solitons are created either by turning off a localized magnetic field or by tipping the local magnetization by 180^0. In particular a scenario of creation of composite solitons in ^{3}He-A runs as follows [25]: 1) by localized magnetic perturbation a large number of $\hat{d}$-soliton-$\hat{d}$-antisoliton pairs are created. (If $\hat{\ell}$ is assumed to be fixed, this problem is solved by the inverse scattering method by ABLOWITZ et al., [26] in the case of a turn-off experiment [27].) 2) Then these solitons and antisolitons run away from their location of creation and stop somewhere due to the spin diffusion within 10^{-1} m sec. 3) Then $\hat{d}$-solitons transform into composite solitons due to the orbital viscosity in the time scale of $10\,(1 - T/T_c)^{\frac{1}{2}}$ m sec. More precisely, if the turned off field is of a rectangular shape,

$$H(z) = H_0\, \theta(a^2 - z^2)$$

with $\theta(x)$ the step function, the total number of created soliton-antisoliton pairs are given by

$$N \cong \frac{1}{2}\,(1 - (\Omega_A/\omega_0)^2)^{\frac{1}{2}}\, N_0, \quad N_0 = \frac{2}{\pi}\,(\omega_0/\Omega_A)\,\frac{a}{\xi} \tag{24}$$

where $\omega_0 = \gamma_0 H_0$. In usual circumstances where $a \sim 1$ cm, solitons are created by hundreds when $\omega_0 > \Omega_A$.

The predicted threshold behavior as well as the number of created composite solitons appear to be consistent with recent experiments [28, 29]. However, details of the scenario may have to be modified. For example, KRUSIUS et al., [30], observed ultrasonically disturbance caused in the $\hat{\ell}$-field associated with creation of the composite solitons after a magnetic field is suddenly turned off. Indeed, they have observed the threshold behavior consistent with the above scenario. However, the time required for appearance of $\hat{\ell}$-disturbances is by a factor of 10^2 larger (i.e., ~ 10^2 m sec) than that expected from the scenario. More recently, BOZLER, et al., [30] recorded free induction decay of magnetization after 180^0 tipping. First of all they have not seen any signal indicating the appearance of $\hat{d}$-solitons (i.e., unshifted Larmor frequency). Furthermore, before the appearance of the splay composite soliton signal with $\lambda_g \simeq 0.7$, a large portion of the system passes through a metastable state with well defined resonance signal (with $\lambda_g \simeq 0.5$) after a lapse of time of the order of 10 m sec. Only about 1 second later a clearly defined composite soliton signal does appear. These results indicate the need for further work on this subject.

Most of the results reported here were obtained in collaboration with P. Kumar and Y. R. Lin-Liu. The present work is supported by the National Science Foundation under grant number DMR76-21032.

References

1. A. J. Leggett, Rev. Mod. Phys. 47, 331 (1975).
2. W. F. Brinkman, H. Smith, D. D. Osheroff, and E. I. Blount, Phys. Rev. Lett. 33, 624 (1974).
3. P. G. de Gennes, Phys. Lett. A44, 271 (1973); in Proc. Nobel Symposium Vol.24 (Academic Press, New York, 1975).
4. V. Ambegaokar, P. G. de Gennes, and D. Rainer, Phys. Rev. A9, 2676 (1974).
5. G. Toulouse and M. Kléman, Journal de Physique 37, L-149 (1976).
6. G. E. Volovik and V. P. Mineev, Pisma ZETF 24, 605 (1976), and ZETF 72, 2256 (1977).
7. M. C. Cross and P. W. Anderson, Proc. of the 14th International Conference on Low Temperature Physics, Otaniemi, Finland, 1975, edited by M. Krusius and M. Vuorio (North Holland, Amsterdam, 1975) Vol.1, p.29.
8. N. D. Mermin and T. L. Ho, Phys. Rev. Lett. 35, 594 (1976).
9. P. W. Anderson and G. Toulouse, Phys. Rev. Lett. 38, 508 (1977).
10. N. D. Mermin, in Quantum Fluids and Solids, edited by S. B. Trickey, E. D. Adams, and J. W. Dufty (Plenum Press, New York, 1977).
11. G. E. Volovik and V. P. Mineev, Phys. Rev. B (to be published).
12. K. Maki and P. Kumar, Phys. Rev. B14, 118 (1976).
13. K. Maki and P. Kumar, Phys. Rev. B16, 4805 (1977).
14. K. Maki and Y. R. Lin-Liu, Phys. Rev. B (in press).
15. D. D. Osheroff, W. van Roosbroeck, H. Smith and W. F. Brinkman, Phys. Rev. Lett. 38, 134 (1977).
16. K. Maki and P. Kumar, Phys. Rev. Lett. 38, 557 (1977); Phys. Rev. B16, 182 (1977).
17. K. Maki and T. Tsuneto, J. Low Temp. Phys. 27, 537 (1977).
18. K. Maki, Phys. Rev. B11, 4262 (1975).
19. K. Maki and P. Kumar, Phys. Rev. B17, 1088 (1978).
20. O. Avenel, M. E. Bernier, E. J. Varoquaux and C. Vibet, in Proc. of the 14th International Conference on Low Temperature Physics, Otaniemi, Finland, 1975, edited by M. Krusius and M. Vuorio. (North Holland, Amsterdam, 1975) Vol.5, p.429.
21. C. M. Gould and D. M. Lee, Phys. Rev. Lett. 37, 1223 (1976).
22. M. C. Cross, J. Low Temp. Phys. 21, 525 (1975).
23. S. Engelsberg, W. P. Brinkman, and D. D. Osheroff, J. Low Temp. Phys. 29, 29 (1977).
24. D. D. Osheroff, in Quantum Fluids and Solids, edited by S. B. Trickey,

E. D. Adams, and J. W. Dufty (Plenum Press, New York, 1977).
25. K. Maki, in Quantum Fluids and Solids edited by S. B. Trickey, E. D. Adams, and J. W. Dufty (Plenum Press, New York, 1977).
26. M. J. Ablowitz, D. J. Kaup, A. C. Newell, and H. Segur, Phys. Rev. Lett. 30, 1262 (1973).
27. K. Maki and P. Kumar, Phys. Rev. B14, 3920 (1976).
28. R. W. Giannetta, C. M. Gould, E. N. Smith, and D. M. Lee, in Quantum Fluids and Solids edited by S. B. Trickey, E. D. Adams, and J. W. Dufty (Plenum Press, New York, 1977).
29. J. Kokko, M. A. Paalanen, R. C. Richardson, and Y. Takano, to be published.
30. M. Krusius, D. N. Paulson, and J. C. Wheatley, preprint.
31. H. M. Bozler, T. Bartolac, and K. Luey, (in preparation).

Creation of Spin Waves in ^{3}HeB

P.W. Kitchenside, R.K. Bullough, and P.J. Caudrey

Department of Mathematics, U.M.I.S.T., P.O. Box 88
Manchester M60 1QD, Great Britain

This article is largely concerned with the following problems: the solution of the *double sine-Gordon* equations [1,2,3]

$$u_{xx} - u_{tt} = \mp (\sin u + \frac{1}{2} \lambda \sin \frac{1}{2} u) \qquad (1)$$

for boundary conditions u, u_x, u_{xx}, etc. $\to 0$, as $|x| \to \infty$, and initial data $u(x,0) = 0$; $u_t(x,0) = a$, $|x| \leq \ell$, $u_t(x,0) = 0$, $|x| > \ell$. Two cases are directly relevant to the spin waves in the superfluid phases of ^{3}He below 2.6 mK: these are the $-$ve sign and $\lambda = 1$, and the $+$ve sign and $\lambda = 0$. The latter is an initial value problem for the sine-Gordon equation and can be solved by an inverse scattering method [4]. The double sine-Gordon equations which arise for $\lambda \neq 0$ are not soluble by any of the techniques presently available for solving nonlinear evolution equations [2,3,5]: there are, for example, apparently only three conservation laws [2] and the systems are not 'integrable'. Evidently only singular perturbation theory [6] and numerical integration are available to solve this problem. Despite success with perturbation theories for the case of the positive sign [6], we do not yet know how to handle similar perturbation theory for the negative sign: $u_{xx} - u_{tt} = -\sin u$ is unstable and its multisoliton solutions are unstable. This paper therefore confines its report to the results of numerical work.

A survey of the physics to which these results are relevant, namely to spin waves in the A-phase of ^{3}He (sine-Gordon case) and the B-phase (double sine-Gordon case) must necessarily be brief. We have elsewhere [3,5,7] introduced an 'adiabatic' Hamiltonian density $\mathscr{H}$ which generalizes that proposed by LEGGETT [8] to describe the unusual NMR behaviours observed. We derive from this the equations of motion

$$\theta_t = \frac{\delta \mathscr{H}}{\delta r_3} = -\gamma B + \gamma^2 \chi^{-1} r_3 \qquad (2)$$

$$r_{3,t} = -\frac{\delta \mathscr{H}}{\delta \theta} = \gamma^{-2} \chi \bar{c}^{2} \theta_{xx} - \frac{\delta H_D}{\delta \theta}$$

and therefore

$$\theta_{tt} - \bar{c}^{2} \theta_{xx} = \gamma^2 \chi^{-1} \delta H_D / \delta \theta \qquad (3)$$

in one space dimension x (we confine discussion to this case). The parameters are: $\gamma = e/m_p c$ = gyromagnetic ratio, χ = static susceptibility, $\bar{c}$

a velocity $\sim v_F$; θ and r_3 are a pair of canonically conjugate variables such that γr_3 describes the magnetization (r_3 is the expectation value of a spin one operator σ_W : σ_W has the eigenvalues $S_W = \pm 1,0$); B is a homogeneous external magnetic field imposed along x. Important results are [7,8] that the dipole interactions H_D are $H_D = -3/5\ g_D(T)\cos^2\theta$ (A-phase) and $H_D = +\ 4/5\ g_D(T)\ (\cos\theta + \cos 2\theta)$ (B-phase). The first case yields the sine-Gordon equation in $u \equiv -2\theta$; the second case yields the double sine-Gordon equation, with negative sign and $\lambda = 1$, in $u \equiv -2\theta$.

A Hamiltonian density for this second case is [7] $\mathcal{H} = 1/2\ u_t^2 + 1/2\ u_x^2 + 2(\cos 1/2\ u + 1/4)^2$. The dipole interaction energy minimum is at either of the two roots δ and $4\pi - \delta$ of $u = 2\cos^{-1}(-1/4)$. There are two solitary wave solutions (kinks) which take u from $u = \delta$ to $4\pi - \delta$ ('$4\pi - 2\delta$' kink) and from $4\pi - \delta$ to $4\pi + \delta = \delta$ (mode 4π) ('2δ' kink). These take the forms

$$u = 2\pi + 4\tan^{-1}\left(\sqrt{3/5}\tanh 1/2\ \Theta\right), \quad u = 4\tan^{-1}\left(\sqrt{5/3}\tanh 1/2\ \Theta\right) \qquad (4)$$

for the $4\pi - 2\delta$ and 2δ kinks respectively: the arguments Θ are $\Theta = \kappa(x - vt) + \Theta_0$ and $\kappa = \sqrt{15/16}\ (1 - v^2)^{-1/2}$. In the rest frame, these kinks have 'masses' 5.1097 and 11.3929 units. One easily realises (and see [7]) (i) that with boundary conditions $u \to \delta$ (mod 4π) the $4\pi - 2\delta$ kink must be followed by the 2δ kink, the order cannot be reversed and such kinks will bump in collision; (ii) with the same boundary conditions a $4\pi - 2\delta$ kink-antikink pair can convert to a -2δ antikink-kink pair if the colliding $4\pi\ \ 2\delta$ pair has enough kinetic energy to produce the 2δ-pair rest mass. The threshold for this is $v = 0.8938$. Below threshold the $4\pi - 2\delta$ kink-antikink pair bumps [3,5,7]. The behaviour just above threshold is particularly interesting [3,5,7] because of the loss of energy by radiation.

We turn to the creation of spin waves and their behaviours. We suppose that an additional inhomogeneous field $\Delta B_0(x)$ is removed at $t = 0$. For $t > 0$ the field B is a constant field B_0. It is easy to see from (2) that there is a jump induced in r_3, Δr_3, and a motion of θ which satisfy

$$\begin{aligned}\theta_t &= \gamma^2\chi^{-1}\Delta r_3 - \gamma\,\Delta B_0(x)\\ \Delta r_{3,t} &= -\ \delta H_D/\delta\theta + \gamma^2\chi c^{-2}\theta_{xx}\end{aligned} \qquad (5)$$

with $\theta = \theta_0$ (a minimum of H_D), $\theta_t = -\gamma\,\Delta B_0(x)$, $\Delta r_3 = 0$, $\Delta r_{3,t} = 0$ at $t = 0$ and $\theta \to \theta_0$, $\theta_{x^{-1}} \to 0$, etc., as $|x| \to \infty$ for all $t >> 0$. This is equivalent to $\theta_{tt} = -\gamma^2\chi\ \ \delta H_D/\delta\theta + c^2\theta_{xx}$ for $t \geq 0$, with $\theta = \theta_0$, $\theta_t = -\gamma\,\Delta B_0(x)$ at $t = 0$, and $\theta \to \theta_0$, $\theta_x \to 0$, etc., as $|x| \to \infty$.

The A-phase (sine-Gordon) problem has a homogeneous case (recall $u \equiv -2\theta$; $\Omega_{\ell A}$ is the longitudinal NMR frequency [7,8])

$$u_{tt} = -\Omega_{\ell A}^2\sin u;\ u = 0,\quad u_t\ \ 2\gamma\,\Delta B_0 \quad \text{at} \quad t = 0. \qquad (6)$$

However, for $u \to 0$, $u_x \to 0$ etc., as $|x| \to \infty$ for all $t \geq 0$, the solution does not remain homogeneous. It does so for the open-ended boundary condition $u_x = 0$, $|x| = \ell$ for all $t \geq 0$. The solutions in this case are $u = \cos^{-1}\mathrm{cn}(wt,k)$, $w = \gamma\,\Delta B_0$ and $0 < k < 1$; $u = \pi + 2\cos^{-1}\mathrm{dn}(wt,k^{-1})$,

$\omega = \Omega_{\ell A}$ and $k > 1$; $k \equiv \Omega_{\ell A}\gamma^{-1}\Delta B_0^{-1}$ is the modulus of the Jacobian elliptic function cn and k^{-1} of dn. When $k = 1$, $u = 2\cos^{-1}\operatorname{sech} wt$. The magnetization therefore rings with the frequencies of the Jacobian elliptic functions. These frequencies are $\pi\gamma\Delta B_0 K^{-1}$ for $\gamma\Delta B_0 > \Omega_{\ell A}$ and $\pi\Omega_{\ell A}(2K)^{-1}$ for $\gamma\Delta B_0 < \Omega_{\ell A}$ in which $K(k)$ is the complete elliptic integral of the first kind $[K(1) = \infty,\ K(0) = \pi/2]$. For small k, linearized theory, the longitudinal NMR has a steady oscillation at the frequency $\Omega_{\ell A}$ which is independent of B_0 and ΔB_0. This is in agreement with the unusual transverse and longitudinal NMR behaviour observed in the A-phase [8,9]. (The transverse theory is not contained in the present analysis.)

Notice that when $k = 1$ the trajectory $u = 2\cos^{-1}\operatorname{sech}\Omega_{\ell A}t$, $u_t = 2\Omega_{\ell A}\operatorname{sech}\Omega_{\ell A}t$, does not ring. This defines a critical field for no ringing $\gamma\Delta B_{0c} = \Omega_{\ell A}$. WHEATLEY [9] has measured ΔB_{0c} following an analysis similar to that given here: agreement between theory and experiment is not perfect. However, the assumption of homogeneity and the boundary condition $u_x = 0$, $|x| = \ell$ were not discussed. Physical boundary conditions certainly involve the effect of $\hat{\underset{\sim}{\ell}}$ vector textures [10] consistent with the condition $\underset{\sim}{\ell}$ normal to the end surfaces. This problem is too hard to consider here. To determine one possible effect of boundary conditions we impose the *fixed ends* condition $u = 0$ at $|x| = \ell$.

The sine-Gordon equation has been solved in a form readily applicable at present only for solutions $u(x,t)$ defined on the *real line* $-\infty < x < \infty$. We therefore approximate the problem of fixed ends by

$$u, u_x, \text{etc.} \to 0, |x|\to\infty;\ u(x,0) = 0; u_t(x,0) = 2\tau, |x| \le \ell,\ u_t(x,0) = 0, |x| > \ell.$$

KAUP [4] has used the inverse scattering method partially to solve this problem. The essential point is that for $\tau < 1$ only stationary breathers are created. Each such breather is of the form

$$u = 4\tan^{-1}\{\cot\mu \sin(t\sin\mu)\operatorname{sech}(x\cos\mu)\}. \tag{7}$$

The number of such breathers is the integral part of $(\ell\tau/\pi) + 1/2$. The value $\tau = 1$ where $\gamma\Delta B_0 = \gamma\Delta B_{0c} = \Omega_{\ell A}$ remains a critical field. For $\tau > 1$ kink-antikink pairs begin to travel out in opposite directions. In terms of the scattering data, pairs of eigenvalues $\zeta, -\zeta^*$ move successively onto the imaginary axis and split-up there. Figure 1, shows four such radiating kink-antikink pairs; two standing wave breathers remain in the central region $|x| \lesssim \ell$.

Corresponding results for the B-phase double sine-Gordon case are more complicated. The homogeneous problem has phase plane trajectories

$$\frac{1}{2}u_t^2 = (\cos u + \cos\frac{1}{2}u) + C. \tag{8}$$

There are two ringing free trajectories namely when the constant $C = 0$ and when $C = 2$. The critical fields are $\gamma\Delta B_{0c_1} = \sqrt{3/5}\,\Omega_{\ell B}$ and $\gamma\Delta B_{0c_2} = \sqrt{5/3}\,\Omega_{\ell B}$. Further details are given in [7]. Preliminary experiments designed to observe the two critical fields are reported by WHEATLEY [9].

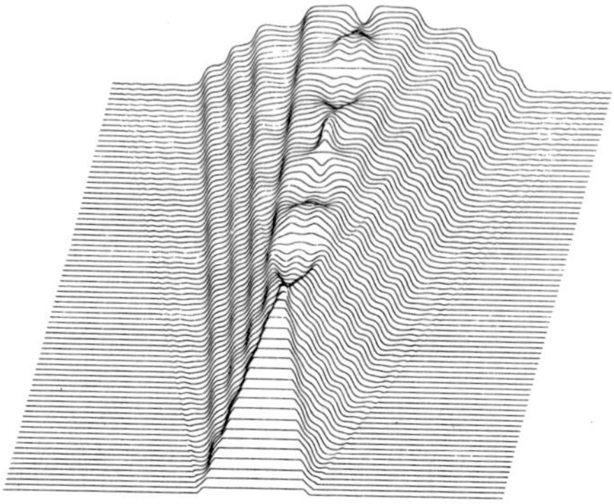

Fig.1 Four outgoing kink-antikink pairs of the sine-Gordon equation generated by initial data $u_t(x,0)= 2.6, |x|<15$. Two standing wave breathers remain in $|x|<15$. Note the initial homogeneous region

The arguments surrounding the boundary conditions for the A-Phase problem apply in similar form to the B-phase problem; but the physical problem of textures is certainly different [10]. The initial value problem (1) (with negative sign and $\lambda = 1$) is not soluble by any known scattering method. Nevertheless, for fields $\Delta B_0 < \Delta B_{0c1}$ we can expect that standing wave breathers may be formed. The Fig. 2 for $u_t = 1$ (less than the critical values $u_t = 3/2$) is evidence supporting this expectation.

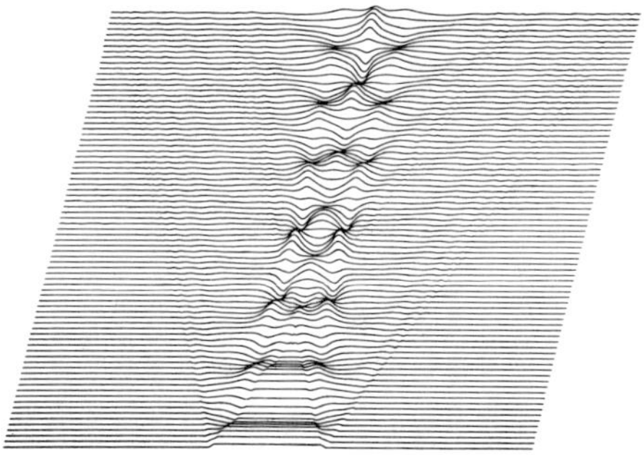

Fig.2 Breather-like solutions of the double sine-Gordon equation generated by initial data $u_t(x,0)= 1, |x|<15$

For boundary conditions $u \to \delta$, $|x| \to \infty$, the $4\pi - 2\delta$ kink cannot be followed by a second such kink. The Fig. 3 for $u_t = 3/2$ shows that two kink-antikink pairs radiate out in opposite directions. The Fig. 3 starts at $t = 40$ in time units and is followed for the longer time shown. The pairs continue to emit radiation but hold together well. This result suggests that for $u_t = 3/2$ at a point $x > \ell$ in a specimen of B-phase ^{3}He, it might be possible to detect magnetically an emitted kink-antikink pair: in contrast in the A-phase a single 2π-kink would be observed for the critical value $u_t = 2$.

The Fig. 4 shows the result of crossing the second critical threshold at $u_t = 5/2$. The boundary condition is $u = \delta, |x| \to \infty$, for the right; $\delta \to 4\pi - \delta \to 4\pi + \delta \to 8\pi - \delta \to 4\pi + \delta \to 8\pi - \delta$. This sequence involves three $4\pi - 2\delta$ kinks, one $(4\pi - 2\delta)$ antikink and one 2δ kink! The conclusion from these figures is that below threshold at $u_t = 3/2$ in the B-phase the homogeneous problem is unstable to break up into breathers; at $u_t = 3/2$ it is unstable to emission of two $4\pi - 2\delta$ kink-antikink pairs; at $u_t = 5/2$ it is unstable to further emission involving one 2δ kink and one 2δ antikink.

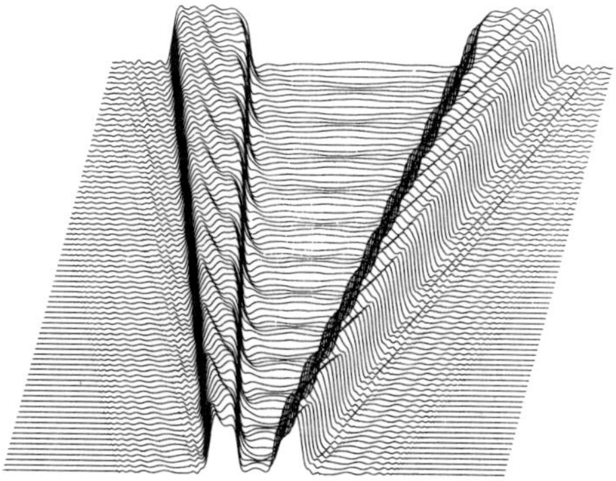

Fig.3 Two 4π-2δ kink antikink pairs of the double sine-Gordon equation generated by the critical initial data $u_t(x,0)=3/2$, $|x|<15$. Time starts at t=40 in this figure

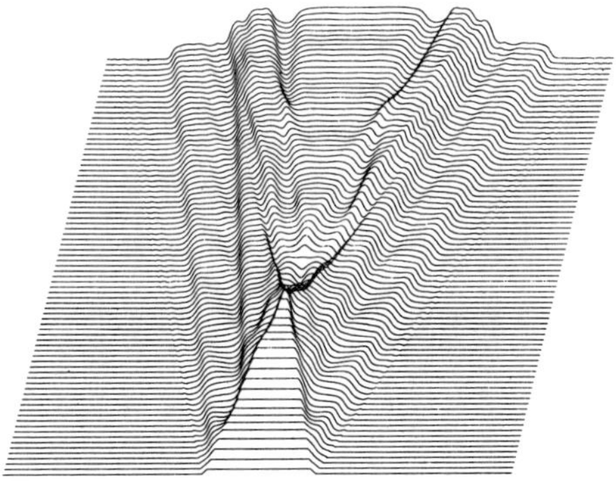

Fig.4 The combination of three 4π-2δ kinks, one $(4\pi$-$2\delta)$ antikink and one 2δ kink of the double sine-Gordon equation travelling to the right, and the corresponding combination of antikinks and kinks travelling to the left, generated by initial data $u_t(x,0)=2.6$. This is above the critical value $u_t=2.5$ for creating 2δ kinks

A conclusion important to the physics is that the critical fields for no ringing NMR remain critical for break up into solitons both in the A-phase and in the B-phase. However, if break up into solitons occurs, as it surely must for most physically realizable boundary conditions, such break up would be accompanied by an apparent ringing signal at least at specific points $x > \ell$ outside the original region of inhomogeneous magnetic field. It is not easy to see that any averaged signal will necessarily be ringing free and this may have some bearing on the imperfect agreement between theory and experiment observed for the A-phase [9].

It seems possible to detect magnetically the different kinks emitted in the A- and B-phases and so obtain further evidence on the correctness of the symmetries assigned to the order parameters in the two different phases.

References

1. S. Duckworth, R.K. Bullough, P.J. Caudrey, J.D. Gibbon: Phys. Lett. 57A, 19 (1976)
2. R.K. Dodd, R.K. Bullough: Proc. Roy. Soc. London A 351, 499 (1976); 352 481 (1977)
3. R.K. Bullough, P.J. Caudrey: "The Multiple sine-Gordon Equations in Non-linear Optics and in Liquid ^{3}He" In *Proceedings of the Symposium on Nonlinear Evolution Equations Solvable by the Inverse Spectral Transform* (Academia dei Lincei, Rome, June 1977) ed. by F. Calogero (in press)
4. D.J. Kaup: "Studies in Appl. Maths.", L1V, 165 (1975) (Massachusetts Inst. of Technology)
5. R.K. Bullough, R.K. Dodd: "Solitons I. Basic Concepts II. Mathematical Structure" In *Synergetics A Workshop*, ed. by H. Haken (Springer-Verlag, Berlin, Heidelberg, New York, 1977) pp.92-119; R.K. Bullough: *Solitons in Physics* (Lectures given at NATO Advanced Study Institute on Nonlinear Equations in Physics and Mathematics, Istanbul, August 1977) ed. by A.O. Barut (in press)
6. P.W. Kitchenside, A.L. Mason, R.K. Bullough, P.J. Caudrey: "Perturbation Theory for the Double sine-Gordon Equation", these Proceedings; A.C. Newell: "Perturbed Soliton Systems", these Proceedings
7. R.K. Bullough, P.J. Caudrey, P.W. Kitchenside: "Bumping Spin Waves in the B-phase of Liquid ^{3}He", to be published in J. Phys. C: Solid State Physics (1978)
8. A.J. Leggett: Rev. Mod. Phys. 47, 331 (1975)
9. J. Wheatley: Rev. Mod. Phys. 47, 415 (1975)
10. K. Maki: "Textures in ^{3}He", these Proceedings

The Interaction of Spin Waves in Liquid He^3 in Several Dimensions

John Gibbon

Department of Mathematics*, U.M.I.S.T., P.O. Box 88
Manchester M60 1QD, Great Britain

Nonlinear Klein-Gordon equations in two or more spatial dimensions of the type

$$\Box\phi = F(\phi) \tag{1a}$$

$$\Box = \frac{\partial^2}{\partial t^2} - \left(\frac{\partial^2}{\partial x^2} + \frac{\partial^2}{\partial y^2} + \frac{\partial^2}{\partial z^2}\right) \tag{1b}$$

persistently occur in theoretical physics as multi-dimensional models of wave propagation in nonlinear media. One important area where equations of this type arise and where multidimensionality arises naturally is in studies of the interaction of spin waves in liquid He^3 [1,2,3]. In the A-phase of He^3, spin waves of magnetization are governed by the sine-Gordon (SG) equation

$$\Box\phi = -\Omega^2_A \sin\phi \tag{2}$$

The constant Ω_A is an NMR frequency and the variable ϕ is an angle which defines the spin orientation [2]. In this case the metric is really (1,-1, -2,-1) for the $\Box$ operator but it is easier to rescale the y variable so that we shall always consider $\Box$ to be in the form given in (1b). Equation (2) also occurs in flux propagation across a Josephson junction [4].

The double sine-Gordon (DSG) equation arises in the B-phase of He^3

$$\Box\phi = \Omega^2_B(\sin\phi + \tfrac{1}{2}\sin\tfrac{1}{2}\phi) \tag{3}$$

Note the change of sign on the r.h.s. of (3) in comparison to (2). For the detailed physics of these equations we refer the reader to refs. [1-3]. A further equation of great interest is the ϕ^4 equation

$$\Box\phi = m^2\phi - \lambda\phi^3 \tag{4}$$

which is of interest in particle physics [5] and studies in lattice dynamics [6].

In 2 dimensions (1 space; 1 time), of all these equations only the SG equation shows perfect soliton behaviour i.e. no ripples are emitted when the solitons collide. The DSG with the opposite sign to (3) governs the propagation of pulses in degenerate SIT in optics [2] and wobbling pulses occur. However for both the bumping spin waves in the He^3 B-phase governed by (3) and for the hyperbolic tangent "kink" solutions of (4) ripples are emitted when one wave collides with another [2,3]. When more spatial dimensions are added as in (1) the situation is no better as far as soliton preservation is concerned. The SG equation has multiple soliton "wavefront" behaviour [7] but special conditions occur on the motion which force the wavefronts to move in triangles whose areas must remain constant in time. Wavefront type solu-

---

*Address after 1st Oc. 1978: Department of Mathematical Physics, University College, Belfield, Dublin 4, Eire.

tions are useless in particle physics but are obviously more useful in solid state physics when one is studying the interaction at differing angles of lines of spins. One of the few equations which have truly four dimensional (Euclidean), analytic, non singular solutions which go to zero everywhere at infinity, is the ϕ^4 equation

$$\Box\phi = C\phi^3 \tag{5}$$

which arises in self-dual Yang-Mills field theories. A solution of (5) is

$$\phi = \sqrt{\frac{-8\lambda}{C}}(x^2 + y^2 + z^2 + t^2 + \lambda)^{-1} \tag{6}$$

This solution has become known as the Polyakov instanton [5].

In order to salvage something from the difficulties which arises in any attempt to find multiple wave solutions of equations such as (1); it is easier to consider solutions of (1) as depending on a single variable $g=g(x,y,z,t)$ [8,9]. Equation (1) is now

$$(\Box g)\phi_g + (\underset{\sim}{\Delta} g)^2\phi_{gg} = F(\phi) \tag{7}$$

where $\underset{\sim}{\Delta}$ plays the role of the gradient operator with an appropriate metric such that $\underset{\sim}{\Delta}.\underset{\sim}{\Delta} = \Box$. There is a result in classical field theory which says that if $\Box g/(\underset{\sim}{\Delta} g)^2 = f(g)$ where f is an arbitrary function then the "surfaces" g are equipotential surfaces. It therefore seems appropriate to take

$$\Box g = A(g) \qquad (\underset{\sim}{\Delta} g)^2 = B(g) \tag{8}$$

As it stands (7) is not integrable but if we take $A = \frac{1}{2}dB/dg$, then (7) becomes

$$\frac{d^2\phi}{dV^2} = -F(\phi) \qquad V = \int |B|^{-\frac{1}{2}} dg \tag{9}$$

The function V plays the role of a potential since $\Box V=0$. We can pick anything for B except (8) may not have any solutions. However it is appropriate to take $B(g) = -g^2\Omega^2$ so $\Omega V = \log g$ and

$$\Box g = -\Omega^2 g \quad ; \qquad (\underset{\sim}{\Delta} g)^2 = -\Omega^2 g^2 \tag{10}$$

where Ω is an appropriate constant (Ω_A or Ω_B etc.).

One set of solutions of (10) are

$$g = \sum_{i=1}^{N} \exp \theta_i \tag{11a}$$

$$\theta_i = p_i x + q_i y + r_i z - w_i t + \delta_i \qquad p_i^{\,2} + q_i^{\,2} + r_i^{\,2} - w_i^{\,2} = \Omega^2 \tag{11b}$$

With ${}^{N}C_2$ conditions on the motion

$$(p_i - p_j)^2 + (q_i - q_j)^2 + (r_i - r_j)^2 - (w_i - w_j)^2 = 0 \tag{11c}$$

The set of equations (11b,c) is overdetermined for $N>7$. As long as (9) can be integrated then solutions can be found trivially. For example the SG kink-type solution is

$$\phi = 4\tan^{-1}(\exp\Omega V) = 4\tan^{-1}g \qquad \Omega = \Omega_A \tag{12}$$

which takes ϕ from 0 to 2π.

It is worth making a comparison at this point with the full multiple soliton solution of the SG equation. Lack of space precludes showing the full formulae but a graph of $\sin\frac{1}{2}\phi$ against x and y at t=0 for three solitons is shown in figure 1. Details are given in [7]. The middle and outer wedge shape areas of the graph are areas in which $\phi=0$ (spin down) and the other areas are $\phi=2\pi$ (spin up). The figure therefore shows areas of differing spins propagating in the (x,y) plane the main result being that the central triangle area os spin propagates without change of form. It was shown in [7] that the area of the triangle formed by the three kinks must remain constant in time. The connection between the full soliton solutions and those given in (11,12) is simple. The conditions on the motion (11c) are exactly those which force all the soliton phase shifts to have infinite value. Miles investigated such solutions for the 2 dimensional KdV equation and called them resonant solutions. It seems therefore that the DSG and ϕ^4 equations have multiple resonant soliton solutions although not full ones. These solutions are essentially very simple and can be thought of as representing a configuration of localised travelling wave solutions. Equation (11c) has a simple meaning. The dispersion relation $p_i^2 + q_i^2 + r_i^2 - w_i^2 = \Omega^2$ expresses the mass energy relationship for a set of solitons of mass Ω. The condition (11c) shows that two solitons of momentum (w_i,p_i,q_i) yield a third of zero mass. This is the classical analogue of Compton scattering.

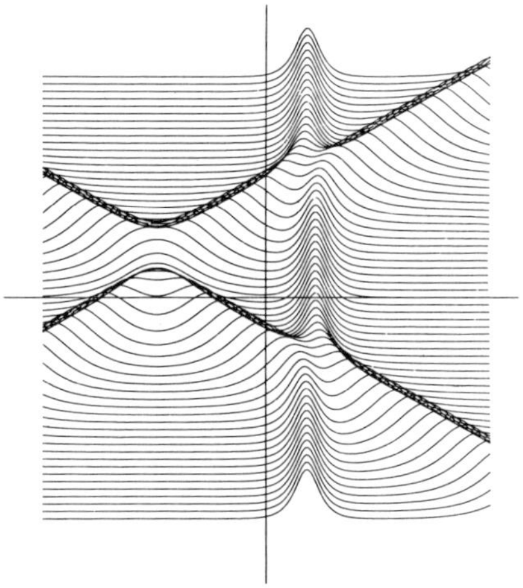

Fig 1

Plot of $\sin\frac{1}{2}\phi$ against x (continuous) and y (discrete) at t=0 for three full solitons (non resonant) for the sine-Gordon equation.

There are two types for the DSG equation [2], which are associated with the zeroes of the r.h.s. of (3) which occur at $\delta = 2\cos^{-1}(-\frac{1}{4})$. There are no real solutions associated with the zeroes which occur at $2N\pi$. One pulse, called the 2δ pulse takes ϕ from $-\delta$ to $=\delta$:

$$\phi = 4\tan^{-1}\left[\sqrt{\frac{5}{3}}\left(\frac{g-1}{g+1}\right)\right] \qquad \Omega=\Omega_B \tag{13}$$

A second solution is the $4\pi-2\delta$ pulse which takes ϕ from δ to $4\pi-\delta$:

$$\phi = 4\tan^{-1}\left[\sqrt{\frac{5}{3}}\left(\frac{g+1}{g-1}\right)\right] \tag{14}$$

Each of these pulses (12,13,14) are spin waves of magnetization where ϕ switches from one value or level to another. Figure 2 represents the 2δ

pulse in the (x,y) plane at $t=0$ for three waves. The configuration is moving down the figure, the upper level representing an area of one spin orientation and the lower level another.

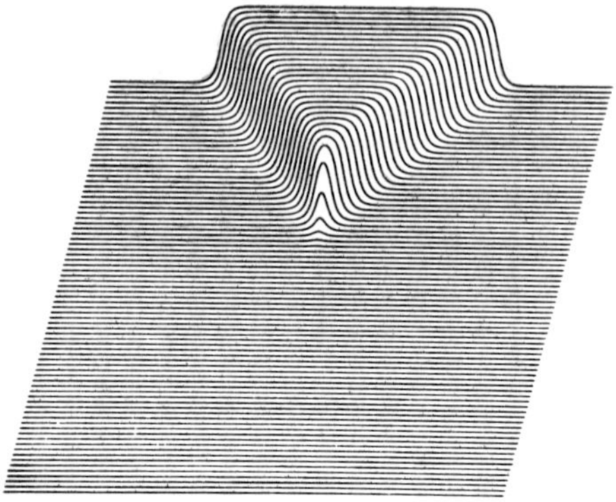

Fig 2

A 2δ pulse solution of the DSG equation $(t=0)$ for two waves $(\Omega_B=1)$. Upper area: $\phi=\delta$ and lower area $\phi=-\delta$

One further set of solutions in 2 dimensions for the SG equation are of interest. If we take $g = \Omega_A{}^2$ and $(\underset{\sim}{\Delta} g)^2 = g\Omega_A{}^2$ in (7) then a solution is

$$g = \tfrac{1}{4}(x^2 - t^2)\Omega_A{}^2 \qquad (15a)$$

where

$$g\phi_{gg} + \phi_g = \sin\phi \qquad (15b)$$

The solution for g is equivalent to a similarity solution. The o.d.e. in (15a) can be transformed into the third Painlevé transcendent which has no essential singularities or branch points. Numerical computation [10] shows that ϕ now moves between 0 and π and has the characteristic of being an oscillating function.

References

1. A.J. Leggett: Rev. Mod. Phys. 47, 331, 1975

2. R.K. Bullough and P.J. Caudrey: Proceedings of the Symposium "Nonlinear Evolution equations solvable by the Inverse Spectral Transform" Pitman Research Notes 1978

3. R.K. Bullough and P.J. Caudrey: "Optical Solitons and their spin wave analogues in He^3" Proc. of 4th Rochester Conference on Quantum Optics: ed. L. Mandel and E. Wolf (Plenum N.Y.) 1978

4. A. Barone, F. Esposito, C. Magee, A. Scott: Riv. Nuovo Cim., 1, 227, 1971

5. R. Jackiw: Rev. Mod. Phys. 49, 681, 1977

6. T. Schneider and E. Stoll: Phys. Rev. Letts, 35, 296, 1975; J.A. Krumhansl and J.R. Schrieffer, Phys. Rev. B11, 3535, 1975

7. J.D. Gibbon and G. Zambotti: Nuovo Cimento, 28B, 1, 1975

8. J.D. Gibbon, N.C. Freeman and R.S. Johnson: Phys. Letts. 65A, 380, 1978

9. J.D. Gibbon, N.C. Freeman and A. Davey, J.Phys.A, 11, L93, 1978

10. G.L. Lamb: Rev. Mod. Phys. 43, 99, 1971

Josephson Transmission Line Oscillators

Alwyn C. Scott

Department of Electrical and Computer Engineering, University of Wisconsin
Madison, WI 53706, USA

1. Introduction

In this paper, the term "Josephson transmission line" (or JTL) will refer to a wave structure composed of two superconducting metal strips separated by an insulating barrier that is thin enough (~ 25Å) to permit transverse (Josephson type) tunneling of electrons in the superconducting state [1]. JTL's as long as 35 cm have been fabricated by a process employing rf sputtering of the metal layers and photoresist definition of all patterns [2,3]. A diagram of the basic structure to be considered is shown in Fig. 1 where a TEM wave is assumed to propagate in the x-direction. The dependent variables of this wave can be viewed as the transverse voltage v(x,t) across the insulating barrier and the logitudinal current i(x,t) flowing parallel to the insulating barrier in the x-direction.

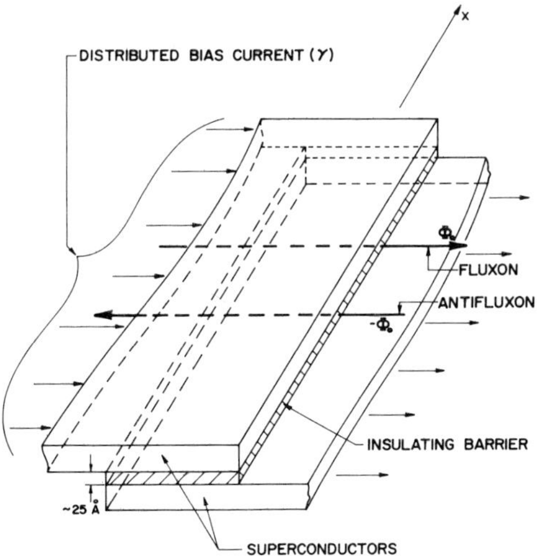

Fig. 1 Sketch of the Josephson transmission line (JTL) considered in this paper

If dissipative effects are neglected, the magnetic flux of the wave

$$\Phi \equiv \int v\, dt$$

is governed by the partial differential equation

$$\Phi_{xx} - (\bar{c})^{-2}\, \Phi_{tt} = (\lambda_J)^{-2} \sin(2\pi\Phi/\Phi_0) \tag{1}$$

where $\bar{c}$ is a characteristic (or wavefront) velocity for the TEM wave, λ_J is a "Josephson penetration length" for magnetic fields into the insulating barrier, and

$$\Phi_0 = h/2e$$

$$\doteq 2.0678 \times 10^{-15} \text{ volt-seconds}$$

is the quantum unit of magnetic flux. Typical values for $\bar{c}$ are 1/15 to 1/20 the velocity of light [3,4] and λ_J ranges from .01 to 1 mm [3]. If x is measured in units of λ_J, time in units of $\lambda_J/\bar{c}$ and magnetic flux (Φ) in units of $\Phi_0/2\pi$, (1) becomes

$$\phi_{xx} - \phi_{tt} - \sin\phi = 0 \tag{2}$$

which is readily recognized as a normalized form of the sine-Gordon equation [5]. The "kink" solutions of this equation

$$\phi_{\pm}(x,t;u,x_0) = 4 \tan^{-1}\left[\exp\left(\pm \frac{x-ut-x_0}{\sqrt{1-u^2}}\right)\right] \tag{3}$$

represent either a fluxon (ϕ_+) traveling in the + x direction or an anti-fluxon (ϕ_-) traveling in the + x direction (see Fig. 1). If the transverse voltage is measured in units of $\Phi_0\bar{c}/2\pi\lambda_J$, then

$$\phi_t = v\,.$$

Also if the longitudinal current is measured in units of $j_0\lambda_J$ (where j_0 is the maximum Josephson current per unit length in the x-direction), then

$$\phi_x = -i\,.$$

Although several interesting applications have been proposed to employ fluxons in information processing systems [6], attention here will be focused upon the fluxon as a generator of electromagnetic radiation. To this end we will be interested in analyzing not the "perfect" sine-Gordon equation given in (2) but structural perturbations of it that permit input and dissipation of energy. Energy input to a fluxon is readily obtained through the distributed bias current (γ) indicated in Fig. 1. The effect of γ is to exert a Lorentz force which accelerates a fluxon in the + x direction and an anti-fluxon in the - x direction. Energy dissipation will be represented by adding phenomenological loss terms of the form $\Gamma|\phi_t|\phi_t$ or $\alpha\phi_t$ to the right hand side of (2). Thus we shall investigate either

$$\phi_{xx} - \phi_{tt} - \sin\phi = \Gamma|\phi_t|\phi_t + \gamma \tag{4a}$$

or

$$\phi_{xx} - \phi_{tt} - \sin\phi = \alpha\phi_t + \gamma \tag{4b}$$

depending upon which is more convenient for the problem at hand.

2. Oscillations on a Finite JTL

In this section we assume our JTL to be of a length L that is long compared with its Josephson penetration length, λ_J, but "finite" so open circuit boundary conditions at the ends influence the oscillations. Thus in the absence of dissipative effects we seek solutions of (2) together with the boundary conditions: $\phi_x(0,t) = \phi_x(L,t) = 0$. The qualitative features of such solutions have been described in a recent survey by Fulton [7]. Analytic expressions can be obtained from the *ansatz* suggested by Lamb [8]. $\phi(x,t) = 4\tan^{-1}[h(x)g(t)]$ where h and g are, in general, Jacobian elliptic functions defined by $(h')^2 = ah^4 + (1+b)h^2 - c$ and $(g')^2 = cg^4 + bg^2 - a$ with a, b and c arbitrary constants. Constabile *et al*. [9] have displayed solutions for the three principal types of oscillatory behavior.

Plasma oscillation. This is an oscillation about $\phi = 0$ described by

$$\phi = 4\tan^{-1}[A\,\mathrm{cn}(\beta x;k_f)\,\mathrm{cn}(\Omega t;k_g)] \tag{5}$$

where

$$k_f^2 = \frac{A^2[\beta^2(1+A^2)+1]}{\beta^2(1+A^2)^2};\quad k_g^2 = \frac{A^2[\Omega^2(1+A^2)-1]}{\Omega^2(1+A^2)^2}$$

and Ω, β and A are related by the *nonlinear dispersion equation*

$$\Omega^2 - \beta^2 = \frac{1-A^2}{1+A^2}\ . \tag{6}$$

Imposition of the open circuit boundary condition at the ends of the JTL fixes the spatial periodicity as

$$\beta_N = \frac{2N}{L}K(k_f) \tag{7}$$

where $N = 1,2,...$ is the number of nodes in the standing wave and $K(k_f)$ is the complete elliptic integral of the first kind.

Bion oscillations. These correspond to bound state oscillations of fluxon-antifluxon pairs (they might also be called "breather" oscillations). They are described by

$$\phi = 4\tan^{-1}\{A\,\mathrm{dn}[\beta(x-x_0);k_f]\,\mathrm{sn}(\Omega t;k_g)\} \tag{8}$$

where

$$k_f^2 = 1 - \left[\frac{1-\beta^2(1+A^2)/A^2}{\beta^2(1+A^2)}\right]; \quad k_g^2 = \frac{A^2[1-\Omega^2(1+A^2)]}{\Omega^2(1+A^2)}$$

and the nonlinear dispersion equation is

$$\beta = \Omega A . \tag{9}$$

The open circuit boundary conditions now require

$$\beta_N = \frac{N}{L} K(k_f) \tag{10}$$

with two possible values for x_0: a) $\beta x_0 = K(k_f)$ and N even (this corresponds to bions located near the center of the JTL) and b) $x_0 = 0$ (this corresponds to fluxons bound to virtual antifluxons at both ends of the JTL).

Fluxon oscillations. Such an oscillation can be qualitatively described as a fluxon that moves to one end of the JTL, is reflected as an antifluxon that moves to the other end, is reflected...etc. It differs from bion and plasma oscillations in that at each point (x) of the JTL, ϕ increases without bound as t increases. Thus a fluxon oscillation can be driven by a distributed bias current (γ) and may serve as a useful ac generator. The generator frequency will be equal to that of the Josephson current, $I_0 \sin(2\pi\Phi/\Phi_0)$. Thus if the dc component of the voltage across the junction is V, the corresponding frequency is

$$f = V/\Phi_0 \tag{11}$$

where Φ_0^{-1} is equal to 483.59 megacycles per microvolt. Fluxon oscillations are described by

$$\phi = 4 \tan^{-1}[A\, \mathrm{dn}(\beta x;k_f)\, \mathrm{tn}(\Omega t;k_g)] \tag{12}$$

where

$$k_f^2 = 1 - \left[\frac{\frac{\beta^2}{A^2}(A^2-1)-1}{\beta^2(A^2-1)}\right]; \quad k_g^2 = 1 - \left[\frac{A^2[\Omega^2(A^2-1)-1]}{\Omega^2(A^2-1)}\right] .$$

The nonlinear dispersion equation is as in (9), and the open circuit boundary conditions again require (10). A monotone increasing behavior of ϕ with t is obtained from (12) by noting that $\phi \to 2\pi$ as $\mathrm{tn} \to \infty$ and $\Omega t \to (2m-1) K(k_g)$ where $m = 0,\pm 1,\ldots$; thus we can switch branches of the arctan function at these instants.

Since fluxon oscillations exhibit a nonzero average value of ϕ_t, a corresponding finite voltage should be measured. The resulting volt-ampere characteristic can be calculated by introducing bias current and dissipation as indicated in (4a). Assuming that Γ and γ are small enough so $\phi(x,t)$ can be approximated by (12), the total input power is the time and space average of $\gamma\phi_t$ while the power dissipation is the time and space average of $\Gamma|\phi_t|\phi_t^2$. Equating these two power terms yields the relation

$$\gamma = 2A^2 \, \Omega^2 \Gamma \left[\frac{6E(k_f)}{K(k_f)} + \frac{1 + (k_g')^2}{A^2} \right] \tag{13a}$$

where $E(k_f)$ is the complete elliptic integral of the second kind, and the time and space average of the normalized voltage is

$$\langle \bar{\phi}_t \rangle = 2\pi\Omega / K(k_g) \ . \tag{13b}$$

From (13) a plot is readily obtained for γ vs. $\langle \Phi_t \rangle$, the volt-ampere characteristic associated with a fluxon oscillation. This relation is plotted in Fig. 2a for an increasing number (N) of fluxons involved in the oscil-

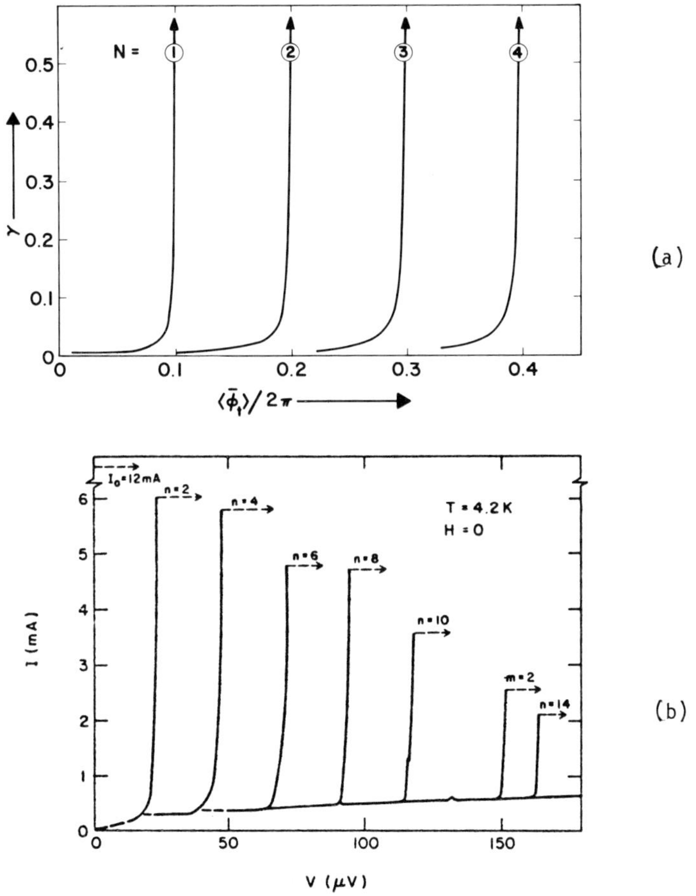

Fig. 2 (a) Plot of current (γ) vs. average voltage from (13). (b) Current-voltage measurements on a lead-oxide-lead Josephson oscillator with dimensions 1.5 mm × 0.2 mm by Chen and Langenberg [10]

lation. For comparison we reproduce, in Fig. 2b, some measurements on a Pb-oxide-Pb rectangular Josephson junction oscillator which have been published by Chen and Langenberg [10]. While the qualitative similarity between Figs. 2a and 2b is encouraging, several reservations and explanatory comments should be made.

a) A major qualitative difference is that the curves of Fig. 2a approach infinity while those of Fig. 2b terminate at some value which is less than the maximum Josephson current. This discrepancy is an artifact of the power balance calculation leading to (13). It was assumed that the structure of the nonlinear mode, i.e. $\phi(x,t)$, remains equal to that given in (12) as γ becomes large. Study of an exact solution of (4a) under periodic boundary conditions by Parmentier and Costabile [11] shows that this is not the case. Their calculations yield branches that terminate at $\gamma = 1$ and display somewhat sharper "elbows".

b) The JTL structure studied in Fig. 2b has dimensions of approximately 1.5 mm × 0.2 mm. Thus the branch labeled "m = 2" corresponds to the lowest transverse oscillation.

c) The branch labeling in Fig. 2b indicates integer multiples of the lowest resonant frequencies of the rectangular Josephson junction cavity which occur at frequencies $f_n = n\bar{c}/2L$. In Fig. 2a, on the other hand, the index N indicates the number of fluxons engaged in the nonlinear oscillation. For one fluxon ($N = 1$) a period is the time required for a fluxon to bounce from one end of the junction to the other at velocity $\bar{c}$. Thus the limiting frequencies to be expected are $f_N = N\bar{c}/L$. The fact that no branches are found for odd values of n in Fig. 2b tends to confirm the theoretical picture presented here.

d) Measurements of power output from the oscillator of Fig. 2b are much less than that indicated by the power balance calculation. Along the $n = 2$ branch, for example, the maximum power output was found to be only 2×10^{-11} watts over a very narrow range (2×10^{-7} volts) of junction voltage. Outside of this narrow range the power output dropped by two orders of magnitude and the volt-ampere characteristic displayed a "fine structure" with hysteresis between adjacent branches. The power balance calculation, on the other hand, indicates that power output should rise as a monotone function of voltage and reach a broad maximum along the vertical segment of the characteristic. It is likely that this discrepancy appears because the "parasitic" effects of plasma and bion modes have not been taken into account. To proceed analytically in this direction will require the more general multiply periodic solutions of the sine-Gordon recently given by Matveev [12].

3. Radiation from a "Wobbling" Fluxon

The oscillator structure to be considered in this section is shown in Fig. 3. Here a JTL of the sort indicated in Fig. 1 is formed as a ring and periodically loaded with localized regions of large Josephson current (the "microshorts") that are separated by an equal spacing a. Means for supplying a distributed current bias (γ) are included so a fluxon can be forced around the circumference of the ring. As the fluxon passes a microshort, it will exchange kinetic energy for locally stored magnetic energy and slow down. On the average, the fluxon will execute a "wobbling" motion at a frequency

$$\omega_o = 2\pi u_m/a \tag{14}$$

where u_m is the mean circumferential velocity of the fluxon. Radiation from the fluxon can be expected at the appropriate Doppler shift of frequency ω_o.

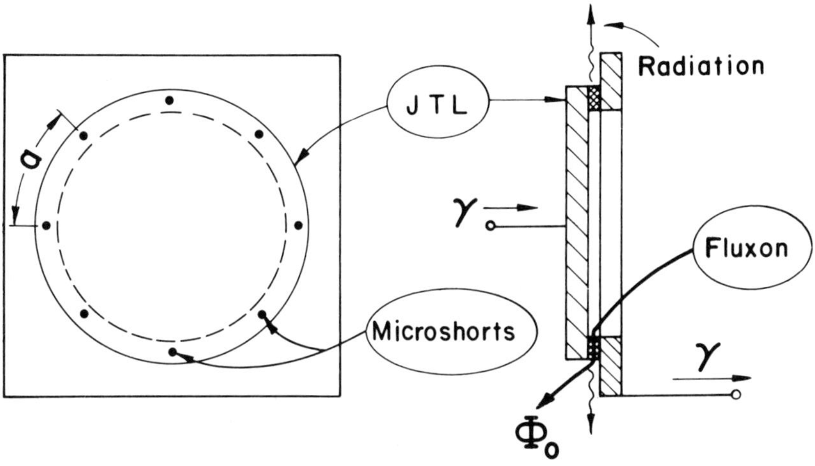

Fig. 3 A "wobbling" fluxon oscillator structure

A convenient analytic tool to study this problem is a soliton perturbation theory in the form recently suggested by Keener and McLaughlin [13]. Only a brief résumé of this approach will be presented here; the reader is referred to [14] for full analytical details. We begin the analysis by writing a structurally perturbed version of the sine-Gordon equation (2) in the form

$$\partial_t \begin{pmatrix} \phi \\ \phi_t \end{pmatrix} + \left(\begin{array}{c|c} 0 & 1 \\ \hline -\partial_{xx} + \sin(\cdot) & 0 \end{array} \right) \begin{pmatrix} \phi \\ \phi_t \end{pmatrix} = \alpha \begin{pmatrix} 0 \\ f(\phi) \end{pmatrix} \tag{15}$$

where the structural perturbation

$$f \equiv -\phi_t - \frac{\gamma}{\alpha} - \frac{\mu}{\alpha} \sum_n \delta(x-na) \sin \phi \ . \tag{16}$$

The first two terms in (16) represent dissipation and current bias as in (4b); the last (summation) term takes account of the increased Josephson current localized at the microshorts. Then a series expansion in the small parameter α takes the form

$$\overline{W} = \overline{W}_0 + \alpha \overline{W}_1 + \dots$$

where $\overline{W} \equiv \text{col}(\phi,\phi_t)$, $\overline{W}_0 \equiv \text{col}(\phi_0,\phi_{0t})$, and ϕ_0 is a solution of the perfect sine-Gordon equation (2). Also $\overline{W}_1 \equiv \text{col}(\phi_1,\phi_{1t})$ is a first order correction governed by the linear equation

$$L\overline{W}_1 = \overline{f} \tag{17}$$

where $\overline{f} \equiv \text{col}(0,f)$ and the operator

$$L \equiv \left(\begin{array}{c|c} \partial_t & 1 \\ \hline -\partial_{xx} + \cos\phi_0 & \partial_t \end{array}\right). \tag{18}$$

If it is assumed that $\phi(x,t)$ is an N-soliton wave at $t = 0$, (17) can be solved as

$$\overline{W}_1 = \int_0^t \int_{-\infty}^{\infty} G(x,t|x',t')\, \overline{f}(x',t')dx'dt' \tag{19}$$

where $G(x,t|x',t')$ is a Green's function that can be explicitly constructed [14]. Secular terms in (19) arise from resonances between soliton contributions to the source $(\overline{f})$ and their corresponding contributions to the Green's function G. These can be avoided by allowing the positions and speeds of the fluxons to modulate slowly with time. To see this, suppose ϕ_0 is a single fluxon

$$\phi_0 = 4\tan^{-1}\left[\exp\left(\frac{x-X}{\sqrt{1-u^2(t)}}\right)\right]$$

where $X \equiv \int_0^t u(t')dt' + x_0(t)$ is the fluxon's position. Then the source (at order α) becomes augmented to

$$\overline{F} = \begin{bmatrix} -\frac{1}{\alpha}\phi_{0,u}\dot{u} + \frac{1}{\alpha}\phi_{0,x'}\dot{x}_0 \\ f - \frac{1}{\alpha}\phi_{0,ut'}\dot{u} + \frac{1}{\alpha}\phi_{0,x't'}\dot{x}_0 \end{bmatrix}. \tag{20}$$

If $\dot{u}$ and $\dot{x}_0$ are chosen so that $\overline{F}$ is orthogonal to the rows of G, secularities are avoided and we have the o.d.e.'s

$$\dot{u} = \frac{\pi\gamma}{4}(1-u^2)^{3/2} - \alpha u(1-u^2)$$

$$+ \frac{\mu}{2}(1-u^2)\sum_n \mathrm{sech}^2\left(\frac{X-na}{\sqrt{1-u^2}}\right)\tanh\left(\frac{X-na}{\sqrt{1-u^2}}\right). \tag{21}$$

$$\dot{x}_0 = -\frac{\mu}{2}\, u \sum_n (X-na)\,\mathrm{sech}^2\left(\frac{X-na}{\sqrt{1-u^2}}\right)\tanh\left(\frac{X-na}{\sqrt{1-u^2}}\right)$$

Integrating these to obtain the periodic motion $u(t) = u(t + 2\pi/\omega_0)$ then determines the wobble frequency and the mean fluxon velocity, u_m. At this point, nonradiated dc power input to the fluxon can be calculated as the product of the Lorentz force (γ) on the fluxon times its mean velocity. Thus

$$P_{nr} = \gamma u_m \quad . \tag{22}$$

Also the augmented source in (20) can be written as a Fourier series in ω_0

$$\begin{aligned}\overline{F}(x',t') = \overline{F}_0(x'-u_m t') &+ \overline{F}_1(x'-u_m t')e^{+i\omega_0 t'} \\ &+ \overline{F}_1^*(x'-u_m t')e^{-i\omega_0 t'} \\ &+ \text{higher harmonics}\end{aligned} \tag{23}$$

and (19) can be rewritten in the form

$$\overline{W}_1 = \int_0^t \int_{-\infty}^{\infty} G_r(x,t|x',t')\,\overline{F}(x',t')dx'dt' \tag{24}$$

where G_r is the "radiative" component of the Green's function which was <u>not</u> made orthogonal to $\overline{F}$. Thus (24) computes the first order radiation from a fluxon as it moves under the influence of (21). The radiative part of the Green's function takes the form*

$$G(x,t|x',t') = \frac{1}{4\pi i}\int_{-\infty}^{\infty} d\lambda \begin{bmatrix} g_{11} & g_{12} \\ g_{21} & g_{22}\end{bmatrix} \frac{\exp\{-i[k(\lambda)(x-x')+\omega(\lambda)\ (t-t')]\}}{\lambda(\zeta^2-\lambda^2)^2} \tag{25}$$

where the $g_{ij} = g_{ij}(x-ut|x'-ut)$, $\zeta = i[(1+u)/(1-u)]^{1/2}$, $k(\lambda) \equiv 2\lambda - 1/8\lambda$,

---

* Explicit expressions for the g_{ij} are given in [14].

and $\omega(\lambda) \equiv 2\lambda + 1/8\lambda$. Thus $\omega^2 = k^2 + 1$ and $d\lambda/\lambda = d\omega/k = dk/\omega$.

The "dc" source term from (23) causes a first order correction to the shape of ϕ_0 [15] which can be neglected when ϕ_0 is used for other first order calculations. The steady state response to the fundamental harmonic in $\overline{F}$ can be defined as

$$\overline{W}_{1ss}(x,t) = \frac{1}{4\pi i} \lim_{t\to\infty} \int_{-\infty}^{\infty} d\lambda \int_{-t}^{t} dt' \int_{-\infty}^{\infty} dx' \, \frac{\exp\{-i[k(\lambda)(x-x')+\omega(\lambda)(t-t')]}{\lambda(\zeta^2-\lambda^2)^2}$$

$$\cdot \, [g_{ij}][\overline{F}_1(x'-u_m t')e^{i\omega_0 t'} + \overline{F}_1^*(x'-u_m t')e^{-i\omega_0 t'}] \, .$$

Integration over x' introduces a factor $\exp\{iku_m t'\}$ so the t' integral involves only

$$\lim_{t\to\infty} \int_{-t}^{t} \exp\{i[\omega(\lambda)+k(\lambda)u_m \pm \omega_0]t'\}dt' = 2\pi\delta(\omega+ku_m \pm\omega_0).$$

This drives radiation at the Doppler shifted wobble frequencies $-(ku_m \pm \omega_0)$. For oscillator applications the higher frequency (ω_+) is of primary interest. Here $\omega_+ = -(\omega_0 + k_+ u_m)$ and $k_+^2 = \omega_+^2 - 1$ so

$$k_+ = \frac{u_m\omega_0 + \sqrt{\omega_0^2+u_m^2-1}}{1 - u_m^2} \quad \text{and} \quad \omega_+ = -\omega_0 \frac{1 + u_m\sqrt{1-(1-u_m^2)/\omega_0^2}}{1 - u_m^2} \tag{26}$$

Note that as $u_m \to 1$, $\omega_+ \to -\omega_0/(1-u_m)$ so a substantial increase in the radiation frequency is obtained through the Doppler shift. Changing the variable of integration from λ to ω then yields

$$\overline{W}_{1ss}^{+} = 128 \, \frac{\exp\{-i[k_+ x + \omega_+ t + \frac{\pi}{2}]\}}{k_+\left[\frac{1+u_m}{1-u_m} + (k_+ + \omega_+)^2\right]^2} \, \cdot$$

$$\cdot \int_{-\infty}^{\infty} [g_{ij}(x-u_m t|x')] \exp(ik_+ x') \, \overline{F}_1(x')dx' \tag{27}$$

$$+ \, cc.$$

$$\equiv col(w,w_t) \, .$$

Away from the fluxon this radiation is governed by the linear Klein-Gordon equation $w_{xx} - w_{tt} = w$ for which the corresponding energy (Hamiltonian) density is $H = \frac{1}{2}(w_x^2 + w_t^2 + w^2)$. Using the dispersion equation ($\omega^2 = k^2 + 1$) and the plane wave relation: $w_x^2 = k_+^2 w^2$ and $w_t^2 = \omega_+^2 w^2$, this can be written $H = w_t^2 = \omega_+^2 w^2$. This energy propagates at the group velocity $d\omega/dk = -k_+/\omega_+$. Thus the radiated power is

$$P_r = \frac{[u_m\omega_0 + \sqrt{\omega_0^2+u_m^2-1}] \cdot [\omega_0+u_m \sqrt{\omega_0^2+u_m^2-1}]}{(1-u_m^2)^2} w^2 . \tag{28}$$

The radiated power will be absorbed by the losses (α) of the JTL and by a useful load. Suppose α is composed of two components: i) α' represents internal losses and ii) α'' represents power absorbed by a shunt load. Then $\alpha = \alpha' + \alpha''$ and the power abosrbed by the load would be

$$P_0 = \frac{\alpha''}{\alpha} P_r . \tag{29}$$

The above discussion has been presented entirely in normalized units. In laboratory units, the limiting velocity of a fluxon ($\bar{c}$) is about 1/15 of the velocity of light or 2×10^7 meters/second. Setting $a = 1$ means that the microshorts are separated by a distance equal to the "Josephson length" $\lambda_J \doteq 5 \times 10^{-4}/\sqrt{J_c}$ microns where J_c is the Josephson current density measured in amperes/meter^2, $\lambda_J \approx 5$ microns and the wobble frequency $f_0 = u_m/\lambda_J$ would be $4 \times 10^{12}(u_m/\bar{c})$ cps. As previously noted, the corresponding Doppler shifted radiation frequency would then approach

$$f_+ \to 4 \times 10^{12}\left(\frac{u_m}{\bar{c}-u_m}\right) \text{ cps}$$

as $u_m \to \bar{c}$, the relativistic limit. For $u_m/\bar{c} = 1/2$, this corresponds to a free space wave length of .075 mm.

References

1. A. C. Scott, Nuovo Cimento B 69, 241(1970).

2. L. S. Hoel, et al., Solid-State Electron. 15, 1167(1972); and P. Rissman and T. Palhomen, ibid. 17, 611 (1974).

3. A. C. Scott, et al., J. Appl. Phys. 47, 3272 (1976).

4. H. T. Yuan and A. C. Scott, Solid-State Electron. 9, 1149 (1966).

5. A. Barone, et al., Rivista del Nuovo Cimento 1, 227 (1971).

6. T. A. Fulton, et al., Proc. IEEE 61, 28 (1973); K. K. Likharev, IEEE Transactions on Magnetics MAG-13, 245 (1977); P. Guéret, ibid. MAG-11, 751 (1975); K. Nakajima et al., J. Appl. Phys. 47, 1620 (1976).

7. T. A. Fulton, in Superconductor Applications: SQIDS and Machines, B. B. Schwartz and S. Foner eds. (Plenum, New York, 1977) p. 125.

8. G. L. Lamb, Jr., Rev. Mod. Phys. 43, 99 (1971).

9. G. Costabile, et al., Appl. Phys. Lett. 32, 587 (1978).

10. J. T. Chen and D. N. Langenberg, in Low Temperature Physics, K. D. Timmerhaus, et al. eds. vol. 3 (Plenum, New York, 1974) p. 289.

11. R. D. Parmentier and G. Costabile, Rocky Mountain J. Math. 8, no. 1 (in Press); R. D. Parmentier, in Solitons in Action, K. E. Lonngren and A. C. Scott, eds. (Academic Press, New York, 1978).

12. V. B. Matveev, Preprint No. 373, Inst. of Theoret. Physics, Univ. of Warsaw, 1976 (unpublished).

13. J. P. Keener and D. W. McLaughlin, Phys. Rev. A 16, 777 (1977); J. Math. Phys. 18, 922 (1977).

14. D. W. McLaughlin and A. C. Scott, Appl. Phys. Lett. 30, 545 (1977); see also Solitons in Action, K. E. Lonngren and A. C. Scott, eds. (Academic Press, New York, 1978).

15. M. B. Fogel, S. E. Trullinger, A. R. Bishop and J. A. Krumhansl, Phys. Rev. Lett. 36, 1411 (1976); Phys. Rev. B 15, 1578 (1977).

Dissipative Structures in Quasi-One-Dimensional Superconductors

A. Baratoff

IBM Zurich Research Laboratory, CH-8803 Rüschlikon, Switzerland

1. Introduction

Consider a superconducting filament of length $d \gg \xi(T)$, the coherence length, close to its transition temperature T_c, and fed by a dc current source. If the filament is connected to massive superconducting electrodes and if its cross-section is sufficiently small, the current density j is uniformly distributed, while the total phase difference ϕ and voltage drop V, if any, are effectively localized within the filament. According to the basic JOSEPHSON relation

$$\dot{\phi} = 2eV/\hbar \ . \qquad (1)$$

Such systems can sustain a time average V_{dc} corresponding to the normal resistance of a segment much shorter than d. This has been attributed [1] to *phase-slip centers*: the increase in ϕ is compensated by -2π jumps in the local phase of the complex order parameter $\psi = fe^{i\chi}$, as the amplitude f periodically vanishes over a section of the filament.

Recently, KRAMER and the author [2] showed that such states appear among one-dimensional solutions of the simplest time-dependent GINZBURG-LANDAU (TDGL) equations. A classification of possible steady states and of threshold solutions signaling global instabilities has been achieved, both for infinite filaments [2] and finite links [3]. In this report, we summarize the main conclusions of those investigations, fit the *kink solutions* first identified by LIKHAREV [4] into that scheme, and illustrate their role in transitions between (globally) different steady states.

2. One-Dimensional Time-Dependent GINZBURG-LANDAU Equations

Although they ignore quasiparticle relaxation processes which are important in real superconductors, the TDGL equations are valid for sufficiently high j [5]. Moreover, they seem to be qualitatively correct at low j and represent the superconducting state quite well close to T_c. We are mainly interested in states with nonzero electrical potential μ and nontrivial x-dependence, and must rely on numerical computations. The complex form of the equations

$$\tau(\dot{\psi} + i\mu\psi) = \psi'' + (1 - |\psi|^2)\psi, \quad \mathrm{Im}\ \psi^*\psi' - \mu' = j \qquad (2)$$

is more convenient in that connection, but the physics involved becomes more obvious if one separates amplitude and phase, i.e.,

$$\tau \dot{f} = f'' + (1 - (\chi)'^2 - f^2)f \tag{3}$$

$$\tau f^2(\dot{\chi} + \mu) = (f^2 \chi')' , \qquad f^2 \chi' - \mu' = j. \tag{4}$$

Thus τ describes the relaxation of f, while (4) implies that the characteristic spatial and temporal scales for μ are $1/\sqrt{\tau}$ and 1, respectively, in terms of the reduced units defined in [2,3]. Finally, the local JOSEPHSON relation, $\mu + \dot{\chi} = 0$, holds in superconducting regions. Contacts to massive electrodes with a possibly different transition temperature T_{c0} are simulated [6] by imposing

$$\mu(0) = V = \dot{\phi} , \qquad \mu(d) = 0, \tag{5}$$

$$\psi(0) = f_0 e^{-i\phi}, \quad \psi(d) = f_0, \quad f_0^2 = (T_{c0} - T)/(T_c - T). \tag{6}$$

The location where $\chi = \mu = 0$, can be chosen to optimize computational accuracy. The results presented here refer to $\tau = 5.79$, a special value singled out by microscopic theory [2-5].

3. Steady-States: Stability Ranges

Unless specified, *stable* henceforth refers to *local stability*. In an *infinite filament* [2] two simple stationary states are known:

- the *normal* (N) state: $\psi = 0$, $\mu' = -j$ (stable for all $j \neq 0$);
- the *superconducting* (S) state: $\psi = f_\infty e^{iqx}$, $q^2 = 1 - f_\infty^2$, $j = f_\infty^2 q$ (stable for $q^2 < 1/3$, i.e., $|j| < j_m = 0.385$).

A particular state is *globally unstable* if at least one stationary or periodic (limit cycle) solution exists which only deviates locally from that state. Stationary solutions representing *critical S-nuclei* in the N-state are found for $j < j_c$, where j_c decreases with τ and approaches j_m for $\tau \sim 2$ (the exact behavior in that range is difficult to determine). A typical example is shown in Fig. 1. As $j \to j_c$ these threshold solutions grow in width and degenerate into two separated, stable, stationary SN-interfaces. According to [4] such *kinks* also exist for $|j| < j_m$; they move towards the N-side if $|j| < j_c$. Once exceeded, a critical S-nucleus actually splits into a pair of such kinks moving in opposite directions.

Besides the unstable stationary *saddle-point* solution [7]

$$\psi = \{\sqrt{2}q + i(1 - 3q^2)^{1/2} \tanh [(1 - 3q^2)^{1/2} (x - x_0)/\sqrt{2}]\}e^{iqx}, \tag{7}$$

which merges with the S-state at $q^2 = 1/3$ and exists for $|j| < j_m$, we also find a *stable oscillatory* solution localized about the S-state in a narrow range $j_c < |j| < j_{min}$, where j_{min} (0.326 for $\tau = 5.79$) also decreases with τ. This is the *phase-slip state* mentioned earlier. Below j_{min}, solution (7) returns to the S-state after one or several phase slips.

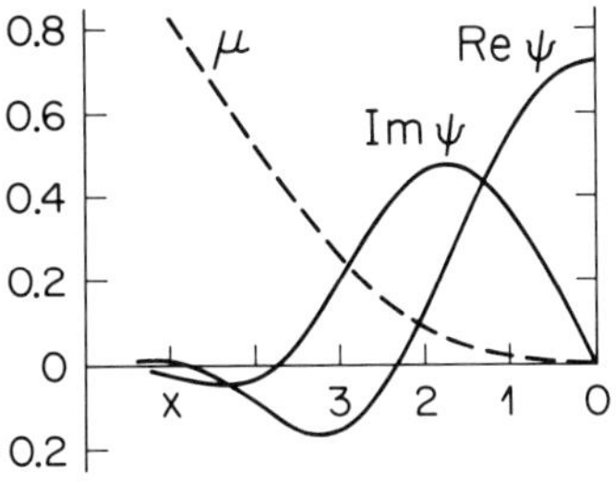

Fig.1 Spatial dependence of the real and imaginary parts of the complex order parameter ψ and of the electrical potential μ for a stationary critical superconducting nucleus at $j=0.30 < j_c$ (0.335 for $\tau=5.79$). Only the left half is shown; $\mu=0$ at the center $x=0$; distances are measured in coherence lengths

The existence of (7) in that range is therefore not connected with a significant global instability.

In the presence of *well-separated boundaries*, (7) or its analog with coth can be matched to $f_0 < f_\infty$ or $> f_\infty$, respectively. Such solutions correspond to *stable* S-states unless x_0 is inside the filament. The N-state is then represented by [5]

$$\psi = g(x) \exp[-i\phi(t)] + g^*(d-x), \qquad (8)$$

both g's being *stationary* solutions *localized near the boundaries* which are stable as long as they do not exhibit maxima, i.e., for $j > j_{sc}$, where $j_{sc} = j_c$ if $f_0 > f_\infty(j_c)$ and $j_{sc} < j_c$ otherwise, (it then corresponds to that threshold solution whose maximum $f^2 = f_0$). Finally, *stable phase-slip* solutions localized *near each boundary* are found if $j_{LA}(q_0) < |j| < j_c$; $j_{LA}(q_0)$ being the current at which (7) has a minimum right at the boundary, i.e., $\sqrt{2}q_0 = f_0$. This occurs for $f_0^2 < 0.34$ if $\tau = 5.79$; no properly matched S-state can then be found if $|j| > j_{LA}(q_0)$!

4. Transitions between Steady States

The situation sketched has all the trappings of first-order phase transitions: globally different states with overlapping ranges of metastability and critical nuclei strongly affected by boundary conditions, as well as interphase boundaries. Moving kinks effect transitions, as illustrated in Fig. 2. A long filament connected to electrodes with $T_{c0} = T_c$ is initially in the N-state ($j = 0.34 > j_c$)(a). In the absence of noise, transitions occur only when the metastability bounds of a particular state are exceeded; thus j is switched to $j = 0.3 < j_c$. Characteristic stages of evolution are shown. First an NS interface detaches itself from each boundary and moves inwards (b). The velocity v of each kink, inferred from the slope of the resulting ramp in $V(t)$ (e) compares well with a direct calculation of $dv/dj|_c$ [4]. Phase slips of growing amplitude develop as the kinks overlap (c); they stop once ψ swings over the saddle-point solution (d). If j is switched into the range between j_{min} and j_c before the subsequent irreversible completion of the S-state takes place, phase slips eventually stabilize as shown on the inset (f). As pointed out by LANDAUER [8] on a simpler analogue of a current-driven transition, noise is only likely to affect the nucleation of kinks during the initial stage. Since the kink velocity vanishes at j_c, transitions and transient readjustments in response to changes in j take a long time close to that value. *Critical slowing-down* is also noticeable in the initial stage of the transition from the state near j_m.

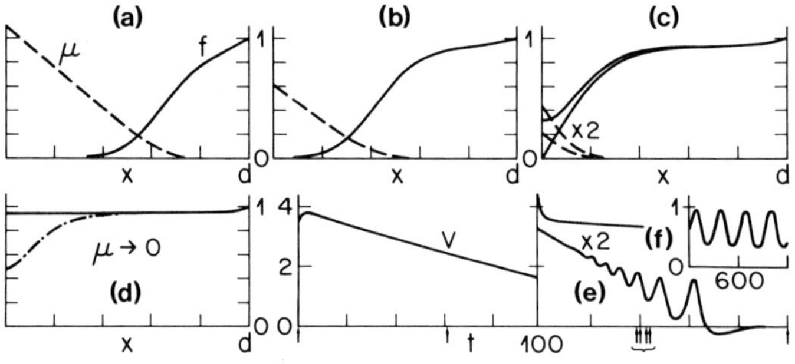

Fig.2 Profiles of f (——) and μ (---) are shown at different stages (a-d) of a transition from the normal state in a filament of length d=20 coherence lengths (only the right half is shown) switched to j=0.3 at t=0; in (c) curves are shown for a typical sequence: phase slip, maximum V, maximum f at center, minimum V; in (d) both the final S-state and the corresponding saddle point (dash-dotted curve) are exhibited. Time dependence of total voltage (e) - times corresponding to (a-d) are indicated by arrows. Stable phase slips localized in the center achieved via timely switching to j=0.33 (f) - the resulting oscillations develop gradually and are not displayed initially

References

1. W.J. Skocpol, M.R. Beasley, M. Tinkham: J.Low Temp. Phys. 16, 145 (1974)
2. L. Kramer, A. Baratoff: Phys. Rev. Lett. 38, 518 (1977)
3. A. Baratoff, L. Kramer: In *Superconducting Quantum Devices and their Applications*, ed. by H.D. Hahlbohm and H. Lübbig (Walter de Gruyter, Berlin 1977) p.62, and to be published
4. K.K. Likharev: ZhETF Pis. Red. (Soviet) 20, 730 (1974); Sov. Phys. JETP Lett. 20, 338 (1974)
5. A. Baratoff, L. Kramer: To be presented at the 15th International Conference on Low-Temperature Physics, Grenoble, August 23-29, 1978
6. K.K. Likharev, L.A. Yakobson: Zh. Eksp. Teor. Fis. (Soviet) 68, 1150 (1975); Sov. Phys. JETP 41, 570 (1975)
7. J.S. Langer, V. Ambegoaker: Phys. Rev. 164, 489 (1967)
8. R. Landauer: Phys. Rev. A15, 2117 (1977); other references cited therein

Solitary Phenomena in Finite Dissipative Discrete Systems*

E. Ben Jacob

Tel Aviv University, Department of Physics and Astronomy
Ramat Aviv, Israel and

Y. Imry

IBM Zurich Research Laboratory, CH-8803 Rüschlikon, Switzerland and
Tel Aviv University, Department of Physics and Astronomy
Ramat Aviv, Israel

1. Introduction

Idealized nonlinear systems which are assumed to be infinite, continuous and nondissipative are well known to exhibit a variety of solitary-wave phenomena [1]. For example, the space-time dependence of the relative phase $\phi(x,t)$ in an extended Josephson junction [2] is governed by a sine-Gordon-type equation. In the idealized case, the sine-Gordon equation is known to possess solitary solutions where a localized change of $\phi(x,t)$ by 2π propagates in an unchanged form with an arbitrary velocity $v \leq c$ where c is the velocity of light in the junction. However, the existence and properties of such solitary solutions for a realistic system is an important unsolved problem [3,4]. Finite dissipation always exists in real systems, presumably balanced by driving forces. Also, finite boundaries may be important and the continuum approximation not always valid for discrete Josephson-junction arrays or for the pendulum systems used for analog simulations. The problem is not only of fundamental interest: a knowledge of the dynamic I-V characteristics and their dependencies on external magnetic fields and external electromagnetic waves is germane to the understanding of Josephson-junction devices [5] and SQUIDS [6]. An interesting case occurs when the driving (e.g., current) source is localized in space. Energy is injected into the system at some point and it has to be fed into the solitary mode at long distances.

We have used qualitative mathematical [7] and numerical [4,8,9] methods to study these problems for models with $N \leq 4$ point junctions. While solitary waves are found to exist in such systems, they do differ in many important respects [10] from the idealized situations, as will be discussed. We also find solitary-like phenomena when the system is not coupled strongly enough to justify a continuum approximation.

2. The Model and Results

We consider a chain of N Josephson point junctions [4,7,11] where nearest-neighbor junctions are coupled inductively and where each of the junctions

*Partially supported by the Commission for Basic Research of the Israeli Academy of Sciences

is described by a simple lumped circuit model where the quasi-particle conductance is approximated as ohmic [12,13]. In addition to the interest in the specific case where $N = 2$, which is the simplest one where the spatial dependence of the phase can exist, the model may be viewed as a numerical simulation of a continuous junction of a length L. This will be discussed later but it should be a reasonable representation of the continuum junction if N is taken as $\gtrsim L/\lambda_J$, where λ_J is the Josephson penetration depth. The equations of the model are (see inset to Fig. 1)

$$\begin{aligned}
&\ddot{\theta}_1 + G\,\dot{\theta}_1 + \sin\theta_1 = I_1 + K\,[\theta_2 - \theta_1 - \theta_{ex}] \\
&\ddot{\theta}_j + G\,\dot{\theta}_j + \sin\theta_j = I_j + K\,[\theta_{j-1} + \theta_{j+1} - 2\theta_j]; \qquad 2 \le j \le N-1 \qquad (1) \\
&\ddot{\theta}_N + G\,\dot{\theta}_N + \sin\theta_N = I_N + K\,[\theta_{N-1} - \theta_N + \theta_{ex}],
\end{aligned}$$

for an open chain. Here θ_j is the phase difference of the jth junction $K \equiv \phi_0/2\pi\tilde{L}\,I_J$, ϕ_0 is the flux quantum, $\tilde{L}$ the inductance and θ_{ex} the flux (in units of $\phi_0/2\pi$) through the circuit between the two consecutive junctions, I_J the Josephson current amplitude, I_j the driving current through the jth junction in units of I_J, and $G = (\omega_p RC)^{-1}$ where $\omega_p = \bar{c}/\lambda_J$ is the Josephson plasma frequency. R is the junction quasiparticle (or shunt) resistance and C its capacitance.

In the case of two junctions [4], three general types of solution exist, resulting in three branches of the dc I-V characteristics: a static solution where $V = \langle\dot{\theta}_1\rangle = \langle\dot{\theta}_2\rangle = 0$, a running solution where θ_1 and θ_2 increase steadily, which has the highest voltage, and a *beating* solution of intermediate voltage where the two junctions interchange static and running roles. The *beating* solution can be viewed as being of a solitary type, a phase change of an integer multiple of 2π (which can also be regarded as n flux units) running across the system. The *beating* solution exists only when K is not too large. For $K \gtrsim 1$ the two junctions move together and the solitary mode does not exist for $N = 2$. More generally, for sufficiently large N, K determines a *correlation length* among the junctions which is equal to $\sqrt{K}$ (in units of the distance between the junctions) and the solitary mode is only visible for $N > \sqrt{K}$. In fact, the continuum approximation for the discrete system should become exact for $K \gg 1$, which will therefore necessitate $N \gg 1$. However, for $K < 1$ the $N \sim 2$-4 case does exhibit solitary modes, which, as long as K is not too small, appear as qualitatively-valid approximations to the continuum case. The discrete case with $K \ll 1$, is very interesting; here, in the *beating mode* one pendulum can make a number, n, of revolutions of the order of $1/K$, while the others are approximately stationary. This is propagated as a function of time along the chain and can be regarded as an n soliton state.

In Fig. 1(a), a bound double soliton-like solution is presented. It is seen that the phase change of 4π is propagated along the chain. The form of this solution is not exactly invariant with the motion and its propagation velocity is not fixed, but a strict periodicity in the motion of the whole system exists. After one soliton crosses the chain, a new one is generated at the first junction and launched along the chain. Other types of solitary solutions are possible. A double (phase change of $2 \times 2\pi$) soliton with a single peak in θ_i, a bound pair of solitons, etc. In the latter case two solitons move at each instant along the chain: see Fig. 1(b).

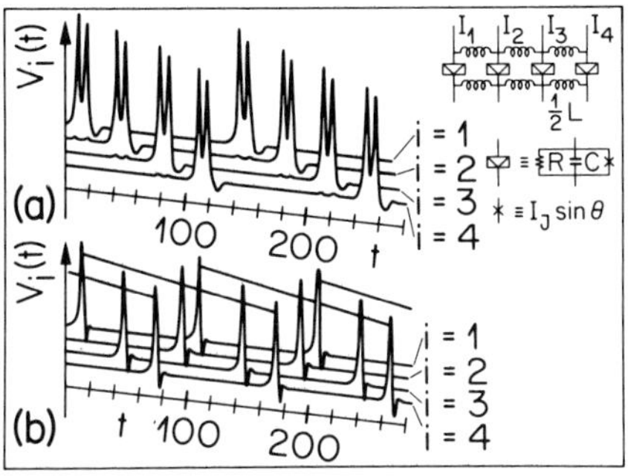

Fig.1 Junction voltages, N=4. (a) A bound two-soliton pair for G=0.8, K=1.8/24π, $I_1=3.2I_J$, $I_2=I_3=I_4=0$; (b) Two correlated solitons, for the same parameters except K=1.8/6π. The straight lines are drawn to guide the eye through each soliton

We emphasize that energy is only fed through the driven junction No. 1 into the system. The solitary modes are launched from junction No. 1 and *leave* the system at the open end - junction No. 4 - with no reflection. While we cannot rule out the possibility that due to energy dissipation, (finite G) the motion will eventually decay for $N \to \infty$, this is apparently not suggested by the fact that within an accuracy of about 1%, no decrease in amplitude is observed at N = 4. It appears that, in fact, the periodic train of solitons is generated at junction No. 1 and that somehow the energy is propagated along the train and thus fed to the leading pulse.

We have also treated a closed ring with N = 4, driven at a point. Here, solitons are launched in two opposite directions with the same sense of rotation. These two modes seem to pass through each other later without any significant changes in shape. Of course, it may also be assumed that they *collide head-on* and reflect backwards, thus not providing proof of the non-interacting nature of these modes.

Each type of solution corresponds, in the Josephson case, to a branch of the I-V characteristics (Fig. 2) where hysteretic jumps among the branches

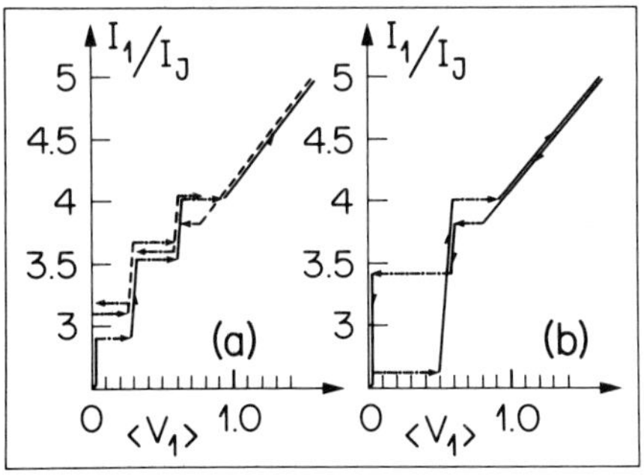

Fig.2 dc I-V characteristics, traced with increasing or decreasing current (see arrows). $I=I_1$, $I_2=I_3=I_4=0$. The smooth lines are for $\phi_{ex}=0$, the dashed one for $\phi_{ex}=\phi_0/4$. (a) Same parameters as in 1(b); (b) G=0.8, K=1.8/12π; the voltage is in units of $\hbar\omega_p/2e$

may occur. From our investigation, an understanding of the conditions for obtaining the various solutions and their limits of stability (including cases where only multiple solitons are possible) are obtained. Also, the dependence on external fields is understood and can be estimated analytically in agreement with the numerical data.

3. Conclusions

Solitary-type solutions are indicated as existing in dissipative driven finite discrete systems. However, they differ in important respects from the ideal case and their behavior does not seem to be just a small modification of the latter one. For example, the propagation velocity is fixed for a given set of parameters of the system [10], the modes are periodically created at the driven end and seem to propagate without any overall dissipation but with changes of shape and velocity, while the discrete system has, for small K, modes that do not exist for the continuous one.

Experimentally, the most direct observation of these effects in the Josephson case would be through the voltage peaks that occur upon the passing of the solitary mode at a given junction. A less direct method to observe these effects would be via the low V branches of the I-V characteristics including their limits of existence and dependence on external fields.

More work is needed on the questions of the interaction among the solitary waves, the finite-temperature noise effects and the evaluation of space-time correlation functions of the phase [14]. The behavior as a function of the dissipation parameter G will be discussed elsewhere. The $G \to 0$ limit does not appear to be a simple regular one.

One of the authors (YI) would like to thank Drs. A. Baratoff, P.M. Marcus, T. Schneider, and E. Stoll for useful discussions and for the hospitality at the IBM Zurich Research Laboratory.

References

1. A.C. Scott, F.Y.F. Chu, D.W. McLaughlin: Proc. IEEE 61, 1443 (1973)
2. B.D. Josephson: Adv. Phys. 14, 419 (1965)
3. T.A. Fulton, R.C. Dynes: Solid State Commun. 12, 57 (1973)
4. Y. Imry, P.M. Marcus: IEEE Trans. MAG-13, 868 (1977)
5. H.H. Zappe: IEEE Trans. MAG-13, 41 (1977)
6. D.C. Tesche, J. Clarke: J. Low Temp. Phys. 29, 301 (1977)
7. V. Imry, L.S. Schulman: J. Appl. Phys. 49, 749 (1978)
8. E. Ben Jacob, Y. Imry: Proc. LT15: To be published
9. E. Ben Jacob, Y. Imry: To be published
10. K. Nakajima, V. Onodera, T. Nakamura, R. Sato: J. Appl. Phys. 45, 4095 (1974)
11. J.A. Blackburn, J.P. Leslie, H.J. Smith: J. Appl. Phys. 42, 1097 (1971)
12. W.C. Stewart: Appl. Phys. Lett. 12, 277 (1968)
13. D.E. McCumber: J. Appl. Phys. 39, 3113 (1968)
14. In quasi-1D magnetic systems, the correlation functions $\cos\theta_i(0)\cos\theta_i(t)$ lead to the observability of soliton-related effects by inelastic neutron scattering. H.J. Mikeska: J. Phys. C11, L29 (1978); J.K. Kjems: Private communication

Stability of Nonuniform States in Systems Exhibiting Continuous Bifurcation

M. Büttiker and H. Thomas

Institut für Physik der Universität Basel, Klingelbergstraße 82
CH-4056 Basel, Switzerland

Many systems which are driven away from equilibrium by pumping or forcing show an instability of a stationary uniform state to a new stationary or travelling-wave state with broken translational symmetry. In continuous systems the loss of stability of the uniform state, which is usually associated with a destabilization of a normal mode, gives rise to a bifurcation of a whole manifold of new solutions. The task is then to select the members of this manifold according to their stability properties, and thus to find the candidates which may be physically realised. The travelling-wave case can be reduced to the stationary case by a transformation to a moving frame. In this paper, we therefore focus attention to stationary states.

Introduction

Consider a system described by a set of fields $\phi(r,t) = \{\phi_1(r,t),\phi_2(r,t), \ldots\}$ and a set of parameters $\alpha = \{\alpha_1,\alpha_2,\ldots\}$ which can be externally controlled. Let the time evolution of the fields $\phi(r,t)$ given by

$$\partial\phi/\partial t = B(\phi,\alpha) \tag{1.1}$$

where B is a nonlinear partial differential operator acting on ϕ and is assumed to be translationally invariant in r and t. A linear stability analysis of a given stationary solution $\phi_s(r)$ leads to the eigenvalue problem

$$L(\phi_s,\alpha)\cdot\delta\phi = \lambda\delta\phi; \quad L = -\nabla_\phi B|_{\phi=\phi_s} \tag{1.2}$$

for the perturbation $\delta\phi(r)\exp(\div\lambda t)$. The state ϕ_s is called (linearly) stable if Re $\lambda(\alpha)\geq 0$ for all eigenvalues, and if the eigenvalues $\lambda = 0$ belong to hydrodynamic modes or to symmetry-restoring modes. If the state is uniform, the eigenfunctions are plane waves, $\delta\phi_q \propto \exp(iqr)$ and the eigenvalues λ_q can be characterized by the wavenumber q.

We are interested in the case that the uniform state becomes unstable for $\alpha=\alpha_c$ against a soft mode with critical wavenumber q_c. In the neighbourhood of the instability we assume a dispersion law

$$\text{Re}\,\lambda(q,\alpha)=-r(\alpha-\alpha_c)+D(q-q_c)^2; \quad \text{Im}\lambda(q,\alpha)=0. \tag{1.3}$$

With increasing α, successively more and more modes become undamped /1,2/, each q mode generating at its point of bifurcation $\alpha_b > \alpha_c$, after splitting off a factor $\exp(iq_c r)$, a small-amplitude periodic solution with period $\Lambda_b = 2\pi|D/r(\alpha_b - \alpha_c|^{1/2}$ and amplitude $A \propto (\alpha - \alpha_b)^{\frac{1}{2}}$. With increasing $\alpha - \alpha_b$, the amplitude will pick up higher harmonics but the states will remain periodic.

Method

The object of the present paper is a general stability analysis of this family of solutions $\phi_s(r, \alpha, \alpha_b)$. Our method is based on the following considerations:

1. Because the nonuniform states $\phi_s(r)$ are states of broken translational symmetry, a set of equivalent states $\phi_s(r+a)$ is generated by an infinitesimal translation. Therefore there always exists a Goldstone mode (GM) $\delta\phi = \partial\phi_s(r)/\partial r$ with egenvalue zero.
2. The linear operator $L(\phi_s)$ has at least the symmetry of the stationary state ϕ_s. If ϕ_s is periodic with period Λ, then $L(\phi_s)$ is invariant under the discrete group T_Λ of translations $n\Lambda$. Therefore the eigenvalues are multivalued functions $\lambda_n(q)$ of the reduced wave vector q over a Brillouin zone (BZ) $-\pi/\Lambda < q \leq \pi/\Lambda$
3. For the zero-amplitude solution $A \to 0$, $\alpha_b = \alpha$, the spectrum is given by (1.3). For the nonuniform states $\alpha_b < \alpha$ (α fixed, α_b decreasing), the width of the BZ changes α $1/\Lambda(\alpha_b)$, and with increasing amplitudes $A(\alpha_b)$ gaps develop at $q = 0, \pi/\Lambda$ separating the bands $\lambda_n(q)$.
4. For a general q in the BZ, no crossing of eigenvalues can occur as A increases.

We show in the following examples that depending on the wavenumber q of the GM, these properties together with k·p perturbation theory around the GM lead to general conclusions about the stability of the family of the periodic states.

Examples

In recent years several methods have been developed to describe systems near the point α_c of instability by time-dependent Ginzburg-Landau equations /3,4/, with only a few order parameter fields varying slowly in space and time being kept to describe the instability.

A. Symmetry-Conserving Instability

The first example is related to current instabilities /5/. The excess electric field ϕ (ϕ real) satisfies the time evolution equation

$$\partial\phi/\partial t = -(1+\phi_{,x})(\alpha\phi - \phi^2) + \phi_{xx} \tag{2.1}$$

which has two stationary uniform states $\phi_{u,1} = 0$, $\phi_{u,2} = \alpha$ with spectrum $\lambda = \alpha + q^2$ and $\lambda = -\alpha + q^2$, respectively. The stationary nonuniform states for $\alpha > 0$ are represented in the $(\phi, \phi_{,x})$-plane in Fig. 1a. The operator L is invariant under T_Λ, and the BZ has therefore width $2\pi/\Lambda$. The GM belongs to q=o and lies in the center of the BZ (Fig. 1b). For $\alpha_b \neq \alpha$, all eigenvalues $\lambda_u(q)$ are nondegenerate and belong to the unit representation of the small group even at q=0. Therefore, the unstable band cannot cross the GM fixed at $\lambda = 0$, q=0, i.e. all periodic states $\Lambda > \Lambda_b$ including the solitary state $\Lambda = \infty$ are unstable.

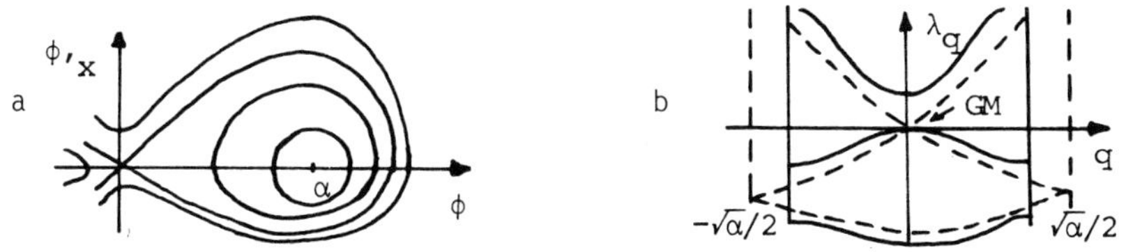

Fig.1a Stationary nonuniform solutions of Eq. (2.1). A solitary solution separates periodic solutions (closed curves) and unbounded solutions (open curves)

Fig.1b Spectrum of the unstable state $\phi=\alpha$ (broken line) and of a periodic solution with period Λ of Eq. (2.1)

B. Symmetry-Breaking Instability

As second example we consider the ϕ^4 potential with a real field ϕ with

$$\partial\phi/\partial t=-\partial V(\phi)/\partial\phi+\phi,_{xx};\quad V(\phi)=-\frac{\alpha}{2}\phi^2+\frac{1}{4}\phi^4. \tag{2.2}$$

The softening of the normal modes $\lambda=-\alpha+q^2$ of the uniform state $\phi=0$ leads to the bifurcation of two uniform states $\phi_u=\pm\alpha^{1/2}$ and the familiy of nonuniform states shown in Fig. 2a.

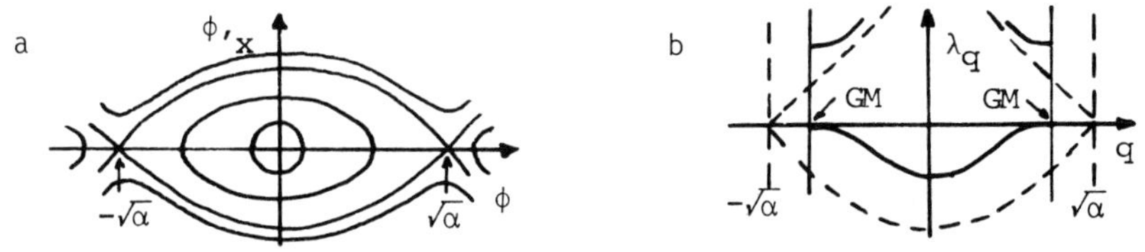

Fig.2a Stationary nonuniform states of Eq. (2.2). Two solitary solutions separate the periodic states (closed curves) from the unbounded states (open curves)

Fig.2b Spectrum of the unstable state $\phi=0$, $\alpha>0$,(broken line) and of a periodic solution with period Λ of Eq. (2.2)

Because a translation by $\Lambda/2$ changes the stationary solution ϕ_s to $-\phi_s$, and L depends only on ϕ_s^2, L is invariant under $T_{\Lambda/2}$. The width of the BZ is thus $4\pi/\Lambda$ and the GM lies at the BZ-boundary (Fig. 2b). Therefore the lowest band $\lambda_o(q,\Lambda)$ has the possibility to becáme positive above a critical Λ. To show that this is not the case, we apply a k·p perturbation around the GM ($\lambda=0$). One finds $\lambda(k)=\{-12\alpha\Lambda\exp(-\Lambda\sqrt{\alpha/2})\}\,k^2+O(|k|^3)<0$, $k=2\pi/\Lambda-q$. Thus for all finite Λ there exist modes with $\lambda(q)<0$, and therefore all periodic solutions are unstable. Only the stationary uniform states $\phi_u=\pm\sqrt{\alpha}$ and the solitary solutions are stable.

As third example we consider the time-dependent Ginzburg-Landau equation for a complex field with ϕ^4 potential

$$\partial\phi/\partial t=-\partial V/\partial\phi^*+\phi,_{xx};\quad V(\phi)=-\frac{\alpha}{2}|\phi|^2+|\phi|^4 \tag{2.3}$$

which describes the onset of convection in the Rayleigh-Bénard instability /3,6,7/,the Laser threshold /7/, as well as chemical instabilities /8/. It is also related to the superconducting phase in thin wires /9,10/. For $\alpha>0$

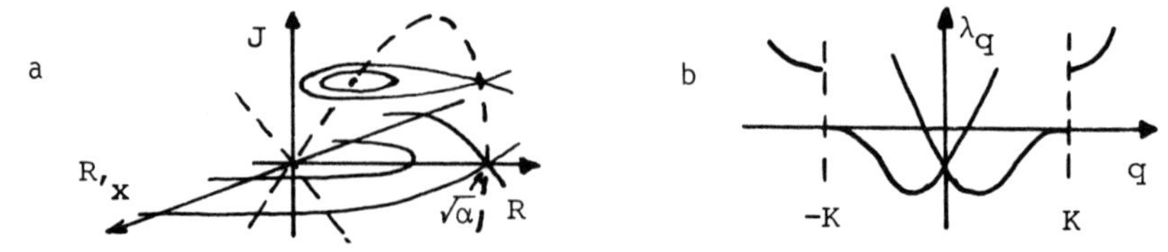

Fig.3a Stationary states of Eq. (2.4). The broken lines represent the "rolls". In each halfplane R>0, J≠0 the unstable roll is surrounded by small amplitude periodic solutions. From the stable roll a solitary solution emerges and returns

Fig.3b Spectrum of an unstable roll in the extended BZ

one has in addition to $\phi=0$, the uniform stationary states $\phi_u=\sqrt{\alpha}\exp(i\Theta_o)$, ($\Theta_o$ arbitrary). The nonuniform states $\phi_s=R\exp(i\Theta)$, R, Θ real, are given by the equations

$$R_{,xx}-\left(\frac{J^2}{R^3}\right)+\alpha R-R^3=0, \quad J=\Theta_{,x}R^2=\text{const} \tag{2.4}$$

and are represented in Fig. 3a. The stability of these solutions is determined by a selfadjoint operator

$$H = \left(-\frac{d^2}{dx^2} - \alpha+2R^2+\frac{J^2}{R^4}\right)\mathbb{1} + i\sigma_y\left(\frac{d}{dx}\left(\frac{J}{R^2}\right)+\frac{J}{R^2}\frac{d}{dx}\right)+\sigma_z R^2 \tag{2.5}$$

acting on a spinor ψ. The uniform stationary states have spectra $\lambda_1=\lambda_2=-\alpha+q^2$ for $\phi_u=0$ and $\lambda_1=2\alpha+q^2$, $\lambda_2=q^2$ for $\phi_u=\sqrt{\alpha}\exp(i\Theta_o)$ ($\alpha>0$). The simplest nonuniform states are the constant-amplitude states ("rolls") $\phi_s=(\alpha-K^2)^{1/2}\exp(iKx)$ (broken line in Fig. 3a) belonging to $J=K(\alpha-K^2)$. For these states, H is still fully translationally invariant, and the spectra are $\lambda=(\alpha-K^2+q^2)\pm(4q^2K^2+(\alpha-K^2)^2)^{1/2}$ /11/ showing that the rolls for $K>\sqrt{\alpha/3}$ are unstable (Fig.3b). Around each unstable "roll" there exist doubly periodic small-amplitude solutions with $\Theta(x)=Qx+\vartheta(x)$, R(x) and $\vartheta(x)$ simply periodic with period Λ. Phase shifts with respect to each of the two periods generate a two-parameter family of equivalent states, giving rise to two $\lambda=0$ modes, $\Psi_1=(R_{,x}, R\vartheta_{,x})$ and $\Psi_2=(0,R)$. $\Psi_1+Q\Psi_2$ corresponds to $\partial\phi_s/\partial x$.

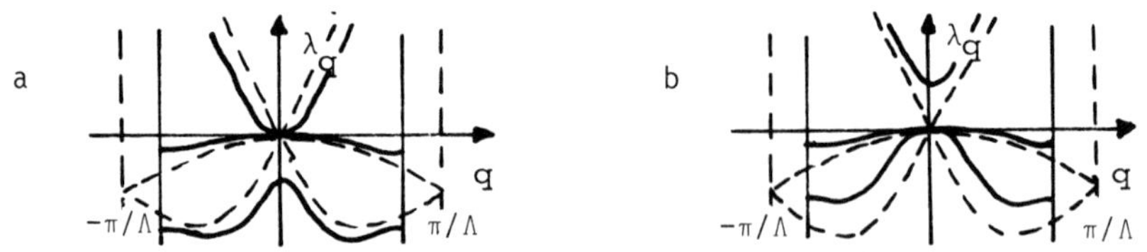

Fig.4a Spectrum of an unstable roll (broken line) and of a (doubly) periodic solution J≠0 (solide line) with gap below $\lambda=0$

Fig.4b Same situation as in Fig. 4a but with gap above $\lambda=0$

For J=0 (2.5) decouples into two independent eigenvalue problems for u and v. The eigenvalue problem for u is identical with that for real ϕ (2.2). The eigenvalue problem for v has a band $\lambda_o(q)<0$. Thus, the periodic solutions are unstable under an additional branch of modes leading in the limit $\Lambda\to\infty$ to an isolated eigenvalue $\lambda_o<0$ of the solitary solution /3,10/.

Whereas for J=0 (2.5) is invariant under $T_{\Lambda/2}$, for J$\neq$0 the translation group is T_Λ. From Fig. 3b it is seen that the unstable rolls have three eigenmodes with eigenvalue λ=0. Two of these are the λ=0 modes discussed above, the third one will have $\lambda_3\neq 0$ as one moves away from the constant-amplitude states, and the stability would depend on the sign of λ_3. However, even if $\lambda_3<0$ initially, the situation of Fig. 4a might change at a critical Λ into that of Fig. 4b, because λ_3 belongs to a different representation and is therefore not prevented from crossing λ=0. A k·p perturbation shows that the λ=0 modes vary as $\lambda=cq^2$, and the stability depends on the sign of c, but this calculation has not been carried through. In Ref. /10/ the solitary solution was shown to be unstable. This suggests that the gap opens downwards(Fig. 4a), rendering all periodic solutions unstable.

References

1. K. Kirchgässner in Synergetics edited by H. Haken, Springer Verlag Heidelberg, New York 1977
2. H. Thomas in the proceedings of the V Internationale Conference on Noise in Physical Systems, Bad Nauheim, March 1978
3. A.C. Newell, J.A. Whitehead, J. Fluid, Mech. 38, 279, (1969)
4. H. Haken, Z. Physik B21, 105 (1975)
5. M. Büttiker, H. Thomas, Phys. Rev. Lett. 38, 78 (1977)
   M. Büttiker, H. Thomas, Solid-State Electronics 21, 95 (1978)
6. L.A. Segel, J. Fluid Mech. 38, 203 (1969)
7. R. Graham, in Fluctuations, Instabilities and Phase Transitions, ed. by T. Riste, Plenum Press, New York and London (1975)
8. Y. Kuramoto, T. Tsuzuki, Prog. Theor. Physics, 54, 687 (1975)
9. J.S. Langer, V. Ambegaokar, Phys. Rev. 164, 498 (1967)
10. D.E. McCumber, B.I. Halperin, Phys. Rev. B1, 1054 (1970)
11. W. Eckhaus, Studies in Non-Linear Stability Theory, Springer Verlag, Berlin, Heidelberg, New York 1965

The Sine-Gordon Chain: Mass Diffusion

T. Schneider and E. Stoll

IBM Zurich Research Laboratory, CH-8803 Rüschlikon, Switzerland

In recent years, there has been growing interest in the theory of Brownian motion of interacting and non-interacting particles in a periodic potential driven by an external field [1-4]. Interesting aspects are: a close analogy between the non-equilibrium phenomena occurring close to the critical field, and continuous phase transitions in thermal equilibrium [3]; the prediction, that the average velocity at small fields and low temperatures might be associated with thermalized sine-Gordon solitons [1].

In this work, we present and discuss some molecular-dynamics results aimed at studying the response on a small field of a sine-Gordon chain in the presence of damping, and subjected to a random force. On the basis of these results we hope to elucidate the relevance of thermalized sine-Gordon solitons in the diffusion process.

The equations of motion of the sine-Gordon chain, using the mechanical analogy, are the Langevin equations

$$M \ddot{X}_i = - \frac{\partial V}{\partial X_i} - \Gamma M \dot{X}_i + \eta_i(t), \quad i = 1,\ldots, N, \tag{1}$$

where

$$V = \sum_i [B(1 - \cos X_i a) + \frac{A}{2} (X_{i+1} - X_i)^2 - E X_i]. \tag{2}$$

The system is a linear array of N-interacting particles of mass M moving in a periodic potential and subjected to an external field. We assume periodic boundary conditions so that

$$X_{N+1} = X_1 . \tag{3}$$

The system is in contact with much lighter particles, treated as a heat bath of temperature T. The collisions with these particles are described by a friction $- \Gamma M \dot{X}_i$ and a random force $\eta_i(t)$ which acts on the particles independently, i.e.,

$$\left.\begin{array}{l} \langle\eta_i(t)\rangle = 0 \\ \langle\eta_i(t)\, \eta_j(t')\rangle = 2\, \Gamma\, M\, k_B T\, \delta_{ij}\, \delta(t - t') \end{array}\right\} . \tag{4}$$

B, A and a are model parameters. The problem is thus equivalent to the Brownian motion of N-particles.

Using a molecular-dynamics technique, as described in [5], we solved the coupled set of Langevin equations (1) numerically. In the present study, we considered systems of 10^3 particles and the model parameters were chosen as follows:

$$\left.\begin{array}{l} M = 1, \quad a = 1, \quad \Gamma = 4, \quad E = 0.1,\ 0.05 \\ B = 1, \quad A = 50,\ 100 \quad k_BT = 15,\ 20,\ 25. \end{array}\right\} \tag{5}$$

This choice matches the regime where the mass transport characterized by the average velocity $\langle\dot{X}\rangle = 1/N \sum_i \langle\dot{X}_i\rangle$ might be dominated by solitons. This regime is defined by

$$E \ll 1, \qquad A \gg B \qquad E_0 = 8\sqrt{A\,B} \gg k_BT, \tag{6}$$

where E_0 is the rest energy of the sine-Gordon (S-G)-kink solution in the undriven system and treated in the continuum approximation. In this regime and the large damping limit, TRULLINGER *et al.* [1] predicted that the mean velocity should be dominated by thermalized S-G-kink solitons. They found in terms of an approximate solution of the Smoluchowski equation,

$$\frac{\Gamma\, M\langle\dot{X}\rangle}{E} \simeq 2\pi\left(\frac{\pi}{2}\right)^{1/2}\left(\frac{E_0}{k_BT}\right)^{3/2} \exp -\frac{E_0}{k_BT}\ , \tag{7}$$

where the kink rest energy is given by (6).

To test these predictions, we calculated the average velocity and followed the time evolution of the system in the steady state. Figure 1 shows snapshots of the instantaneous particle positions at two times.

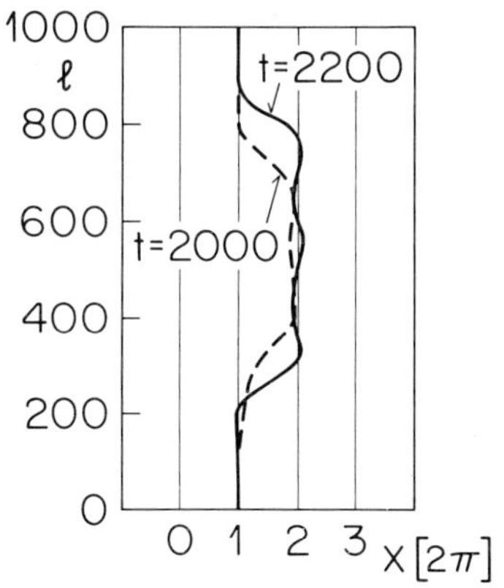

Fig.1 Snapshots of the instantaneous particle positions at two instances of time for B=1, A=100, E=0.1, k_BT=20 and Γ=4. The particles are labeled by ℓ and X denotes the magnitude of their displacement

These results clearly demonstrate that the mass transport in the regime where (7) is expected to hold is governed by moving kinks and antikinks. For a more detailed pictorial presentation of the kink dynamics, we refer to our motion picture. In particular, it illustrates the creation of kink and antikink pairs by thermal fluctuations. Subsequently, both partners move in opposite directions under the action of the applied field. Moreover, a kink may annihilate an antikink created at a later time, or vice versa. This picture is consistent with that of LANDAUER [6]. In Table 1, we compare our results for the mean velocity with (7). From the results for $A = 100$ and $k_BT = 20$ at different fields, it is seen that we are in, or close to, the linear response regime. Accordingly, comparison with (7) makes sense. For this purpose we divided our estimates for $\Gamma\langle\dot{X}\rangle/E$ by the temperature-dependent terms of (7). The resulting constants are listed in the last but one column of Table 1. In the last column, we used an alternative expression where the prefactor $(E_0/k_BT)^{3/2}$ in (7) is replaced by $(E_0/k_BT)^2$.

Table 1 Numerical estimates for the average velocity for different temperatures and fields. The other model parameters are given by (5)

A	k_BT	E	$\Gamma\langle\dot{X}\rangle$	$\frac{\Gamma\langle\dot{X}\rangle}{E}\left(\frac{E_0}{k_BT}\right)^{-3/2}\exp\left(\frac{E_0}{k_BT}\right)$	$\frac{\Gamma\langle\dot{X}\rangle}{E}\left(\frac{E_0}{k_BT}\right)^{-2}\exp\left(\frac{E_0}{k_BT}\right)$
100	15	0.1	0.023 ± 0.002	3.86 ± 0.37	1.67 ± 0.17
100	20	0.1	0.049 ± 0.007	3.34 ± 0.45	1.67 ± 0.17
100	25	0.1	0.071 ± 0.005	3.04 ± 0.21	1.70 ± 0.17
100	20	0.05	0.029 ± 0.003	3.96 ± 0.42	1.98 ± 0.20
50	15	0.1	0.051 ± 0.007	3.02 ± 0.40	1.56 ± 0.16

The quite good agreement between these constants supports the functional relationship (7), as obtained from an approximate solution of the Smoluchowski equation. The modified prefactor leads, however, to a better agreement.

To summarize, we have shown that the average velocity or the diffusion coefficient is, in the appropriate regime (6), dominated by kink and antikink solitons of the sine-Gordon equation.

References

1. S.E. Trullinger, M.D. Miller, R.A. Guyer, A.R. Bishop, F. Palmer, J.A. Krumhansl: Phys. Rev. Lett. 40, 206 (1978); Phys. Rev. Lett. 40, 1603 (1978)
2. R. Festa, E. Galleani d'Agliano: Physica 90A, 229 (1978)
3. T. Schneider, E. Stoll, R. Morf: To be published, Phys. Rev. B 18, (1978)
4. R.A. Guyer, M.D. Miller: Phys. Rev. A 17, 1205 (1978); Phys. Rev. A 17, 1774 (1978)
5. T. Schneider, E. Stoll: Phys. Rev. B 17, 1302 (1978)
6. R. Landauer: Private communication

Solitary Wave Propagation as a Model for Poling in PVF_2

A.J. Hopfinger, A.J. Lewanski, T.J. Sluckin, and P.L. Taylor[1]

Case Western Reserve University, Cleveland, OH 44106, USA

1. Introduction

A piezoelectric and pyroelectric material of great technological importance is the phase-I, or β-phase, form of poly(vinylidene fluoride), also known as PVF_2. In this phase the crystal consists of a parallel array of planar zigzag macromolecules, each constructed of alternating and oppositely oriented CF_2 and CH_2 units. The large electric dipole moment of 7×10^{-30} Cm of each CF_2 unit and their parallel orientation in the unit cell make this material piezoelectric. The orthorhombic unit cell, shown in Fig. 1, has a dipole moment of 1.4×10^{-29} Cm.

The β-phase material is most commonly produced by stretching films of the more readily crystallizable α phase. Because the chain backbone follows a more tortuous path through the unit cell of the α phase than the planar zig-zag of the β phase, a longitudinal stress favors formation of the β phase. In such films the chain direction is generally aligned with the direction of strain but no net macroscopic electric polarization is exhibited; presumably the polarization of individual crystallites lies in random directions perpendicular to the strain direction. Electrically active material may, however, be produced by annealing a sample in a symmetry-breaking electric field of sufficient strength, a process known as poling. The microscopic description of the poling process and of the mechanism of electrical activity remains a matter of some controversy at the time of writing[1].

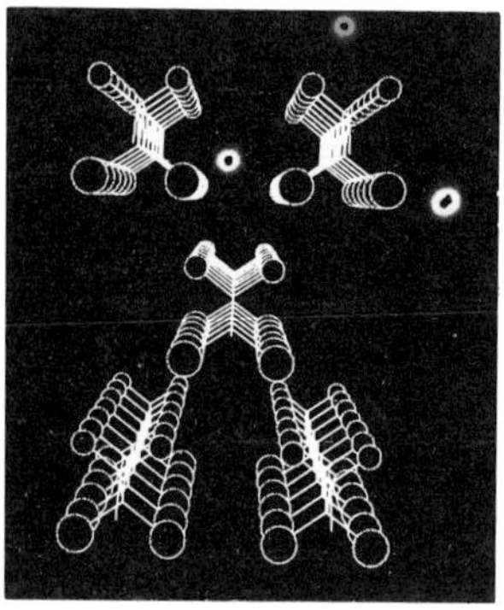

Fig.1

The present paper describes one aspect of a research program directed towards an elucidation of the microscopic behavior of β PVF_2. We here

1. Work supported by the U.S. Army Research Office and in part by the U.S. National Science Foundation.

explore a model in which the poling process involves a reversal of the direction of polarization of the chains within a crystallite through the propagation of a 180° twist along the chain length.

2. The Polymer in Equilibrium

It is, in principle, possible to treat crystalline polymers merely as lattices with a rather large number of atoms per unit cell. In practice, on the other hand, it is natural to take advantage of the relative strength and rigidity of the intrachain covalent forces in comparison with the much weaker van-der-Waals forces of the interchain interaction. This can be accomplished by first calculating the properties of an isolated chain and then adding the effects of interchain forces in a mean-field formalism. Such a program has been applied[2] to systems consisting of only CH_2 units (polymethylene) or only CF_2 units (polytetrafluoroethylene), and is currently being applied[3] to PVF_2. The central quantities in these calculations are the distribution $n(\theta)$ of torsional angles along a chain, and its temperature variation in equilibrium. The idealized Hamiltonian that we use is

$$H = \sum_i U(\theta_i,\theta_{i+1}) + W[n(\theta)] \tag{1}$$

where U is an intrachain interaction and W is an interchain interaction. These interactions were computed using conformational analysis[4]. The free energy f of this system per monomer is determined from the smallest eigenvalue of the transfer-integral equation

$$\exp(-\beta f)\psi(\theta) = (2\pi)^{-1}\int_0^{2\pi} \exp-\beta[U(\theta,\theta')+\tfrac{1}{2}h(\theta)+\tfrac{1}{2}h(\theta')]\psi(\theta')d\theta' \tag{2}$$

where the interchain forces act through the mean field $h = \delta W/\delta n(\theta)$ and where $\beta = (k_B T)^{-1}$. The distribution of torsional angles is given by

$$n(\theta) = \psi(\theta)\chi(\theta) \tag{3}$$

with $\chi(\theta)$ the eigenfunction of the equation adjoint to (2). This formalism allows the prediction of the limits of stability and metastability of various crystalline and melted phases.

3. The Poling Process

Under conditions in which β-PVF_2 is stable or metastable, the interchain potential energy was found to favor a parallel or antiparallel alignment of each CF_2 unit with the local crystalline b-axis. Upon taking some liberties with the geometry of the permissible excursions of torsional angles we are led to consider a Hamiltonian of the form

$$H = \sum_i [\tfrac{1}{2}I\dot{\theta}_i^2+\tfrac{1}{2}K(\theta_i-\theta_{i+1})^2-A_2\cos 2\theta_i-A_1\cos\theta_i] \tag{4}$$

Here θ_i represents the orientation of the ith monomer unit relative to the crystalline b-axis, I is its moment of inertia, K is a torsional parameter of the intrachain interaction, $0<A_1 = pE + \Delta << A_2$ with Δ and A_2 interchain energy constants and pE the product of monomeric electric dipole moment and electric field. The addition of temperature-dependent Brownian-motion forces $F_i(t)$ and a phenomenological damping mechanism

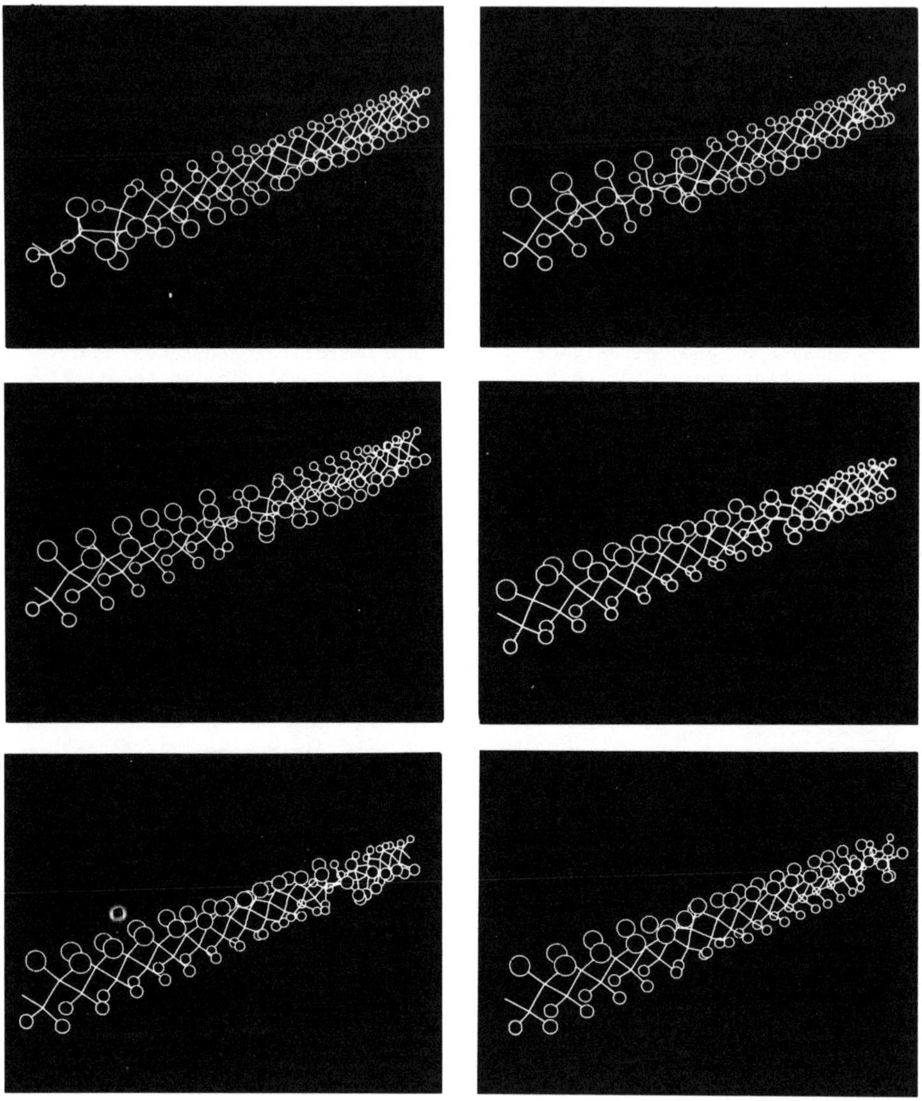

Fig.2 In this sequence a wave of polarization is seen to propagate as the chain rotates 180° about its axis. The electric field is applied downwards to a sample at intermediate temperature

characterized by a constant η then result in an equation of motion of the form

$$I\ddot{\theta}_i = -2A_2\sin 2\theta_i - A_1\sin\theta_i + K(\theta_{i+1} - 2\theta_i + \theta_{i-1}) - I\eta\dot{\theta}_i + F_i(t) \tag{5}$$

Similarities between this system and those studied by other authors [5] are immediately apparent.

In the poling mechanism envisaged for this model, the high temperature and applied field induce the rotation of a chain from a conformation in which all θ_i are in the vicinity of π to one in which all θ_i are close to zero. The birth and propagation of a solitary wave of polarization are then predicted by the solution of (5).

In the continuum limit, for low T, the velocity v, and thickness x_o, of the solitary wave can be determined analytically. We obtain

$$\frac{x_o}{a} = \left[\frac{K}{4A_2 + A_1^2/I\eta^2}\right]^{\frac{1}{2}}, \quad \frac{v}{a} = \left[\frac{K}{I(1+4A_2 I\eta^2/A_1^2)}\right]^{\frac{1}{2}} \tag{6}$$

where a is the polymer repeat unit. However, the low damping and discrete nature of the system limits the applicability of (6), and so a numerical study was made of this system. The results were displayed as a computer-generated ciné-film, sample frames of which are shown in Fig. 2.

More details of this work will be given in a future publication.

References

1. A review of the properties of PVF_2 is given by Y. Wada and R. Hayakawa in Japan. J. Appl. Phys, 15 2041 (1976).
2. F. P. Boyle, P. L. Taylor and A. J. Hopfinger, J. App. Phys. 46 4218 (1975), J. Chem. Phys., 67 353 (1977), J. Chem. Phys. to be published.
3. F. P. Boyle, A. J. Hopfinger, T. J. Sluckin and P. L. Taylor, Bull. Am. Phys. Soc., 23 379 (1978).
4. A. J. Hopfinger, Conformational Properties of Macromolecules (Academic, New York 1973).
5. S. E. Trullinger *et al*. Phys. Rev. Lett., 40 206 (1978) and references therein. In the undamped, zero-temperature, continuum limit one finds the double sine-Gordon equation discussed by S. Duckworth *et al*. (Phys. Lett. 58A, 19, 1976) and J. D. Gibbon *et al*. (Phys. Lett. 65A, 380, 1978).

Theory of One-Dimensional Ionic and Solitary Wave Conduction in Potassium Hollandite

J.B. Sokoloff and A. Widom

Northeastern University, Boston, MA 02115, USA

Potassium hollandite is an ionic conductor which exhibits one dimensional character in the sense that the mobile ions move in one dimensional channels provided by the remaining ions in the crystal [1]. There is a theorem which states that the conductivity of a one dimensional conductor should be zero if it has impurities, because all the states of the current carrier are localised [2]. The reason for this can be roughly understood on the basis of the following simple picture: if the distribution of impurities is represented by a row of n potential barriers, each with reflection coefficient R_1, the transmission coefficient for a particle incident on this n barrier system approaches zero in the thermodynamic limit when all coherence effects between barriers are neglected (which is good for random impurities). In fact, the mean free path (i.e. the distance through the system of barriers over which the wave function of a particle incident on the barriers drops to a small fraction of its incident value) is $(C\,R_1)^{-1}$ where C is the number of impurities per unit length and R_1 is the transmission coefficient of one barrier. For the potential barrier

$$Vo\ (1 + e^{\chi/b})^{-1}$$

where Vo is the barrier height and b its width, the reflection coefficient is given by

$$R_1 = \left[\frac{\sinh\pi(k_1-k_2)b}{\sinh\pi(k_1+k_2)b}\right]^2 \approx e^{-4\pi k_2 b}$$

for k_1b and $k_2b >> 1$, where

$$k_1 = (2mE)^{\frac{1}{2}}/h$$

$$k_2 = [2m(E-Vo)]^{\frac{1}{2}}/h$$

where E and m are the incident particle energy and mass [3]. For m equal to the potassium ion mass, the mean free path is already ≈0.1 mm when $E-Vo = 10^{-4}$eV. That is, the potassium ions are essentially classical and travel over the barriers without reflection. The apparent resistenceless current flow implied by this result does not occur for the high concentrations of potassium ions actually found in potassium hollandite because the ions can dissipate momentum by particle-particle scattering in which momentum is not conserved because of the lattice potential due to the other ions.

For high ionic concentrations (i.e. most channel sites occupied by potassium ions), conduction is best described by the solitary wave like motion of vacancies in the potassium ion lattice. For applied electric fields small compared with the value necessary to push a vacancy over the potential barrier which must be overcome to move the vacancy to the next lattice site, electrical conduction proceeds by thermally activated hopping, in which thermal equilibrium is re-established between hops. In this case, the method of Weiner and Sanders [4], originally proposed for describing one dimensional dislocation motion, can be applied because the vacancy is a type of dislocation in one dimension. The vacancy hopping rate is given by

$$f_{\pm} = \frac{1}{2\pi} \left(\frac{\alpha_2}{m}\right)^{\frac{1}{2}} \nu \exp \left[- \frac{\Delta E_a \pm \frac{1}{2} eaE}{kT}\right]$$

where α_2 is the force constant of the potential well in which a potassium ion sits, ΔE_a is the activation energy, a is a lattice constant of the channel which holds the potassium ions, E is the applied field and ν is a dimensionless quantity defined in reference 4. Then, the current density and conductivity are found from

$$J = \sigma E = nea\,(f_+ - f_-) = \left[\frac{ne^2b^2}{kT} \frac{1}{2\pi} \left(\frac{\alpha_2}{m}\right)^{\frac{1}{2}} \nu\, e^{-\Delta E_a/kT}\right] E,$$

where b is the mean potassium ion spacing and n is the number of vacancies per unit volume. The quantity ΔE_a is found by solving the equilibrium equations for the ground state of the one vacancy case and with the vacancy in the activated position and taking the difference of the energies for both configurations. The result is an activation energy much smaller than the observed 0.2 eV [5]. In reality, there are always breaks in the channels in which the potassium ions reside and the actually observed activation energy is the energy to get around such a break [6]. The calculated activation energy and conductivity should be found, however, in the higher frequency A.C. conductivity.

We would like to thank the U.S. Department of Energy (Contract Number BG-77-5-02-4432) and the National Science Foundation (Grant Number (DMR 75-06789) for financial support.

References

1. H.U. Beyeler, Phys. Rev. Lett. 37, 1557 (1976); H.U. Beyeler, L. Pietronero, S. Strässler and H.J. Wiesmann, Phys. Rev. Lett. 38, 1532 (1977).
2. N.F. Mott, Advances in Physics 16, 49 (1967).
3. L.D. Landau and I. Lifshitz, Quantum Mechanics - Nonrelativistic Theory, 2nd edition (Pergamon Press, Oxford, 1965), p.78.
4. J.H. Weiner and W.T. Sanders, Phys. Rev. 134, A1007 (1964).
5. J. Singer, H.E. Kautz, W.L. Fielder, J.S. Fordyce, "Fast Ion Transport in Solids", ed. W. van Gool (American Elsevier Publishing Company Inc. New York, 1973), p.653.
6. D. Kuse and H.R. Zeller, Phys. Rev. Lett. 27, 1060 (1971).

IV. Summary

Summary: Where Do Solitons Go from Here?

S.E. Trullinger

Department of Physics, University of Southern California
Los Angeles, CA 90007, USA

When Alan Bishop asked me to help organize this symposium, he gave me two very difficult tasks. The first was to obtain travel support money from the National Science Foundation to assist the U.S. participants. The completion of this task was made possible largely due to the efforts of John Connolly and Lew Nosanow in the Division of Materials Science at NSF. These gentlemen realized the importance of the symposium and worked very hard to insure that a decision on the travel proposal would be made in time. We are all indebted to them and the National Science Foundation without whose generous assistance this symposium would not have been possible.

The second task which Alan gave to me was to give a summarizing talk at the end of this symposium. One obvious purpose of such a talk is to remind the participants of the key points and important findings that have been discussed during the symposium. I have spent several hours trying to prepare an adequate summary and this has proven to be impossible. Although I attribute this failure mostly to my own inability, I would like to place at least part of the blame on the enormous success of the symposium. We have listened to almost forty talks in the last three days and each one has contained very important results and concepts. The talks have exhibited in a dramatic way the breadth of applications of solitons and other nonlinear structures in condensed matter.

Although it does seem impossible to do justice to forty talks in the forty minutes I am allotted, I think it is important for all of us to at least briefly review what has transpired. For this purpose I would like to ask each of you for your help, by relying on your own memories of the last three days, rather than my imposing on you my own recollections. In order to stimulate your memories, I shall recite a list of some of the key words or phrases that have cropped up (perhaps more than once) during the symposium. After each item on the list I will pause for a second or two to allow you to reflect on those features connected with the word or phrase that impressed <u>you</u> the most. Let us begin:

Integrable Systems
Conservation Laws
Inverse Scattering Transform
Canals, Barges, Horses
Solitons (aristocratic and plebian)
Phonons, Mesons, Radiation (or whatever you want to call them)
Kinks
Bions
Breathers
Pulsons
Wobblers
Shelves
Fluxons
Parasitic Modes
Twists
Cylindrical Solitons
Phase Slips
Vortices
Phase Boundaries
Domain Walls
Clusters

Mexican Hats
Microdomains
Soft Modes
Instantons
Droplets
"Langer did it all ten years ago"
Spin Density Waves
Charge Density Waves
Mass Density Waves
Discommensurations
Dislocations
Etch Pits
Polariton Solitons
Pendulum Chains
Periodic Distortions
Bifurcations
Instabilities
A Poem
ϕ-particles
Schottky Defects
Operator Democracy
Universality
Fermions/Bosons
Electron-Hole Pairs
Incommensurate to Commensurate Transitions
Kohn Anomalies
Precursors
Longitudinal Lock-in
Fermi Surface Nesting
$\sqrt{13}$ Superlattice
2 k_F and 4 k_F
Thermalized Solitons
Poling
Lunch Bags
Help-wanted ads
Transfer Integral
Partition Function
Fokker-Planck
Smoluchowski
Ionic Conduction
Vacancy Motion
Soliton Lattices
Stacking Solitons
The ω-phase
Epitaxy
Frustration
Clouds of Erratic Trajectories
Fixed Points
Cantor Sets
Devils' Staircase
Textures
Spin Triplet p-Waves
Goldstone Modes
Homotopy Theory
Linear Textures
Pointlike Solitons
Vortex Lines
Disgyrations
Boojums
Snarks
d-Solitons
Twist Solitons
Splay Solitons
Composite Solitons
π's and δ's
Flash Bulbs

I hope this list has reminded you of the amazing explosion which has taken place in the physics community and condensed matter physics in particular. Such diversity of applications of solitons and related objects is a sure sign of the fundamental importance of a new pattern of thought or paradigm, to use Jim Krumhansl's phrase. Once the uninitiated can bring himself to revolt from his "linear" upbringing (via the harmonic oscillator and Schrödinger's equation) and accept the fact that large-amplitude localized objects can exist which do not spread in time, then he stands on the threshold of developing that wonderfully simple and beautiful pattern of thought that emphasizes the remarkable stability of such objects, their very natural use as elementary excitations or as fundamental objects present in the ground state, their coexistence in many cases with extended linearized solutions, their striking particle-like behavior in the presence of perturbations, and perhaps most important of all, their essential role in describing so many physical phenomena that cannot be explained in any other way. Now that we have all been exposed to this paradigm, who among us will be able to ignore the admonishment by G. B. WHITHAM [1] in the last sentence of his book on Linear and Nonlinear Waves: "...one should not always turn too quickly to a search for the ε."?

Now we turn to the question, "Where do solitons go from here?" Let me preface my retreat from this question by remarking that no area of physics seems to have been left untouched by solitons, and as we have witnessed here, condensed matter physics is no exception. It seems folly to try to predict the specific areas where solitons will crop up in the years ahead. Who could have predicted three years ago the items on the above list? Indeed, if such predictions could be made for three years hence, it would be a sure sign that the renaissance of nonlinearity is complete. This symposium augurs the contrary.

Nevertheless, we can be certain of some aspects of the soliton's future. It won't be long before the soliton becomes as natural and familiar a tool in nonlinear physics as the harmonic oscillator has become in linear physics. Indeed, because of its richness of structure and ubiquity in applications, the soliton will become one of the cornerstones of modern physical science. For this reason, solitons will also find their way into the classroom and textbooks as deliciously simple and beautiful examples of what nonlinear physics has to <u>offer</u> and of what it <u>requires</u>.

Never before has a single concept brought together physicists of such varied backgrounds and enabled them to communicate foreign ideas and applications using a common language. The student of solitonics has an unparalled opportunity to venture from his own field of interest and rekindle his love for <u>all</u> areas of physics. This opportunity is one of the major legacies of this symposium and others like it, and I have no doubt that solitons will periodically converge at future symposia where we can learn where solitons <u>have gone</u>, and gain insights and intuition to further refine and enrich the emerging pattern of thought.

In closing, I would like to ask the participants to join with me in asking Chris Eilbeck and Al Scott to organize a conference in 1984 at Heriot-Watt University in Edinburgh to commemorate the 150th anniversary of Scott-Russell's first observation of the soliton. I sincerely hope that they will be able to commandeer a flat-prowed barge and some horses in order to recreate the momentous event and I'm looking forward to joining all of you in a horseback ride along that famous canal in Scotland to observe that "singular and beautiful phenomenon".

References

1. G. B. Whitham, <u>Linear and Nonlinear Waves</u> (Wiley and Sons, New York, 1974).

Index of Contributors

H. Haken

Synergetics

An Introduction

Nonequilibrium Phase Transitions and Self-Organization in Physics, Chemistry and Biology

2nd enlarged edition. 1978. 153 figures. Approx. 360 pages
(Springer Series in Synergetics, Volume I)
ISBN 3-540-08866-0

Synergetics, deals with profound and striking analogies recently discovered between the self-organized behavior of seemingly quite different systems in physics, chemistry, biology, sociology and other fields. The cooperation of many subsystems such as atoms, molecules, cells, animals, or humans may produce spatial, temporal or functional structures. Their spontaneous formation out of chaos is often strongly reminiscent of phase transitions.

This book, written by the founder of synergetics, provides an elementary introduction into the basic concepts and mathematical tools. Numerous exercises, figures and simple examples greatly facilitate the understanding. The basic analogies are demonstrated by various realistic examples from fluid dynamics, lasers, mechanics, engineering, chemical and biochemical systems, ecology, sociology and theories of evolution and morphogenesis.

The second edition differs from the first by an additional chapter on chaotic motion, a rapidly growing field, and by new sections on laser pulses and on morphogenesis.

Springer-Verlag
Berlin
Heidelberg
New York

Springer Tracts in Modern Physics

Ergebnisse der exakten Naturwissenschaften

Editor: G. Höhler
Associate Editor:
E. A. Niekisch

Springer-Verlag
Berlin
Heidelberg
New York

Volume 81
G. Leibfried, N. Breuer

Point Defects in Metals I

Introduction to the Theory
1978. 138 figures, 22 tables. XIV, 342 pages
ISBN 3-540-08375-8

Contents: Introduction and Survey. – Harmonic Aapproximation and Linear Response (Green's Function) of an Arbitary System. – Lattice Theory. – Continuum Theory. – Transition from Lattice to Continuum Theory. – Statics and Dynamics of Simple Single Point Defects. – Scattering of Neutrons and X-Rays by Crystals. – Probability, Distributions and Statistics. – Properties of Crystals with Defects in Small Concentration. – Appendix.

Volume 82

Electronic Structure of Noble Metals and Polariton-Mediated Light Scattering

With contributions by B. Bendow, B. Lengeler
1978. 42 figures, 20 tables. VI, 114 pages
ISBN 3-540-08814-8

Contents: B. Lengeler: de Haas-Van Alphen Studies of the Electronic Structure of the Noble Metals and Their Dilute Alloys. – B. Bendow: Polariton Theory of Resonance Raman Scattering in Solids.

Volume 83
E. Amaldi, S. Fubini, G. Furlan

Pion-Electroproduction

Electroproduction at Low Energy and Hadron Form Factors
1978. 47 figures. Approx. 180 pages
ISBN 3-540-08998-5

Contents: Quantities of Physical Interest. – Theoretical Approaches. – Main Features of the Experiments, Preliminary Tests and Measurements. – Hadron Form Factors from Electroproduction. – Other Developments. – Appendices.

Volume 84

Collective Ion Acceleration

With contributions by C. L. Oison, U. Schumacher
1978. 63 figures. Approx. 240 pages
ISBN 3-540-09066-5

Contents: U. Schumacher: Collective Ion Acceleration With Electron Rings. – C.L. Olson: Collective Ion Acceleration With Linear Electron Beams.